Olfaction in Animal Behaviour and Welfare

FSC
www.fsc.org
MIX
Paper from responsible sources
FSC® C013604

Olfaction in Animal Behaviour and Welfare

Edited by

Birte Lindstrøm Nielsen

INRA, France

CABI

CABI is a trading name of CAB International

CABI
Nosworthy Way
Wallingford
Oxfordshire OX10 8DE
UK

Tel: +44 (0)1491 832111
Fax: +44 (0)1491 833508
E-mail: info@cabi.org
Website: www.cabi.org

CABI
745 Atlantic Avenue
8th Floor
Boston, MA 02111
USA

Tel: +1 (617) 682 9015
E-mail: cabi-nao@cabi.org

A catalogue record for this book is available from the British Library, London, UK.

Library of Congress Cataloging-in-Publication Data

Names: Nielsen, Birte Lindstrøm, editor.
Title: Olfaction in animal behaviour and welfare / edited by Birte Lindstrøm Nielsen, INRA, France.
Description: Boston, MA : CABI, 2017. | Includes bibliographical references and index.
Identifiers: LCCN 2017002054 (print) | LCCN 2017026487 (ebook) | ISBN 9781786391605 (epdf) | ISBN 9781786391612 (epub) | ISBN 9781786391599 (pbk. : alk. paper)
Subjects: LCSH: Smell. | Animal behavior.
Classification: LCC QP458 (ebook) | LCC QP458 .O4316 2017 (print) | DDC 591.5--dc23
LC record available at https://lccn.loc.gov/2017002054

ISBN-13: 978 1 78639 159 9

Commissioning editor: Caroline Makepeace
Associate editor: Alexandra Lainsbury
Production editor: Tim Kapp

Typeset by SPi, Pondicherry, India
Printed and bound in the UK by CPI Group (UK) Ltd, Croydon, CR0 4YY

Contents

List of Contributors

Anders Ågmo, Institute of Psychology, University of Tromsø, N-9037 Tromsø, Norway.
Luisa Amo, Department of Evolutionary Ecology, Museo Nacional de Ciencias Naturales (MNCN-CSIC), C/ José Gutiérrez Abascal, 2, E-28006 Madrid, Spain.
Raimund Apfelbach, Neurobiology/Animal Physiology, University of Tübingen, Auf der Morgenstelle 28, D-72076 Tübingen, Germany.
J. Elizabeth (Liesbeth) Bolhuis, Adaptation Physiology Group, Department of Animal Sciences, Wageningen University and Research, De Elst 1, NL-6708 WD Wageningen, The Netherlands.
Vincent Bombail, NeuroBiologie de l'Olfaction, INRA, Université Paris-Saclay, Domaine de Vilvert, F-78350 Jouy en Josas, France.
Roisin Campbell-Palmer, Faculty of Technology, Natural Sciences and Maritime Sciences, Department of Natural Sciences and Environmental Health, University College of Southeast Norway, N-3800 Bø i Telemark, Norway.
Caroline Clouard, Adaptation Physiology Group, Department of Animal Sciences, Wageningen University and Research, De Elst 1, NL-6708 WD Wageningen, The Netherlands.
Laurent Dormont, CNRS Centre d'Ecologie Fonctionnelle et Evolutive, UMR 5175, 1919 Route de Mende, F-34293 Montpellier Cedex 5, France.
Markus Fendt, Medical Faculty, Institute for Pharmacology and Toxicology, Otto-von-Guericke-Universität Magdeburg, Leipziger Strasse 44, D-39120 Magdeburg, Germany.
Peter G. Hepper, Animal Behaviour Centre, School of Psychology, Queen's University Belfast, Belfast BT7 1NN, Northern Ireland, UK.
Tadeusz Jezierski, Institute of Genetics and Animal Breeding, Polish Academy of Sciences, Department of Animal Behaviour, Jastrzebiec, Poland.
Olivia Le Moëne, Institute of Psychology, University of Tromsø, 9037 Tromsø, Norway.
Frédéric Lévy, Physiologie de la Reproduction et des Comportements, INRA UMR 85, CNRS UMR 7247, Université F. Rabelais, IFCE, F-37380 Nouzilly, France.
Nicolas Meunier, NeuroBiologie de l'Olfaction, INRA, Université Paris-Saclay, Domaine de Vilvert, F-78350 Jouy en Josas, France.
Birte L. Nielsen, NeuroBiologie de l'Olfaction, INRA, Université Paris-Saclay, Domaine de Vilvert, F-78350 Jouy en Josas, France.
Raymond Nowak, Physiologie de la Reproduction et des Comportements, INRA UMR 85, CNRS UMR 7247, Université F. Rabelais, IFCE, F-37380 Nouzilly, France.
Marije Oostindjer, Norwegian University of Life Sciences (NMBU), PO Box 5003, N-1432 Ås, Norway.
Olivier Rampin, NeuroBiologie de l'Olfaction, INRA, Université Paris-Saclay, Domaine de Vilvert, F-78350 Jouy en Josas, France.
Frank Rosell, Faculty of Technology, Natural Sciences and Maritime Sciences, Department of Natural Sciences and Environmental Health, University College of Southeast Norway, N-3800 Bø i Telemark, Norway.
Roland Salesse, NeuroBiologie de l'Olfaction, INRA, Université Paris-Saclay, Domaine de Vilvert, F-78350 Jouy en Josas, France.
Burton Slotnick, Department of Psychology, American University, 4400 Massachusetts Ave NW, Washington, DC 20016, USA.
Deborah L. Wells, Animal Behaviour Centre, School of Psychology, Queen's University Belfast, Belfast BT7 1NN, Northern Ireland, UK.
Tristram D. Wyatt, Department of Zoology, University of Oxford, South Parks Rd, Oxford OX1 3PS, UK.

Prologue

As part of his famous collection of case histories *The Man Who Mistook His Wife For A Hat*, Dr Oliver Sacks wrote an article titled, 'The Dog Beneath the Skin' about a patient who had experimented with drugs. They ended up greatly enhancing his sense of smell. He described smelling in detail he had never experienced before. In a perfume store, he could distinguish all the different scents. Each slight difference in odour was now attached to an emotion. Each person he met now had a unique 'smell-face' and he could smell their emotional state. He described entering a whole new perceptual world. There are a few other humans that can distinguish a myriad of smells: the best wine stewards can identify hundreds of wines by smell alone, for example.

This story fascinated me because I am a visualizer: my thoughts are in pictures instead of words. I had to learn that most people think in words. The first step to understanding is realizing that other people may perceive and think differently. This same principle needs to be applied to our understanding of animals.

A dog's social life revolves around smell. Time and time again I have seen dogs being walked and being jerked away from the things they are smelling. When a person walking a dog impatiently pulls his dog away from the bush where other dogs have left their mark, he is not allowing the dog to go into its smell universe. Dogs seem to like my luggage because it has been to many different smell places such as farms. At the airport, the security dog was in love with my well-worn handmade leather briefcase. These dogs are trained to sit when they smell contraband, but since it was obvious that there was nothing bad in my bag, the officer yanked his dog away.

As I wrote this Prologue, pictures associated with smells came into my head. Since I am a visualizer, I see the pictures first, and then smell the odour associated with them. I see a cookie store that I went to when I was a graduate student. To entice customers, they had a fan that blew cookie smells into the street. Installation of the fan was a great business move. It increased sales.

In my work with handling cattle, I have emphasized their reactions to what they see. Small distractions such as a reflection on a wet floor may make the animals balk and refuse to move through a race. Cattle will also react to smells. Handling in brand new facilities is often difficult due to the smell of fresh paint. The animals often refuse to enter the freshly painted area until I have dirtied it up.

There is a smell of fear, and I have seen the reaction of cattle to it. At one place, a steer flipped over onto his back in a single file race. It took about 10 min to get the animal back on his feet. When he was stuck, he had slobbered on the floor. The next animal in line refused to walk over his 'fear spit'. One animal balked and backed up several times before walking over the floor where the downed animal had slobbered. To get the other cattle to move easily, the race had to be washed down. It appears that for such fear odours to be secreted, the animal has to have a period of 5–20 min of severe stress. The presence of a fear pheromone has been documented by scientific studies discussed in this book.

Since during my career I have worked on improving slaughter plants, I am often asked, 'Are animals afraid of slaughter?' In answer I would say that I have witnessed hundreds of cattle calmly walking into facilities that had blood on them – the most important factor seems to be that there has been no prolonged obvious aversive events with a duration of over 5–10 min. Cattle behaviour and cortisol levels are similar during movement through races both on the farm and at the abattoir.

To end this Prologue, I will leave you with something to ponder. I once saw a group of intently curious heifers forming a circle around a groundhog that was on their pasture. Their behaviour was definitely motivated by curiosity, because the young heifers had never seen a groundhog before. On another occasion I witnessed a group of beef cows on a pasture congregated at a place where a herd-mate had died. The grass looked normal, but they all put their noses to this patch of ground and sniffed. They did not bellow or become agitated but just sniffed and sniffed. They seemed intently interested. Were they distressed or just curious? What were the cows experiencing when they sniffed the place where their herd-mate had died? Learning more about smell may help answer these questions.

Temple Grandin
February 2017

Preface

If you are reading this, you're either considering buying or have already acquired this book, and now you are contemplating whether to read the whole thing or just dip into the chapters of relevance to you (or, indeed, put it back on the shelf). I would encourage you to do the former (but then again, I would, wouldn't I?). Here's why:

This book is one of the most accessible books on olfaction that you will ever come across. The chapters are structured so as to give you an overview of the subject, written in a style aimed to be easily comprehensible. The book is not meant to review all the literature on the subject but to give you a taste of the type of research which has been carried out on olfaction. You will find nuggets of knowledge in each chapter for potential use in your project report, your thesis, your teaching or your next dinner conversation.

Because the world of smells is fascinating.

As humans, we use our nose much more than we think. From knowing when to change the baby's nappy to choosing a perfume, we smell our surroundings almost unknowingly. Is there a gas leak? This fish is off! And if we use our noses to such an extent, how complex is the olfactory world of animals, where the sense of smell is a much more important sensory modality?

The idea for this book came about as a continuation of an opinion paper co-written by a group of researchers from across Europe. The subject of the short paper (Nielsen *et al.*, 2015) was the absence of olfaction and odours in most scientific studies relating to applied ethology and animal behaviour. Most behavioural studies involving odours are carried out by neuroscientists working on rodents and interested in the mechanics of the brain with particular reference to olfaction. The information available on other animal species managed by humans, such as farm, zoo or companion animals, was relatively scarce and often difficult to find. One of the first and most cited papers on environmental enrichment of animal housing mentions odours only once ('*Greater attention to the sounds and odours coming from the external environment is also warranted*', Newberry, 1995), and gives no examples or suggestions of the use of olfactory stimulation as a form of enrichment for captive animals. How could we entice researchers, who do not have olfaction as their primary subject area, to at least consider the role of odours in their experiments? Information on olfaction and behaviour was not always adequate or easily accessible, either being non-scientific or otherwise buried in articles heavy in neurological jargon. An evidence-based yet accessible book on olfactory behaviour was missing. Many of the authors of the original paper agreed to contribute to a book on the subject, and other experts on olfaction joined in. The book you're reading now is the result.

Odorant receptors and the organization of the olfactory system was uncovered as recently as 26 years ago, which led to the Nobel Prize in physiology or medicine 2004 being awarded jointly to the two scientists behind the discovery (Buck and Axel, 1991). Incidentally, the 2005 Ig Nobel Prize in biology (a prize intended to celebrate the unusual, and spur people's interest in science) was awarded to four scientists for painstakingly smelling and cataloguing the peculiar odours produced by 131 different species of frogs when the frogs were feeling stressed (Smith *et al.*, 2003). Hopefully this book will motivate its readers to consider the effects of odours and the role of olfaction in their work.

The book is divided into three sections: the first section (Chapters 1–5) will introduce you to some of the terminology and give you some basic information about olfaction, how to carry out behavioural tests using odours, as well as the relative roles of the species, the environment, previous experience and – for humans – the cultural background. The second section (Chapters 6–11) deals with the role of olfaction in different contexts and across species, including feeding behaviour, reproduction, disease detection and animal housing. The final section (Chapters 12–15) focuses on the olfactory behaviour of animals managed by humans, in other words farm, zoo, lab and companion species. In many of the chapters, the chemical names of different odorants are included. However, knowledge of the exact nature of the chemicals used is often not essential for the understanding of the study described, but they may as well be properly named in each example, as opposed to being referred to as odorants A and B.

I started this introduction by encouraging you to read the whole book. That's still the case, but if your specific interest is, say, foraging in wombats, you should pay particular attention to Chapters 6 and 13, which between them cover olfaction in feeding behaviour and zoo animals, respectively. Likewise, any combination of two chapters will provide a quick overview of the importance of olfaction for a subject area – from reproductive behaviour in sheep to stress responses in rats.

Happy reading!

Birte L. Nielsen, Editor
Jouy-en-Josas, France
November 2016

References

Buck, L. and Axel, R. (1991) A novel multigene family may encode odorant receptors – a molecular-basis for odor recognition. *Cell* 65, 175–187. DOI: 10.1016/0092-8674(91)90418-X

Newberry, R.C. (1995) Environmental enrichment: increasing the biological relevance of captive environments. *Applied Animal Behaviour Science* 44, 229–243. DOI: 10.1016/0168-1591(95)00616-Z

Nielsen, B.L., Jezierski, T., Bolhuis, J.E., Amo, L., Rosell, F., Oostindjer, M., Christensen, J.W., McKeegan, D., Wells, D. and Hepper, P. (2015) Olfaction: an overlooked sensory modality in applied ethology and animal welfare. *Frontiers in Veterinary Science* 2, 69. DOI: 10.3389/fvets.2015.00069.

Smith, B.P.C., Tyler, M.J., Williams, B.D. and Hayasaka, Y. (2003) Chemical and olfactory characterization of odorous compounds and their precursors in the parotoid gland secretion of the green tree frog, *Litoria caerulea*. *Journal of Chemical Ecology* 29, 2085–2100. DOI: 10.1023/A:1025686418909

1 The Nuts and Bolts of Olfaction

Nicolas Meunier[1,2] and Olivier Rampin[1]

[1]Neurobiology of Olfaction, INRA, Université Paris-Saclay, Jouy-en-Josas, France; [2]Université de Versailles Saint-Quentin, Versailles, France

A living being does plenty of things with the information conveyed by the presence of an odour. One of the most basic needs is food, which can be recognized through chemical signature cues. Odours also play a major role when looking for a mate. Many animals will rely mainly on olfaction to initiate reproduction; this is the case in most domesticated species. Animals will also use olfaction to orientate themselves, to create social bonds as well as to avoid predators and imminent threats such as fire. In this chapter, we will present the basic information to understand how olfaction works and an overview of its importance in animal behaviour.

What is an Odour?

Living beings need to detect information in their environment to interact with it. Many different types of information are accessible, such as those conveyed by concentration of molecules, pressure, light, heat, and electric or magnetic fields. The simplest and probably the first system that developed during the course of evolution is the detection of molecules in the environment. While this ability appears obvious in animals, it is shared by all living beings. Indeed, most bacteria are able to detect a new source of energy such as lactose in the absence of glucose and can adapt their metabolism to use it. Some bacteria can even perform complex communication through molecule exchanges (Waters and Bassler, 2005). Similarly, plants detect molecules released by neighbouring conspecifics eaten by herbivores and produce in response various compounds to reduce their own attractiveness to herbivores (Karban *et al.*, 2013). However, in the following we will concentrate on odours and olfaction in relation to the animal kingdom.

Can all volatile molecules be defined as an odour? An odour is first of all a mixture of molecules (each referred to as an odorant) that is different from its surrounding (air and water do not smell) and that triggers a sensation in the animal detecting it. An odour is defined by the nature and the concentration of odorants that are present in it. Natural odours like those released by animal fluids (such as sweat, urine, saliva and tears) contain hundreds of odorants with very different chemical properties: acids, aldehydes, ketones, etc. An odour allows the identification of its source based on the variety of odorants present. As odours are conveyed in the form of a complex airflow (referred to as plumes) in the air or currents in water, it also contains information about the distance to the source, and time elapsed since it has been released. The concentration of a single odorant changes at each time point and at each location point of a given milieu. The complexity of this information suggests that the olfactory system is adapted to this spatiotemporal dynamic.

In a laboratory setting, it is difficult to control this heterogeneity of smells originating from the variety of odorants, their concentration and their spatiotemporal dynamic. Therefore, research labs often use a small number of odorants, prepared in a limited number of concentrations to study the response of an organism. In humans, one odorant (i.e. a single molecular compound) is often sufficient to be perceived as an odour, e.g. isoamyl acetate is recognized as banana odour. We use this property in laboratories to simplify experiments assuming that it holds true also for non-human animals.

There are puzzling data about odorants and odours. Experiments in humans reveal that the smell of a molecule is very different depending on its concentration, ranging from pleasant to unpleasant and finally irritant. For example, heptanoic acid is pleasant at low concentrations around 10^5 diluted in mineral oil, reminiscent of cheese, whereas at higher concentration (around 10^3 dilution) the smell will resemble your socks after you have just finished jogging on a hot day. Finally, heptanoic acid is impossible to smell when only 10 times diluted as it will immediately cause you to block your respiration. Furthermore, while some molecules with different chemical structures are perceived as the same odour, some molecules that are mirror images of each other (enantiomers) are distinguished after some training even for those not blessed with a good nose. Although correlations are found between molecular features of an odorant and its perceived intensity, pleasantness and familiarity (Keller and Vosshall, 2016), there is no unique relationship between an odorant and the odour it is associated with. Finally, both in humans (Livermore and Laing, 1996) and rodents, odorant discrimination within an odour remains poor. In other words, neither rats nor humans are good at answering the question 'How many odorants are there in this odour?'

The detection of an odour is the job of olfaction, one of the five senses of humans, and it goes along with taste (or gustation), which is more directed towards the detection of food qualities during ingestion. Both are commonly grouped under the term of chemical senses because they allow the detection of chemicals in the environment. Their distinction can be tricky especially for species living in water. In the following, we will focus our attention mainly on the terrestrial animal world.

How is an Odour Detected and Processed by the Nervous System?

The molecular basis of odour detection

To start at the very beginning: animals are built from elementary structures called cells. The frontier between a cell and the outer world is the membrane maintaining the integrity of the cell. This protection leads to isolation from the environment, which is far from optimal when you need to interact with it. Thus, all cells produce on the surface of their membrane transporters to exchange molecules and receptors to interact with the environment. Part of each receptor is exposed to the extracellular compartment. It is this part that will bind, reversibly, the molecules of the environment such as odorants. By binding molecules to membrane receptors, cells obtain information from their environment. Some molecules bind to the receptor even when they are present at low concentration. These molecules have a high affinity for the receptor. They are referred to as ligands. Through binding, the shape of the receptor changes, which drives a signal into the cell. Figure 1.1A describes this mechanism, which is called signal transduction because it transduces (converts) an external signal (here a molecule at a given concentration) into an internal signal.

The binding of an odorant to a receptor is facilitated by the movement of the molecule in the environment and the behaviour of the animal. A sniff, which changes the flux of air that enters the nostrils, and redirects air in the nasal cavity towards the olfactory epithelium, improves odorant detection. In animals, the olfactory receptors that bind odorants are present in the membrane of specialized cells called olfactory neurons. Olfactory receptors and olfactory neurons refer thus to two different things that are easy to confuse. The olfactory receptors are the membrane proteins binding odorants. The olfactory neurons are the first order neurons of the olfactory system.

The largest family of receptors in the genome

Industrial chemistry has synthetized new molecules that were previously absent on earth, and among these are thus potentially new odorants. Astonishingly, as long as these molecules are volatile, they can activate the olfactory system, despite never having existed during the evolution of this system. The combinatory range of organic molecules is huge, indicating that the olfactory system discriminates among tens of thousands of different odorants. Furthermore, this discrimination is very subtle as we are able to distinguish very similar molecules (Fig. 1.1B), even between enantiomers (the same molecule but arranged differently in space, just like our two hands). This capacity is very similar to the immune system's ability to fight pathogens. Indeed, the immune system needs to recognize an immense diversity of pathogens, some of them never encountered before. Similarly to the immune system, the olfactory system deals with molecular diversity and specificity of recognition by having a great variety of olfactory receptors and a combinatory system.

In mammals, the genome contains approximately 30,000 genes. Around 5% encode for olfactory receptors, which is huge, making olfactory receptors the largest receptor gene family (Buck and Axel, 1991). Among terrestrial vertebrates, elephants, horses and cows have the greatest number of olfactory receptors, while birds and primates have the lowest (Fig. 1.1C). This diversity appears to be consistent with the olfactory

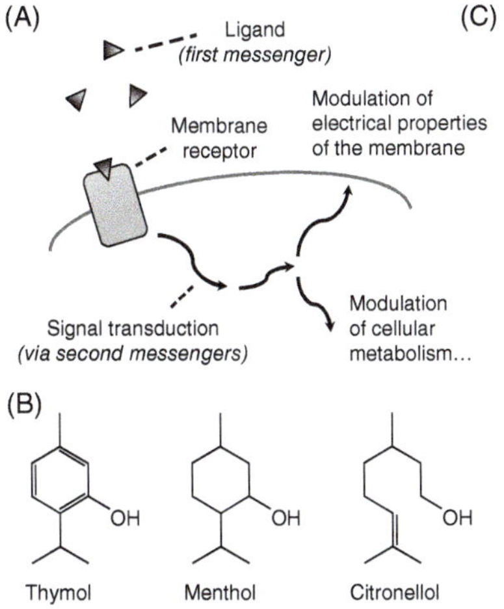

(C)

Species	OR number
Elephant	1948
Cow	1186
Rat	1207
Pig	1200
Horse	1066
Mouse	1130
Dog	811
Human	396
Orangutan	296
Chicken	283

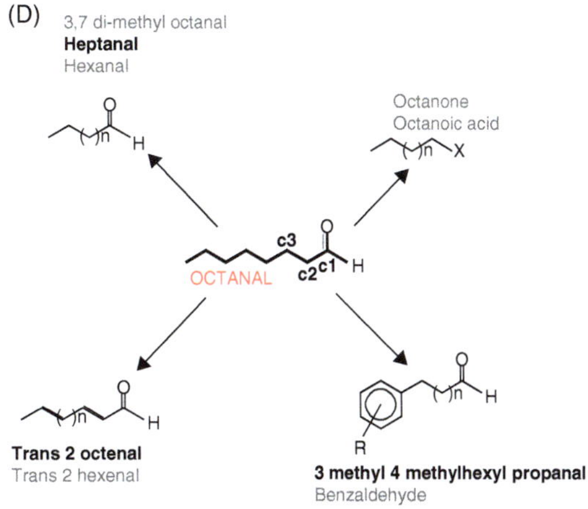

Fig. 1.1. Olfactory receptors. (A) A cell gathers information from its environment through receptors present at its membrane. The binding of a molecule (usually at low concentration to be considered a ligand) to the receptor triggers a cellular response. In the olfactory system, the odorants bind to olfactory receptors (OR) present in the membrane of the olfactory sensory neurons. (B) Molecules with comparable chemical structures elicit different odour perception in humans (thymol: herbal, thyme, spicy; menthol: minty; citronellol: floral, rosy, citrus). (C) Number of estimated functional olfactory receptors (OR) in various species. (D) Mouse olfactory receptor I7 is strongly activated by octanal. Molecules with related chemical structure (in bold) also activate this receptor but not molecules with slightly different chemical groups (in grey). It indicates that the chemical group is more important for binding to a given olfactory receptor than slight adjustments in molecular structure, which does not prevent the molecule from binding to the receptor. (Modified from Araneda *et al.*, 2000, and Niimura *et al.*, 2014.)

capacities of a given species, as this number decreases according to the importance of olfaction among the different senses (Niimura *et al.*, 2014). Elephants, with their 2000 olfactory receptors, rely heavily on olfaction especially for sexual behaviour. They are among the few known mammals in which odorant molecules drive behavioural responses before and during mating (Rasmussen and Greenwood, 2003). Nevertheless, primates with their great visual system or platypuses endowed with a keen electric field sense still have about 400 functional olfactory receptors. It is also worth noticing that birds in the past have been wrongly considered as having poor olfaction, whereas they are in fact no worse than many mammals, with almost 300 olfactory receptors for the chicken. However, no clear relationship can be found between olfactory ability to detect low concentration of odorants and olfactory gene numbers. Furthermore, olfactory capacities and sensitivity also depend on training as humans can develop a very acute sense of smell (e.g. perfumers, sommeliers and oenologists).

Only a few olfactory receptors have been characterized in terms of which molecules they recognize, and most studies have been performed in mice due to the genetic tools available. These results show that, overall, some olfactory receptors can be activated by molecules sharing structural similarities (Fig. 1.1D), while others very specific, being activated by one particular molecule (Araneda *et al.*, 2000; Saito *et al.*, 2009). A key point to understanding how olfaction distinguishes among tens of thousands of different molecular compounds is the fact that an olfactory neuron expresses only one of the hundreds of olfactory receptors available in the genome.

First step … the olfactory epithelia

Olfactory neurons are distributed over the whole olfactory epithelium in the nasal cavity (Fig. 1.2A–C). In vertebrates, two major epithelia are implicated in odour detection: the main olfactory epithelium and the vomeronasal epithelium (often referred to as the vomeronasal organ). The latter has long been thought to have appeared with terrestrial life but recent studies indicate that there was already segregation of those two systems in early forms of vertebrates such as lamprey (Ubeda-Bañon *et al.*, 2011). Although the vomeronasal organ exists in all mammals, it regresses at the end of the embryonic development in bats, cetaceans and most primates including humans (Mucignat-Caretta, 2010). In terrestrial animals, the vomeronasal organ is specialized in the detection of molecules present in body secretions that are usually not very volatile. Thus, the animal must be in direct contact with the source in order to detect these molecules. For a long time, the two systems have been considered complementary: the detection of odours carrying general information was performed by the main olfactory epithelium, whereas the vomeronasal organ was thought to be specialized in the detection of odours, the famous pheromones, released by individuals from the same species and eliciting innate behaviours (see Chapter 3). Recent studies show that both epithelia participate in pheromone detection (Leypold *et al.*, 2002; Mandiyan *et al.*, 2005), and both epithelia also share the same structure. They are composed of olfactory neurons surrounded by supporting cells ensuring a stabilized ionic environment. Olfactory neurons are in direct contact with the external milieu, which is rich in pathogens and has high levels of oxygen, so they regularly undergo apoptosis (programmed cell death) and are renewed from basal cells (multipotent cells present in the olfactory epithelium). Olfactory neurons bear cilia containing the olfactory receptors (Fig. 1.2A) and these cilia are surrounded by mucus (Fig. 1.2B), which limits the toxicity of the environment. In terrestrial animals, this mucus is also important to make the lipophilic odorant molecules more soluble in this hydrophilic gel, probably thanks to carrier proteins called odorant binding proteins. It also contains a large number of enzymes implicated in the degradation of these odorants. Indeed, when an animal relies on olfaction to follow scent-trails, it needs to analyse swiftly the changes in odorant concentration while it moves. Odorants need thus to be cleared speedily from the mucus in order to be able to perform such a task.

In terrestrial animals, the nose contains olfactory and respiratory epithelia involved in three tasks: odour detection, and conservation of body heat and water. The respiratory epithelium is very similar to the epithelium along the airways; it is thinner than the olfactory epithelium and contains mainly cells bearing motile cilia, which transport particles and pathogens trapped in the mucus towards the digestive tract. One distinctive feature, which is highly variable between species, is the percentage of the nasal epithelium area devoted to olfaction. It is usually higher in small animals (e.g. a grey squirrel has a higher percentage of its epithelium area in the nose devoted to olfaction than a white-tailed deer). For animals with similar body mass, this will be limited by the snout form, as some animals with a great visual system have a smaller nose to improve binocular vision, but also by the animals' diet and habitat. Typically, within animals of the same body mass, canids have the highest percentage of nasal epithelium devoted to olfaction (the coyote being the champion; Fig. 1.2C), a percentage that can be linked to their enhanced olfactory ability (Yee *et al.*, 2016).

Odorant detection is based on combinatory recognition

We mentioned earlier that one olfactory neuron expresses only one type of the hundreds of olfactory receptors available in the genome. One olfactory receptor is activated by a few odorants with chemical structures that share common traits. When odorants flow into the nasal cavity up to the olfactory epithelium, a population of olfactory neurons that bear receptors binding to those odorants is activated. The combination of odorants within an odour is reflected in the olfactory epithelium by the combination of receptors/olfactory neurons activated. Another very important feature of the olfactory system is that olfactory neurons that bear identical olfactory receptors send their axons onto a limited number of specialized spherical structures, the glomeruli, within the olfactory bulb (Fig. 1.2D). In the mouse, there are about two glomeruli per olfactory receptor and thus approximately 2000 glomeruli. It is therefore the combination of olfactory neurons activated by an odorant that in turn will activate a population of glomeruli near the surface of the olfactory bulb (Fig. 1.2E). This creates a form of glomerular map related to a specific odorant at a given concentration. If considering a two-way state of activation of these glomeruli (on or off), this system can represent at least 2^N odours, where N is the number of olfactory receptors. Even with species having a relatively small number of olfactory receptors (~300), this can potentially lead to 2^{300} different combinations of glomeruli activation states. This corresponds to a huge number of different odours that can be distinguished, far more than the potential combination of atoms present in organic molecules. Figure 1.2F shows three different representations of real activation states of glomeruli in response to different odorant stimulation, representing different glomerular maps.

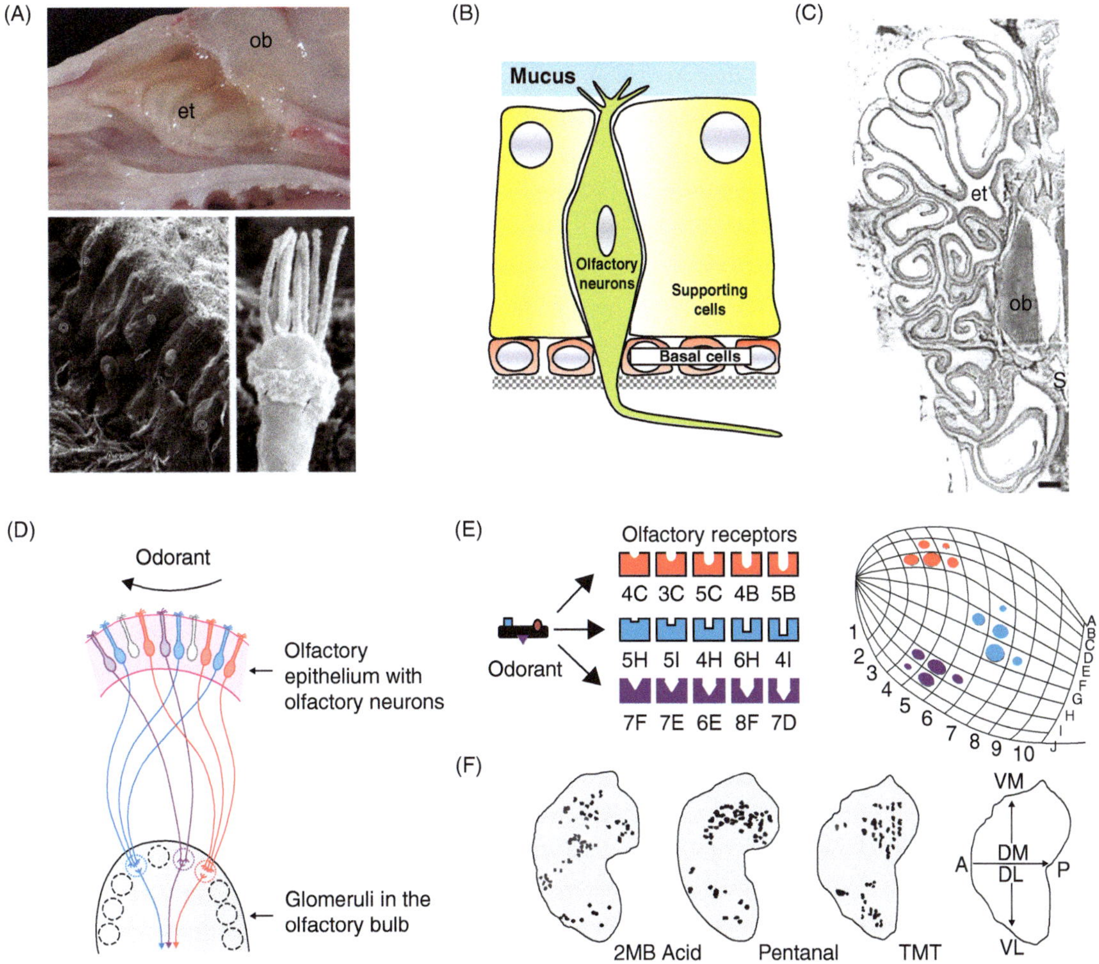

Fig. 1.2. From olfactory neurons to olfactory bulb maps. (A) Top: View of a rat's nasal cavity showing the endoturbinates (et) covered by the main olfactory epithelium and lying next to the olfactory bulb (ob). Below: Electron microscopy of human olfactory epithelium showing the organization of supporting cells surrounding the olfactory neurons bearing cilia (magnified on the right) on the surface of the epithelium and sending axons to the olfactory bulb. Adapted from Morrison and Costanzo (1990). (B) Schematic representation of the olfactory epithelium. (C) Sagittal view of the main olfactory epithelium from a coyote. The endoturbinates form a complex network to increase the surface of the olfactory epithelium. Adapted from Yee *et al*. (2016). (D) Odorants activate olfactory sensory neurons (OSNs) in the olfactory epithelium. OSNs of the same colour bear the same olfactory receptor. Axons from neurons expressing these identical receptors converge into specific glomeruli in the olfactory bulb. (E) A given odorant bears different chemical groups represented here by a circle, triangle and square. By means of these different structures, the odorant activates different populations of OSNs expressing different receptors with more or less affinity (4C; 5H and 7F each fitting perfectly to the chemical groups depicted, and with similar receptors having less affinity). Due to the convergence of OSNs expressing the same olfactory receptor onto a single glomerulus, the detection of one odorant by the population of different OSNs will in turn activate a population of glomeruli. It yields a representation of the odorant identity on the surface of the olfactory bulb (mapped on the right), with glomeruli receiving signals with a strength depending of the affinity of the receptors for the odorant (coordinates on the map relate to olfactory receptor number). (F) Three examples of activation maps for various odorants in mice in the olfactory bulb (orientation of the olfactory bulb is presented on the right (anterior/posterior/lateral/medial/dorsal/ventral). For different odorants there are thus different maps of activation in the olfactory bulb. (Adapted from Mori and Sakano, 2011.)

What is the use of having more than 300 olfactory receptors then? Recent studies point out that some olfactory receptors are specialized in recognizing specific molecules (Sato-Akuhara *et al.*, 2016). One possibility is therefore that species with an important olfactory receptor repertoire possess more specialized olfactory receptors that could help to recognize innately relevant odours in the environment. As the ligands are only partially known for olfactory receptors in mice and almost unknown for other species, the importance of the olfactory repertoire can currently only be speculated on.

Interestingly, while the olfactory receptors in insects are very different from those of vertebrates, the same organization of olfactory neuron projections by glomerular maps is present in their brain. This indicates an evolutionary convergence to solve the discrimination problem of molecules with very different structures (Su *et al.*, 2009). While an odorant elicits a specific activation pattern in the olfactory bulb, this representation evolves with the meaning associated with this odorant. For example, in associative reward learning, an odour is associated with a reward. At subsequent exposure to the odour, olfactory bulb neurons that are activated by the odour increase their activity (Nuñez-Parra *et al.*, 2014). Similarly, an odour associated with a danger will have an enhanced representation in the olfactory bulb compared with a passive exposure to the same odorant (McGann, 2015). Furthermore, studies on mice have shown that the olfactory bulb processes the olfactory information along two pathways: one deals with the innateness of an odorant such as molecules emanating from predators, while the other is specialized in learning (Fig. 1.3).

Many brain structures are involved in the processing of odours

Olfactory bulb outputs reach many different structures in the brain and most are grouped under the term of the olfactory cortex (Fig. 1.4). This includes among others the olfactory tubercle, the piriform cortex and parts of the amygdala. The olfactory cortex is well developed in rodents, in which it occupies the ventral tier of the brain. In other mammals its size appears smaller compared with the development of the neocortex. Anatomical studies reveal that axons of neurons connected to a single glomerulus leave the olfactory bulb and richly innervate different areas of the olfactory cortex (Sosulski *et al.*, 2011; Igarashi *et al.*, 2012). As each glomerulus receives information from only one type of olfactory neuron responding to few odorants, it suggests a parallel processing of the same odorant information by the different structures of the olfactory cortex. Within a structure, different characteristics of an odour may be processed in different regions. In humans, the anterior region of the piriform cortex encodes the composition of an odour, identifying what odorants are present in the mix, such as 'there is isoamyl acetate here' (Gottfried *et al.*, 2006). In contrast, the posterior region of the piriform cortex encodes perceived quality but not structure, hence would provide a synthetic representation of the odour, such as 'smells like banana, doesn't it?', an example of a perceptual object (Wilson and Sullivan, 2011). The piriform cortex is likely to be involved in other tasks that enrich the decoding of odorants (e.g. perception of odour similarity between odorant molecules, imagining odours, categorizing odours with a hedonic value, and memorization), and its activity is modulated by attention. The piriform cortex, along with the olfactory tubercle, not only responds to odours, but also to other sensory inputs, among them taste, sound and touch. Convergence of different sensory inputs would increase the amount of information available to elaborate a perceptual object. Finally, in some of these brain structures, the response to an odour is modulated by the expected value of a reward (be it positive or negative) that the odour has been associated with at previous exposures (Setlow *et al.*, 2003).

As well as the olfactory cortex, other structures are also innervated by the olfactory bulbs' outputs including the hippocampus, which is involved in memory. This is peculiar to the processing of olfactory information as all other senses are first processed by the thalamus before going to other brain areas. The direct linkage between the olfactory bulbs and the hippocampus contributes to the strong capacity of odours to evoke vivid memories (Mainland *et al.*, 2014). So it is likely that brain structures not only encode the complexity of an odour (the variety and concentration of odorants in it), but also other aspects such as hedonic value, prior experience and motivational associations. Is it possible that the processing of these different characteristics is carried out by separate brain structures? Figure 1.5 illustrates this question, displaying odorant properties in a two-dimensional space. There, the odour of an oestrous vixen presented to a male rat carries quite different information (Rampin *et al.*, 2006). On the one hand the predator odour is

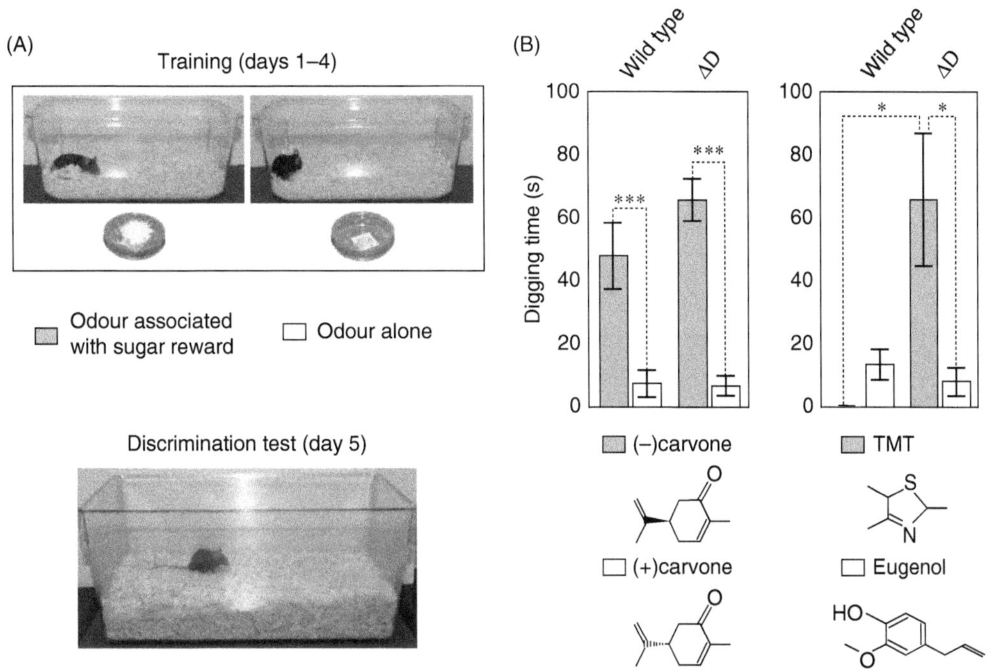

Fig. 1.3. Mice cannot learn to associate an odorant inducing innate fear with a reward except if part of their olfactory bulb is inactivated. (A) Mice were trained for 4 days to associate a sugar reward with an odour. The odorants used were either (–)carvone, which is considered neutral for the animal, or the predator-related odorant trimethylthiazoline (TMT). (B) On the fifth day, the time spent digging by the mice was measured when in the presence of the odorant previously used during training or in the presence of a similar odorant. Two strains of mice were used: one unaltered (wild type) and one with half of the olfactory bulbs genetically inactivated (ΔD). After the training, most mice spent far more time digging in the presence of the odour associated with sugar than when the related odorant was present. Most mice could learn to get a reward from the neutral (–)carvone odorant, which they could discriminate from its enantiomer (+)carvone. However, only the ΔD mice could associate TMT with the reward, whereas normal mice were unable to associate TMT, which innately induces fear, with a reward. This would indicate that part of the olfactory bulb most likely deals with the innateness of the odorant while the other part (still active in ΔD mice) allows the processing of odorants in terms of learning. (Adapted from Kobayakawa *et al.*, 2007.)

frightening; on the other hand the oestrous odour is sexually arousing. There are indications that it is, at least in part, possible to search for specialized processing of odour labelling in distinct areas of the brain. An example is the amygdala, a collection of different nuclei, recognized as an important brain centre for the processing of emotions, including fear. In rodents, amygdala neurons respond to the odour of a predator, and amygdala lesions abolish fear evoked by this predator odour. The amygdala can further transmit information to the hypothalamus, a brain structure that controls the secretion of stress hormones (see Chapter 10). The orbitofrontal cortex is involved in adaptive behaviour and decision-making. Within the orbitofrontal cortex, pleasant and unpleasant odours activate different regions (Gottfried *et al.*, 2002), and by integrating information from other structures such as the amygdala, this would help select and execute an appropriate behavioural strategy (Schoenbaum *et al.*, 1999).

As a conclusion, the activation map of an odour in the brain corresponds to a network encompassing several brain structures. These structures contribute to the production of a perceptual object based on a first step of analytical approach, then by exchanging information. Modulation of this encoding is made by the convergence of other sensory modalities and the status of the animal, e.g. fed or fasted, healthy or sick, calm or stressed.

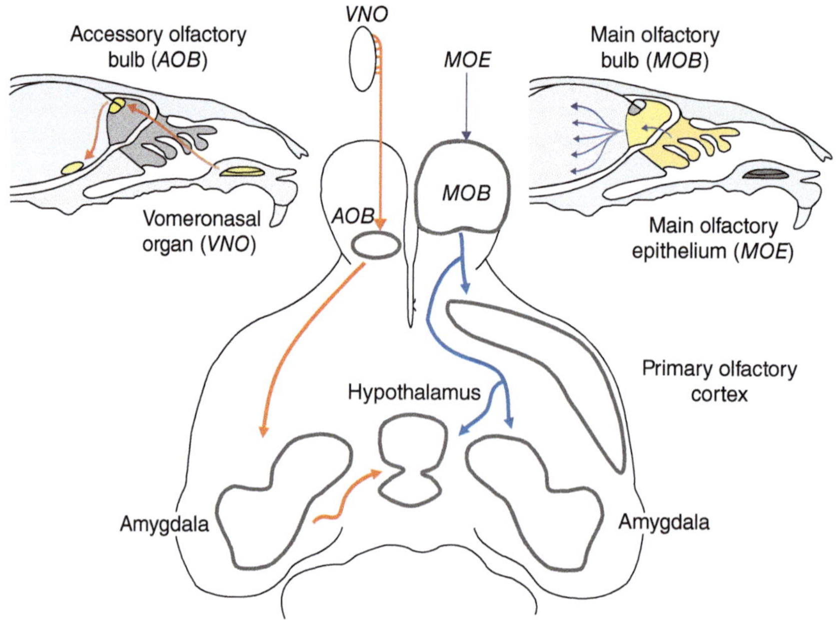

Fig. 1.4. Olfactory pathways in the nose and brain of rodents. Top left and right: representation of a rodent head sectioned in the sagittal plane (rostral to the right). Odorants present in the nasal cavity activate two sets of olfactory neurons: those in the vomeronasal organ (VNO, top left), and those in the main olfactory epithelium (MOE, top right). Axons of these neurons travel through two anatomically distinct pathways. The VNO efferent pathways reach the accessory olfactory bulb (AOB), where they relay to neurons that in turn inform the amygdala, a collection of brain nuclei involved in emotions (shown in the middle figure, a horizontal view of the rodent brain). Further, connections are established with the hypothalamus, another collection of brain nuclei involved in fundamental behaviours such as sex, fighting and food intake. The MOE efferent pathways travel to and relay through the main olfactory bulb (MOB) to inform the primary olfactory cortex, the amygdala and the hypothalamus. (Adapted from Dulac and Wagner, 2006.)

The Behavioural Consequences of Odour Detection

Odours elicit a variety of behavioural responses in animals. The next chapters report detailed information on the contribution of odours to food intake including prey detection (Chapter 6), sex and reproduction (Chapter 7), intergenerational interactions (Chapter 8), as well as disease and stress including predator detection (Chapters 9 and 10). To put into context the information given so far, we will present here some translational aspects of odour detection and their consequences for behaviour.

Me and the others

Odours released by an individual represent a very rich identity card, and therefore contribute to the definition of the self. Everybody possesses an individual odour. Only fiction has speculated on the deleterious effects of lacking one, e.g. the fictional character Jean-Baptiste Grenouille, the man with no odour in *Perfume: The Story of a Murderer* by German writer Patrick Süskind. Animals distinguish among individuals of their own or of a different species using olfactory cues. For example, trained mice can distinguish between the odours of individual red junglefowl (Karlsson *et al.*, 2010) and horses discriminate between body odour samples from unfamiliar conspecifics (Peron *et al.*, 2014). Through detection and analysis of olfactory cues animals will adapt their behaviour towards another animal, either known or unknown. Hence Antarctic prions (*Pachyptila desolata*) that live in dense colonies, find their nest burrows when they return from long overseas foraging trips by recognizing their partner's odour (Bonadonna *et al.*, 2007). African elephants use odours and other cues to answer the question 'Who is my enemy?', as they show more fear responses when they smell garments

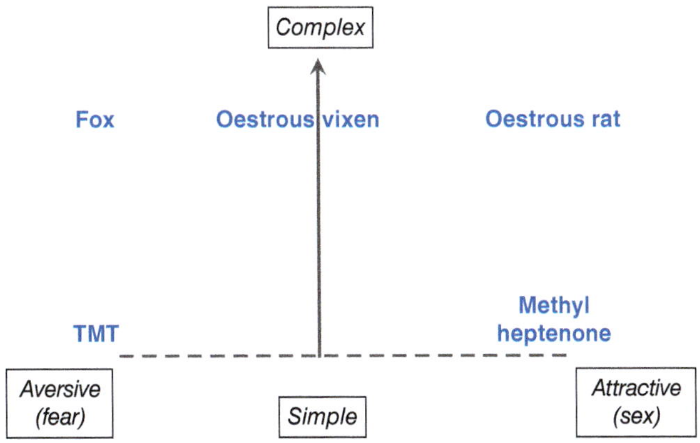

Fig. 1.5. Odour representations in a 2D space. The *x*-axis represents, in a male rodent's world, valence, from negative on the left (such as the odour of a predator eliciting fear) to positive on the right (the odour of an oestrous female is appealing). The *y*-axis represents complexity of an odour, from simple (a single odorant, bottom) to complex (natural odours, which are blends of hundreds to thousands of odorants). Fox odour frightens rats, and a single molecule, trimethylthiazoline (TMT), mimics the frightening effect of fox odour (Vernet-Maury *et al*., 1968). Odours from oestrous female rats sexually arouse male rats (Sachs *et al*., 1994), and a single molecule, methylheptenone, mimics the arousal effect of oestrous females (Nielsen *et al*., 2013). Male rats exposed to the odour of oestrous vixens display fear episodes intermingled with sexual arousal episodes (Rampin *et al*., 2006). Other axes may be represented, e.g. familiarity (with innate reactions to unknown odours at one end and conditioned reactions to odours linked to a reward at the other).

worn by men from hunter-breeder tribes than garments from agriculturalists (Bates *et al.*, 2007).

So, if you were a rat sniffing the animal next to you, you would probably be able to tell whether or not the animal (i) is a threat or a prey, (ii) is male or female, (iii) is immature or an adult, (iv) is a potential sexual partner now/later or never, (v) belongs to your family, and (vi) is in good health. Metabolites of proteins involved in the immune system may contribute to the odorant signature of an individual (Burger *et al.*, 2015), and selection of a partner for sex and reproduction may be influenced by the detection of his/her immune identity based on such olfactory cues (Wedekind and Furi, 1997). However, while those studies are almost 40 years old, they have been recently questioned and the individual identity of one animal may be more related to the variability of the protein it expresses. Indeed, fragments of many proteins not related to the immune system are present in urine and those fragments, detected by the VNO in mice, reflect the genetic uniqueness of one individual (Overath *et al.*, 2014). In addition, metabolites of sex steroids are likely to carry information on the sex and hormonal status of the releaser.

One or many odorants?

Is the importance of odours strong enough to initiate behaviour solely based on odour information? There are examples showing that an odour, even a single odorant, can be the unique trigger of behaviour. Bombykol, a pheromone released by the female silkworm moth *Bombyx mori*, elicits a complex series of behavioural and physiological responses of the male moth, including flight direction and genital activation (Butenandt *et al.*, 1961). The finding that one odorant alone can elicit a behavioural response suggests that it recruits a reflex-like, hard-wired and genetically programmed neural network. This neural network is very selective: bombykol is efficient but not bombykal, a molecule that is structurally very close. It is species dependent: bombykol attracts male moths from the bombyx species only. It is also sex oriented, as only males are attracted. And once the neural network is activated, it is not disrupted by other olfactory cues: the natural background odour bouquet of the environment does not influence the flight of male noctuid moths towards the female (Badeke *et al.*, 2016). The rabbit mammary pheromone, 2-methylbut-2-enal, elicits suction behaviour in the new-born rabbit (Schaal *et al.*, 2003), this example being described in more detail in Chapter 8. Both in the male moth and the new-born rabbit, the neural network is short-lived. In the male moth, this is due to the short life span of the animal, whereas in the new-born rabbit, the growth and development of the animal augments the panel of other senses, including hearing and vision. As a consequence, there is a shift in the balance of information processed by the brain, with the importance of olfactory inputs being challenged by other sensory inputs, creating a need to integrate and rank these different types of information.

In other species there are examples that not a single but a series of odorants is required to coordinate adapted behavioural responses within a given context. For example, in reproductive behaviour, females and males exchange olfactory cues that influence the nature of subsequent sexual interactions. In fish, a precise temporal pattern of sex pheromones released by females and males drives the production of gametes and sexual behaviour through interactions between the sexes (Sorensen *et al.*, 1998). Sexually active elephant bulls release odorants such as frontalin that are an attractant to females (Rasmussen and Greenwood, 2003) and, in turn, female elephants release a pre-ovulatory pheromone that is an attractant to males (Rasmussen, 2001).

In contrast to when a single odorant induces behaviour, several odorants may activate different olfactory information pathways in the brain, affecting the same or different brain nuclei, so as to elicit one or several components of behaviour. A simple interpretation is that different odorants recruit many, parallel reflex pathways. Nevertheless, for the sake of integration and coordination, the neural networks underlying these reflex pathways need to be interconnected. As evidenced by the examples above, some of these pathways are sex specific. It suggests that their organization and their activation have been influenced either genetically or through sex hormones. Studies of fish reproduction suggest that these neural pathways may be recruited sequentially during the course of reproductive behaviour. So in contrast to the moth model, sexual odours with multiple odorants should trigger a memorization of past information so that the animals progress in their behaviour, and do not stay at the same point or repeat indefinitely the same behavioural sequence. Such neural networks, which are more complex, may be less fixed than those found in moths and thus be more prone to external adjustments: European male starlings use milfoil (commonly known as yarrow) for courtship and nest building, and their attractiveness to this plant odour exists only during the breeding season (De Groof *et al.*, 2010).

We have seen that the brain of an individual has the capacity to integrate information from different odours gathered over time, so as to elaborate an adaptive behavioural response. A single source can release different odorants as a function of time, with each odorant eliciting behavioural responses in different species. An example is decomposition of a body, which releases a variety of odorants, and is accompanied by colonization by arthropods. The temporal pattern of arthropod colonization follows the changes of the odorant bouquet released by the corpse and/or by previous arthropods. This could be viewed as a model of a 'meta-brain' made from the sum of all the brains of the arthropods that have sequentially moved towards the decomposing body, and this model could illustrate the series of reflex pathways activated by different odorants within a single brain of, say, a fish or an elephant.

One can use the odour of others for personal benefit (so beware of perfumes!). Bolas spiders feed on male moths from different species that they attract by releasing odorants that mimic the female moth's sex pheromones (Haynes *et al.*, 2002). A single spider feeds on different moth species, and thus releases different odorants. However, some odorants that are attractants for males of one moth species may be repulsive for male moths of another species. So spiders are not only able to synthesize odorants found in several moth species, but also to decrease the production of some of these odorants aversive to potential future victims. Remember that the male moth responds in an unchangeable manner: once he has smelt the female pheromone, his neural network is activated onto a one-way road, driving the male moth towards the spider, and no alternative is possible. In contrast, the spider appears to have a much more accommodating neural network. Another example of using the odours of others for personal benefit comes from beetles of the *Neopyrochroa flabellata* species. The males feed on Spanish flies and concentrate their victims' noxious chemical, cantharidin, in cephalic glands. During courtship, female beetles sample the male's gland, and preferentially mate with cantharidin-positive males (Eisner *et al.*, 1996). In this case, the animals do not synthesize the odorant of interest, but extract it and concentrate it from their environment. If males predate on Spanish flies because they are attracted by cantharidin, should one consider that females in fact predate on males? Finally, sea birds can locate krill to feed on by picking up a trail of dimethylsulfide, an odorous gas emitted by phytoplankton, which is the main feed source for krill (Nevitt, 2000). In this case, the animals use an indirect olfactory cue to target food. Other fascinating examples of the olfactory capability of pelagic sea birds can be found in Chapter 4.

There are contexts in which a single odour source releases several odorants that convey conflicting information. Laboratory rats not only display fear, but also sexual arousal when exposed to the odour of oestrus vixen faeces (Rampin *et al.*, 2006; Fig. 1.5). In this model there is no apparent reciprocal inhibition of fear and sexual arousal, suggesting that the neural networks that support the two behavioural responses are mainly independent. In rodents, the processing of conflicting stimuli (e.g. fear and sex) by parallel pathways has been demonstrated in the amygdala (Choi *et al.*, 2005), a brain structure that receives olfactory inputs. Parallel pathways remain separate in the amygdala, but then converge onto another structure, the ventromedial hypothalamus, which may represent the integrative and decision centre. This can also give rise to conflicting information. For example, cows in oestrus are mounted by other cows in the herd, whereas cows in di-oestrus are not. If a di-oestrous cow is rubbed with her own vaginal mucus collected while she was in oestrus, then other cows in the herd mount her. However, if she is rubbed with the oestrous vaginal mucus from a different cow, then her herd-mates do not mount her (Nishimura *et al.*, 1991). This would indicate that in the brains of these cows, there is a conflict between oestrus-related odorants and individual distinctive odours.

Odours and what else?

The importance of odours in the information world of animals depends on where they live. Animals that are nocturnal or living in burrows or in the deepest of oceans with no access to light mainly rely upon and use their chemical senses, as well as touch and hearing, to interact with the environment. This is the case for many fishes (Nosal *et al.*, 2016) and typical of moles, which are nearly blind and use the variation in odour arriving at each nostril to hunt worms (Catania, 2013).

Most of the time, animals will use available information through a variety of senses to adapt their behaviour. The use of those senses can be balanced as a function of space or time. An example is sharks hunting for prey. They use a combination of senses according to their distance to the prey (Fig. 1.6), and the spatiotemporal sequence of the stimuli they use is first odours integrated with water turbulence sensed through their lateral line, then vision, hearing, electric field sensitivity and, finally, touch (Gardiner *et al.*, 2014). Another example comes from sea turtles. While at sea, they use geomagnetic and chemical senses (Endres *et al.*, 2016) to reach the tiny islands they will nest on. Geomagnetic information leads turtles in the vicinity of the islands and airborne and waterborne odours allow the retrieval of the exact nest location. Mice interact with other mice mostly through olfaction, and when their

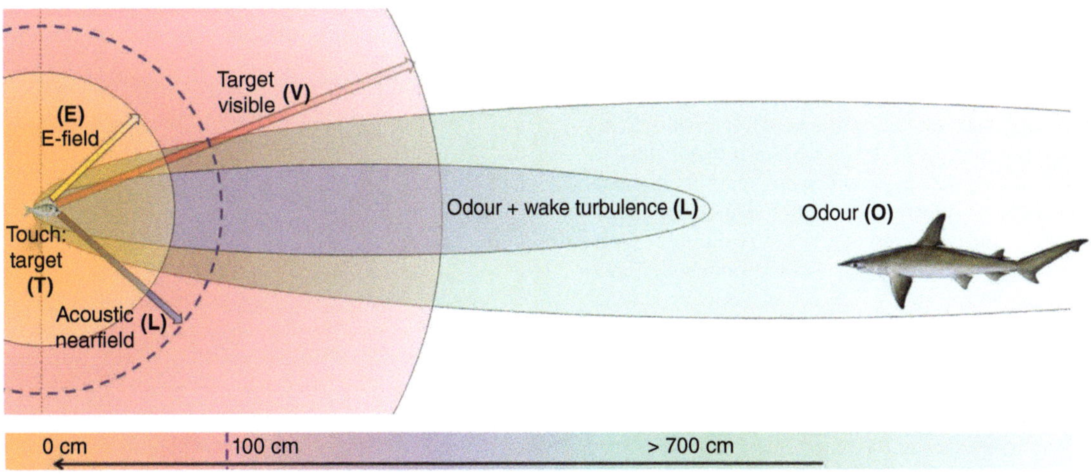

Fig. 1.6. A shark uses different types of information from the environment to catch prey. First, the shark tracks the prey from long distance based on the prey odour carried by the water current that the shark swims against. About 10 m from its prey, the shark uses visual and acoustic clues to approach the prey. It finally catches it, guided by the electric signal released by all living animals (E-field) and its sense of touch. (Adapted from Gardiner *et al.*, 2014.)

olfaction is hindered through genetic manipulation they will no longer initiate social contact (Leypold *et al.*, 2002; Mandiyan *et al.*, 2005). This does not mean that social interactions in mice rely only on olfaction, as it is known that they communicate through ultrasonic vocalization. Notably, a recent study shows that females use vocalization to induce paternal parental care from their male partner (Liu *et al.*, 2013). On the other hand, animals with one particularly acute sense will favour this sense over olfaction, such as seen in raptors, which have a highly developed visual system and therefore rely much less on other senses to hunt their prey.

Does this mean that animals with a highly developed visual system such as humans have poor olfactory capacities? Olfactory loss in humans as well as olfactory perturbation in mice (Fleischmann *et al.*, 2008) are correlated with signs of depression (Croy *et al.*, 2014), demonstrating its importance (see also Chapter 14). A recent study carried out with brave students shows that humans, if they abandon their bipedalism for a while, are capable of tracking scents successfully on the ground using strategies similar to those used by dogs (Porter *et al.*, 2007). Thus, even if olfaction is not the dominant sense of a given animal species, it still contributes to information upon which different behaviours are initiated.

Perspectives: Building Biosensors Using the Specificity of Olfactory Receptors

The high affinity and specificity of interactions between olfactory receptors and odorants make the olfactory receptor family an attractive candidate to build biosensors. The coupling of electronic devices and grafted olfactory receptors can potentially make detectors with lower sensitivity than existing chemical methods such as gas chromatography mass spectrometry (GC-MS). Furthermore, biosensors could be cheaper and with fewer constraints allowing direct analyses of samples collected in the field. The extent of application of biosensors is huge (Glatz and Bailey-Hill, 2010) and one area is the detection of odours linked to a particular animal state such as its reproductive or health status (see also Chapter 9). In farm animals, the detection of oestrus governs the timing of artificial insemination. In some species, oestrus detection is not easy, and methods have been developed that use a variety of devices to measure changes related to reproductive state, including vaginal electrical resistance, hormones in milk (Friggens and Chagunda, 2005) as well as locomotion. The detection of odorants that would characterize the oestrus stage using biosensors would be a promising alternative and some advances have been made already (Nielsen *et al.*, 2011, 2013). These odorants still need to be characterized better and their specific olfactory receptors identified. This information is likely to be acquired via fundamental research in the field of olfaction.

References

Araneda, R.C., Kini, A.D. and Firestein, S. (2000) The molecular receptive range of an odorant receptor. *Nature Neuroscience* 3, 1248–1255. DOI: 10.1038/81774

Badeke, E., Haverkamp, A., Hansson, B.S. and Sachse, S. (2016) A challenge for a male noctuid moth? Discerning the female sex pheromone against the background of plant volatiles. *Frontiers in Physiology* 7, 143. DOI: 10.3389/fphys.2016.00143

Bates, L.A., Sayialel, K.N., Njiraini, N.W., Moss, C.J., Poole, J.H. and Byrne, R.W. (2007) Elephants classify human ethnic groups by odor and garment color. *Current Biology* 17, 1938–1942. DOI: 10.1016/j.cub.2007.09.060

Bonadonna, F., Miguel, E., Grosbois, V., Jouventin, P. and Bessiere, J.M. (2007) Individual odor recognition in birds: an endogenous olfactory signature on petrels' feathers? *Journal of Chemical Ecology* 33, 1819–1829. DOI: 10.1007/s10886-007-9345-7

Buck, L. and Axel, R. (1991) A novel multigene family may encode odorant receptors: a molecular basis for odor recognition. *Cell* 65, 175–187. DOI: 10.1016/0092-8674(91)90418-X

Burger, D., Dolivo, G., Marti, E., Sieme, H. and Wedekind, C. (2015) Female major histocompatibility complex type affects male testosterone levels and sperm number in the horse (*Equus caballus*). *Proceedings of the Royal Society B: Biological Sciences* 282, 20150407. DOI: 10.1098/rspb.2015.0407

Butenandt, A., Beckmann, R. and Hecker, E. (1961) [On the sex attractant of silk-moths. I. The biological test and the isolation of the pure sex-attractant bombykol]. *Hoppe-Seyler's Zeitschrift fur physiologische Chemie* 324, 71–83.

Catania, K.C. (2013) Stereo and serial sniffing guide navigation to an odour source in a mammal. *Nature Communications* 4, 1441. DOI: 10.1038/ncomms2444

Choi, G.B., Dong, H.W., Murphy, A.J., Valenzuela, D.M., Yancopoulos, G.D., Swanson, L.W. and Anderson, D.J. (2005) Lhx6 delineates a pathway mediating innate reproductive behaviors from the amygdala to the hypothalamus. *Neuron* 46, 647–660. DOI: 10.1016/j.neuron.2005.04.011

Croy, I., Nordin, S. and Hummel, T. (2014) Olfactory disorders and quality of life – an updated review. *Chemical Senses*, 39, 185–194. DOI: 10.1093/chemse/bjt072

De Groof, G., Gwinner, H., Steiger, S., Kempenaers, B. and Van der Linden, A. (2010) Neural correlates of behavioural olfactory sensitivity changes seasonally in European starlings. *PLoS ONE* 5, e14337. DOI: 10.1371/journal.pone.0014337

Dulac, C. and Wagner, S. (2006) Genetic analysis of brain circuits underlying pheromone signaling. *Annual Review of Genetics* 40, 449–467. DOI: 10.1146/annurev.genet.39.073003.093937

Eisner, T., Smedley, S.R., Young, D.K., Eisner, M., Roach, B. and Meinwald, J. (1996) Chemical basis of courtship in a beetle (*Neopyrochroa flabellata*): cantharidin as precopulatory 'enticing' agent. *Proceedings of the National Academy of Sciences USA* 93, 6494–6498. DOI: 10.1073/pnas.93.13.6494

Endres, C.S., Putman, N.F., Ernst, D.A., Kurth, J.A., Lohmann, C.M. and Lohmann, K.J. (2016) Multi-modal homing in sea turtles: modeling dual use of geomagnetic and chemical cues in island-finding. *Frontiers in Behavioral Neuroscience* 10, 19. DOI: 10.3389/fnbeh.2016.00019

Fleischmann, A., Shykind, B.M., Sosulski, D.L., Franks, K.M., Glinka, M.E., Mei, D.F., Sun, Y., Kirkland, J., Mendelsohn, M., Albers, M.W. and Axel, R. (2008) Mice with a 'monoclonal nose': perturbations in an olfactory map impair odor discrimination. *Neuron* 60, 1068–1081. DOI: 10.1016/j.neuron.2008.10.046

Friggens, N.C. and Chagunda, M.G. (2005) Prediction of the reproductive status of cattle on the basis of milk progesterone measures: model description. *Theriogenology* 64, 155–190. DOI: 10.1016/j.theriogenology.2004.11.014

Gardiner, J.M., Atema, J., Hueter, R.E. and Motta, P.J. (2014) Multisensory integration and behavioural plasticity in sharks from different ecological niches. *PLoS ONE* 9, e93036. DOI: 10.1371/journal.pone.0093036

Glatz, R. and Bailey-Hill, K. (2011) Mimicking nature's noses: from receptor deorphaning to olfactory biosensing. *Progress in Neurobiology* 93, 270–296. DOI: 10.1016/j.pneurobio.2010.11.004

Gottfried, J.A., O'Doherty, J. and Dolan, R.J. (2002) Appetitive and aversive olfactory learning in humans studied using event-related functional magnetic resonance imaging. *Journal of Neuroscience* 22, 10829–10837.

Gottfried, J.A., Winston, J.S. and Dolan, R.J. (2006) Dissociable codes of odor quality and odorant structure in human piriform cortex. *Neuron* 49, 467–479. DOI: 10.1016/j.neuron.2006.01.007

Haynes, K.F., Gemeno, C., Yeargan, K.V., Millar, J.G. and Johnson, K.M. (2002) Aggressive chemical mimicry of moth pheromones by a bolas spider: how does this specialist predator attract more than one species of prey? *Chemoecology* 12, 99–105. DOI: 10.1007/s00049-002-8332-2

Igarashi, K.M., Ieki, N., An, M., Yamaguchi, Y., Nagayama, S., Kobayakawa, K., Kobayakawa, R., Tanifuji, M., Sakano, H., Chen, W.R. and Mori, K. (2012) Parallel mitral and tufted cell pathways route distinct odor information to different targets in the olfactory cortex. *Journal of Neuroscience* 32, 7970–7985. DOI: 10.1523/jneurosci.0154-12.2012

Karban, R., Shiojiri, K., Ishizaki, S., Wetzel, W.C. and Evans, R.Y. (2013) Kin recognition affects plant communication and defence. *Proceedings of the Royal Society B: Biological Sciences* 280, 20123062. DOI: 10.1098/rspb.2012.3062

Karlsson, A.C., Jensen, P., Elgland, M., Laur, K., Fyrner, T., Konradsson, P. and Laska, M. (2010) Red junglefowl have individual body odors. *Journal of Experimental Biology* 213, 1619–1624. DOI: 10.1242/jeb.040279

Keller, A. and Vosshall, L.B. (2016) Olfactory perception of chemically diverse molecules. *BMC Neuroscience* 17, 55. DOI: 10.1186/s12868-016-0287-2

Kobayakawa, K., Kobayakawa, R., Matsumoto, H., Oka, Y., Imai, T., Ikawa, M., Okabe, M., Ikeda, T., Itohara, S., Kikusui, T., Mori, K. and Sakano, H. (2007) Innate versus learned odour processing in the mouse olfactory bulb. *Nature* 450, 503–508. DOI: 10.1038/nature06281

Leypold, B.G., Yu, C.R., Leinders-Zufall, T., Kim, M.M., Zufall, F. and Axel, R. (2002) Altered sexual and social behaviors in trp2 mutant mice. *Proceedings of the National Academy of Sciences USA* 99, 6376–6381. DOI: 10.1073/pnas.082127599

Liu, H.X., Lopatina, O., Higashida, C., Fujimoto, H., Akther, S. *et al.* (2013) Displays of paternal mouse pup retrieval following communicative interaction with maternal mates. *Nature Communications* 4, 1346. DOI: 10.1038/ncomms2336

Livermore, A. and Laing, D.G. (1996) Influence of training and experience on the perception of multicomponent odor mixtures. *Journal of Experimental Psychology: Human Perception and Performance* 22, 267–277. DOI: 10.1037/0096-1523.22.2.267

Mainland, J.D., Lundstrom, J.N., Reisert, J. and Lowe, G. (2014) From molecule to mind: an integrative perspective on odor intensity. *Trends in Neuroscience* 37, 443–454. DOI: 10.1016/j.tins.2014.05.005

Mandiyan, V.S., Coats, J.K. and Shah, N.M. (2005) Deficits in sexual and aggressive behaviors in Cnga2 mutant mice. *Nature Neuroscience* 8, 1660–1662. DOI: 10.1038/nn1589

McGann, J.P. (2015) Associative learning and sensory neuroplasticity: how does it happen and what is it good for? *Learning and Memory* 22, 567–576. DOI: 10.1101/lm.039636.115

Mori, K. and Sakano, H. (2011) How is the olfactory map formed and interpreted in the mammalian brain? *Annual Review of Neuroscience* 34, 467–499. DOI: 10.1146/annurev-neuro-112210-112917

Morrison, E.E. and Costanzo, R.M. (1990) Morphology of the human olfactory epithelium. *The Journal of Comparative Neurology* 297, 1–13. DOI: 10.1002/cne.902970102

Mucignat-Caretta, C. (2010) The rodent accessory olfactory system. *Journal of Comparative Physiology A.* 196, 767–777. DOI: 10.1007/s00359-010-0555-z

Nevitt, G.A. (2000) Olfactory foraging by Antarctic procellariiform seabirds: life at high Reynolds numbers. *The Biological Bulletin* 198, 245–253. DOI: 10.2307/1542527

Nielsen, B.L., Jerôme, N., Saint-Albin, A., Thonat, C., Briant, C., Boue, F., Rampin, O. and Maurin, Y. (2011) A mixture of odorant molecules potentially indicating oestrus in mammals elicits penile erections in male rats. *Behavioral Brain Research* 225, 584–589. DOI: 10.1016/j.bbr.2011.08.026

Nielsen, B.L., Jerôme, N., Saint-Albin, A., Rampin, O. and Maurin, Y. (2013) Behavioural response of sexually naive and experienced male rats to the smell of 6-methyl-5-hepten-2-one and female rat faeces. *Physiology and Behavior* 120, 150–155. DOI: 10.1016/j.physbeh.2013.07.012

Niimura, Y., Matsui, A. and Touhara, K. (2014) Extreme expansion of the olfactory receptor gene repertoire in African elephants and evolutionary dynamics of orthologous gene groups in 13 placental mammals. *Genome Research* 24, 1485–1496. DOI: 10.1101/gr.169532.113

Nishimura, K., Utsumi, K., Okano, T. and Iritani, A. (1991) Separation of mounting-inducing pheromones of vaginal mucus from estrual heifers. *Journal of Animal Science* 69, 3343–3347.

Nosal, A.P., Chao, Y., Farrara, J.D., Chai, F. and Hastings, P.A. (2016) Olfaction contributes to pelagic navigation in a coastal shark. *PLoS ONE* 11, e0143758. DOI: 10.1371/journal.pone.0143758

Nuñez-Parra, A., Li, A. and Restrepo, D. (2014) Coding odor identity and odor value in awake rodents. *Progress in Brain Research* 208, 205–222. DOI: 10.1016/B978-0-444-63350-7.00008-5

Overath, P., Sturm, T. and Rammensee, H.G. (2014) Of volatiles and peptides: in search for MHC-dependent olfactory signals in social communication. *Cellular and Molecular Life Sciences* 71, 2429-2442. DOI: 10.1007/s00018-014-1559-6

Peron, F., Ward, R. and Burman, O. (2014) Horses (*Equus caballus*) discriminate body odour cues from conspecifics. *Animal Cognition* 17, 1007–1011. DOI: 10.1007/s10071-013-0717-9

Porter, J., Craven, B., Khan, R.M., Chang, S.J., Kang, I., Judkewitz, B., Volpe, J., Settles, G. and Sobel, N. (2007) Mechanisms of scent-tracking in humans. *Nature Neuroscience* 10, 27–29. DOI: 10.1038/nn1819

Rampin, O., Jerôme, N., Briant, C., Boué, F. and Maurin, Y. (2006) Are oestrus odours species specific? *Behavioural Brain Research* 172, 169–172. DOI: 10.1016/j.bbr.2006.04.005

Rasmussen, L.E. (2001) Source and cyclic release pattern of (Z)-7-dodecenyl acetate, the pre-ovulatory pheromone of the female Asian elephant. *Chemical Senses* 26, 611–623. DOI: 10.1093/chemse/26.6.611

Rasmussen, L.E. and Greenwood, D.R. (2003) Frontalin: a chemical message of musth in Asian elephants (*Elephas maximus*). *Chemical Senses* 28, 433–446. DOI: 10.1093/chemse/28.5.433

Sachs, B.D., Akasofu, K., Citron, J.H., Daniels, S.B. and Natoli, J.H. (1994) Noncontact stimulation from estrous females evokes penile erection in rats. *Physiology and Behavior* 55, 1073–1079. DOI: 10.1016/0031-9384(94)90390-5

Saito, H., Chi, Q., Zhuang, H., Matsunami, H. and Mainland, J.D. (2009) Odor coding by a mammalian receptor repertoire. *Science Signaling* 2, ra9. DOI: 10.1126/scisignal.2000016

Sato-Akuhara, N., Horio, N., Kato-Namba, A., Yoshikawa, K., Niimura, Y., Ihara, S., Shirasu, M. and Touhara, K. (2016) Ligand specificity and evolution of mammalian musk odor receptors: effect of single receptor deletion on odor detection. *Journal of Neuroscience* 36, 4482–4491. DOI: 10.1523/jneurosci.3259-15.2016

Schaal, B., Coureaud, G., Langlois, D., Ginies, C., Semon, E. and Perrier, G. (2003) Chemical and behavioural characterization of the rabbit mammary pheromone. *Nature* 424, 68–72. DOI: 10.1038/nature01739

Schoenbaum, G., Chiba, A.A. and Gallagher, M. (1999) Neural encoding in orbitofrontal cortex and basolateral amygdala during olfactory discrimination learning. *Journal of Neuroscience* 19, 1876–1884.

Setlow, B., Schoenbaum, G. and Gallagher, M. (2003) Neural encoding in ventral striatum during olfactory discrimination learning. *Neuron* 38, 625–636. DOI: 10.1016/S0896-6273(03)00264-2

Sorensen, P.W., Christensen, T.A. and Stacey, N.E. (1998) Discrimination of pheromonal cues in fish: emerging parallels with insects. Current Opinion in Neurobiology 8, 458–467. DOI: 10.1016/S0959-4388(98)80032-9

Sosulski, D.L., Bloom, M.L., Cutforth, T., Axel, R. and Datta, S.R. (2011) Distinct representations of olfactory information in different cortical centres. *Nature* 472, 213–216. DOI: 10.1038/nature09868

Su, C.Y., Menuz, K. and Carlson, J.R. (2009) Olfactory perception: receptors, cells, and circuits. *Cell* 139, 45–59. DOI: 10.1016/j.cell.2009.09.015

Ubeda-Bañon, I., Pro-Sistiaga, P., Mohedano-Moriano, A., Saiz-Sanchez, D., de la Rosa-Prieto, C., Gutierrez-Castellanos, N., Lanuza, E., Martinez-Garcia, F. and Martinez-Marcos, A. (2011) Cladistic analysis of olfactory and vomeronasal systems. *Frontiers in Neuroanatomy* 5, 3. DOI: 10.3389/fnana.2011.00003

Vernet-Maury, E., Le Magnen, J. and Chanel, J. (1968) [Emotional behavior of rats; influence of the odor of a predator and a non-predator]. *Comptes rendus hebdomadaires des séances de l'Académie des sciences. Série D: Sciences naturelles* 267, 331–334.

Waters, C.M. and Bassler, B.L. (2005) Quorum sensing: cell-to-cell communication in bacteria. *Annual Review of Cell and Developmental Biology*, 21, 319–346. DOI: 10.1146/annurev.cellbio.21.012704.131001

Wedekind, C. and Furi, S. (1997) Body odour preferences in men and women: do they aim for specific MHC combinations or simply heterozygosity? *Proceedings of the Royal Society B: Biological Sciences* 264, 1471–1479. DOI: 10.1098/rspb.1997.0204

Wilson, D.A. and Sullivan, R.M. (2011) Cortical processing of odor objects. *Neuron* 72, 506–519. DOI: 10.1016/j.neuron.2011.10.027

Yee, K.K., Craven, B.A., Wysocki, C.J. and Van Valkenburgh, B. (2016) Comparative morphology and histology of the nasal fossa in four mammals: gray squirrel, bobcat, coyote and white-tailed deer. *The Anatomical Record (Hoboken)* 299, 840–852. DOI: 10.1002/ar.23352

2 Innateness and Learning in Olfactory Behaviour and Odour Perception

Birte L. Nielsen

Neurobiology of Olfaction, INRA, Université Paris-Saclay, Jouy-en-Josas, France

Olfaction and odours influence the behaviour of most animals, even in species not usually famous for their ability to smell. Sheep, for example, are able to discriminate between conspecifics based on their odour alone (Baldwin and Meese, 1977). Also, we rarely think of birds as olfactory beings, but odours can be quite important for avian behaviour, and there is even evidence to suggest that olfaction is involved in the migratory behaviour of certain bird species (Wikelski *et al.*, 2015). As it happens, an interesting example of odour learning and use of smell comes from colony roosting sea birds: Bonadonna *et al.* (2004) found that blue petrels can find their nesting burrow based on smell alone, without acoustic or visual cues. This arctic bird species nests in burrows underground, and is not known for its visual acuity; occasionally colliding with upright obstacles on the shore. In addition, they prefer to forage on moonless nights to avoid predation. So an individual petrel returning from a night of foraging is guided by the smell of its own burrow to locate its entrance hole among the thousands of such holes in close proximity, which make up the petrel colony (Bonadonna *et al.*, 2004). As also mentioned in Chapter 1, this ability appears to be based on recognizing the odour of their partner (Bonadonna *et al.*, 2007).

The blue petrel identifying the signature odour of its own nest is an example of how learning about certain odours can influence their meaning. We learned in Chapter 1 that different species have different olfactory capacities, consequently leading to differences in the importance of chemical communication and information. There is also variation between animals in their ability to learn, both between but also within species, and for most odours, their meaning (i.e. the biologically relevant information carried by the odour) is rarely inherent, but depends on the context in which the odour is encountered and may change as the animal develops. Some examples of this are presented and discussed in this chapter.

Innate Response to an Odour

As humans, when encountering a smell, we may categorize it as being (more or less) pleasant, (more or less) aversive or simply neutral. For most odours we do not remember the first time we smelled it, and we probably think that we have always considered a given odour as good smelling or bad smelling. However, when they are first encountered, surprisingly few odours give rise to an immediate response of like and dislike, and in Chapter 5 some of the cultural aspects of odour perception in humans will be presented. In non-human animals, the most well-known odours giving rise to an immediate, unconditioned (i.e. not learned) response are those produced by predators. The innateness of this response makes sense in evolutionary terms because inability to react appropriately the first time a predator approaches could likely result in death. Rodents are the most commonly used model for this type of research, and the predator smells most widely studied are those emitted by cats and foxes (but see Apfelbach *et al.*, 2015 for other prey and predator species). These predator odours originate from urine, faeces and fur (e.g. Christensen and Rundgren, 2008; Wernecke *et al.*, 2015). The first volatile odorant compound to be identified as an innate predator odour was trimethylthiazoline (better known as TMT), found in fox faeces by Évelyne Vernet-Maury (1980). This molecule is detectable by rats in very dilute concentrations ($4.6*10^{-15}$ mol/l; Laska *et al.*, 2005), and has been used in a large number of studies (see the review by Fendt and Endres, 2008). The fear response of rodents induced by TMT is distinguishable from the aversive responses seen when exposed to the odour of spoiled food, thus illustrating differences in the innate and learned processing of these odours in the brain (Fendt *et al.*, 2003).

One study investigating the innateness of predator odours in the wild was carried out by Chamaillé-Jammes *et al.* (2014). They chose a remote island off the west coast of Canada where a population of black-tailed (sitka) deer had been living for over 90 years. Wolves had never been present in this group of islands, whereas brown bears lived in the area. The researchers wanted to know if the deer showed innate threat-sensitive foraging with respect to these two predators. They therefore set up 98 bait stations, where a circle (1 m diameter) of the flat ground was sprayed with one of four different odours, including urine from bears and wolves, with water being used as the control treatment. Pieces of apple were spread on top of the odorized area and small open-top containers holding the odour were pushed into the ground to ensure a continuous perfusion of the smell. Following odorization and baiting, the first visit to each station by a deer was monitored using motion-detector cameras recording videos at 1 frame/s. Figure 2.1 shows the percentage of sniffing observed at these visits and the proportion of apple pieces eaten by the deer. It is clear that the response of the deer to the wolf odour is greater than that seen with water and bear urine. Two additional odours tested (cologne and petrol) did not yield results different from those obtained with water and bear urine, indicating that it is not the novelty of wolf odour that gives rise to the observed behaviour. One explanation for the stronger response is that wolves are a more dangerous threat than black bears to the deer, and despite the long absence of wolves, their odour has remained innately fear-inducing for the deer. The authors add that the different responses to the two predator odours are not related to risk of encounter as wolves range more than black bears, and thus are less likely to be found close to the urination sites (Chamaillé-Jammes *et al.*, 2014).

Changes to Innate Odour Valence via Infection

An innate behavioural response to an odour, such as avoidance of some predator odours as described above, can still be modulated. For example, in

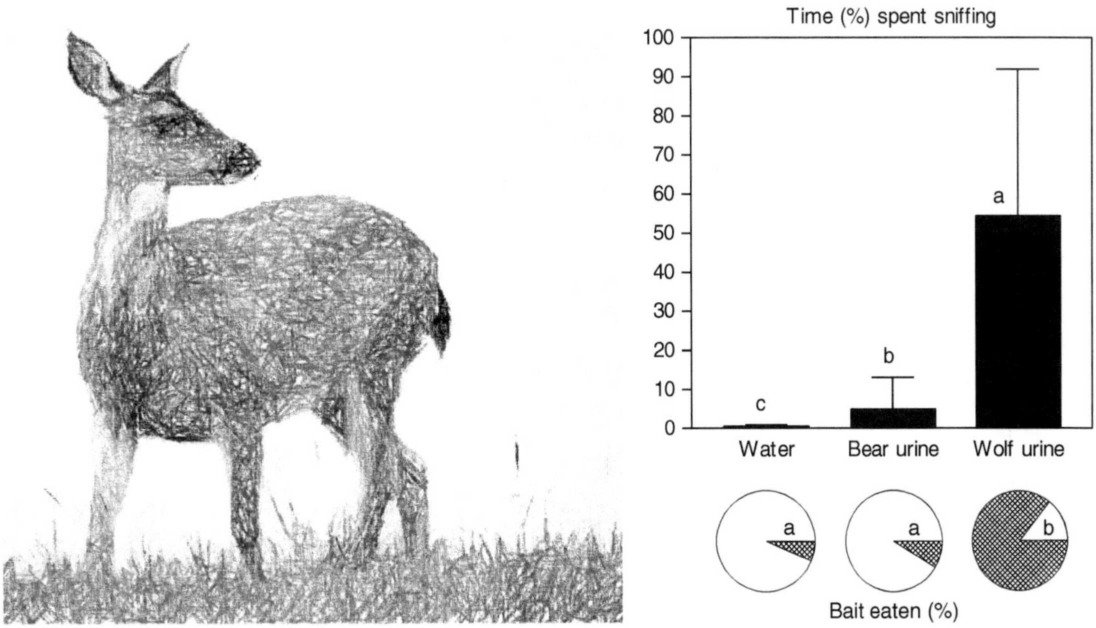

Fig. 2.1. Behavioural response of black-tailed deer when visiting feeding stations sprayed with water (control), bear urine or wolf urine, and baited with apple pieces. The bar chart show the percentage of time the deer spent sniffing, and the pie charts show the percentage of bait eaten (in white) for the three odour treatments, respectively. Letters indicate significant differences. (Based on data from Chamaillé-Jammes *et al.*, 2014, and a photo by Steve Hillebrand.)

contrast to mice reared in standard laboratory rodent cages, those raised in a complex and enriched environment show an attenuated response to TMT (Sotnikov *et al.*, 2014). We now know a lot about how changes in the perception of fear-inducing odours are processed in the brain (Takahashi, 2014). Changes have been found in the behavioural response of mice to cat odour when the rodents have been infected with the protozoan parasite *Toxoplasma gondii* (Vyas *et al.*, 2007). Although sexual reproduction of this parasite occurs only in cats, it can infect many warm-blooded animals, including sheep, poultry and humans (Worth *et al.*, 2013). When rodents are the intermediate host, the parasite induces behavioural changes that make the mouse or rat more likely to be eaten by the ultimate host by reducing their anxiety and fear of cats, including cat odours (although this has recently been challenged, see Worth *et al.*, 2013). Such parasite-induced changes in host behaviour are called adaptive host manipulation, and another example of this can be found in the liver fluke (Badie, 1975), where one of the intermediate hosts is a species of ant. Upon infection by the fluke, the ant begins to spend the nights at the top of blades of grass, thereby increasing the likelihood of being eaten by ruminants, the ultimate host for this fluke.

In the case of *T. gondii*, the infected animal undergoes epigenetic changes at the brain level leading to reduced fear of cat odour possibly through changes in odour perception (Hari Dass and Vyas, 2014). Poirotte *et al.* (2016) wanted to investigate if similar behavioural changes occurred in primates. They used adult male and female chimpanzees kept in five enclosures with two to nine animals per group. The infection status of all 33 animals was tested and within each group, between one and four animals were found to be infected with *T. gondii*. Samples of three types of urine were presented to the chimpanzees on the fence of their enclosure on three occasions and in random order. The urine samples used were from leopards, human males and from either lions or tigers, neither of which are natural predators of chimpanzees. The approaches of the chimpanzees towards the odours were recorded, together with their sniffing and licking of the odour source, as well as touching it and sniffing their hand. The observer was blind to the infection status of the animals, and data were analysed taking into account sex and dominance rank of the individual chimpanzee together with order of odour presentation. Chimpanzees infected with *T. gondii* were found to approach the leopard urine more than their uninfected group mates, whereas no differences were found for human and tiger or lion urine (Fig. 2.2). Infected chimpanzees also approached leopard urine more than the urine of the other felids, altogether indicating that the *T. gondii* protozoan may have induced an altered perception of the leopard urine, which would increase the likelihood of the chimpanzee being successfully predated by this member of the Felidae family.

Innate or Learned Response to an Odour?

As stated earlier, only a few odours give rise to an innate behavioural response, and these are mainly associated with predator avoidance. Another innate aspect of olfaction relates to specific anosmia, where certain individuals are genetically unable to smell a certain odour. An example of this is the distinctive smell of our urine after consumption of asparagus: Pelchat *et al.* (2011) found that roughly 6% of the population were unable to detect this particular odour; this was not associated with the ability to produce the smelly urine, which we all do, but linked to a single nucleotide polymorphism (SNP; a variation that occurs at a specific position in the genome). When a response to an odour *is* observed, it can be difficult to ascertain if the response is truly innate. When 3-day old human infants are presented with different odours, and tested for measures of like and dislike (based on changes in, for example, facial expressions and respiration rate), some odours may be inherently pleasant (vanillin) and aversive (butyric acid) at high concentrations (Soussignan *et al.*, 1997). There are, however, few odours that babies reject, and the results may be an artefact of butyric acid stimulating the trigeminal nerve, which provides sensory input to the face and mouth, at high concentrations (Alaoui-lsmaïli *et al.*, 1997). A clear divergent distinction was made by bottle-fed and breast-fed babies, respectively, when tested with these milk odours, and the authors state that 'olfactory preferences are largely due to previous exposure, even at this early age'.

In an elegantly constructed experiment, Logan *et al.* (2012) investigated whether the first suckling performed by new-born mice was odour-induced and if so, whether the behavioural response to the odour was innate. This study was inspired by the finding that new-born anosmic mice often did not initiate suckling and thus failed to survive (e.g. Wong

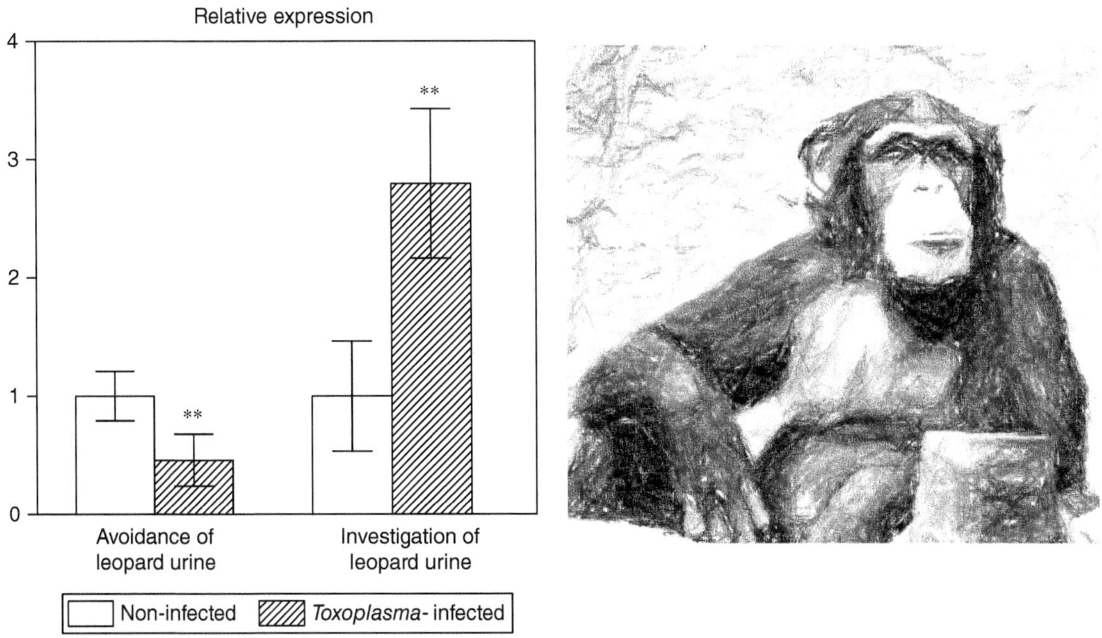

Fig. 2.2. Chimpanzees infected with the parasitic protozoan *Toxoplasma gondii* display less avoidance and more investigation towards the smell of urine from leopards, a natural predator of chimpanzees, than their non-infected conspecifics (**, $P < 0.01$). (Based on data from Poirotte *et al.*, 2016.)

et al., 2000), as well as the relatively recent identification of a specific suckling pheromone in rabbits (Schaal *et al.*, 2003; see also Chapter 3). Using normal, wild-type strains of laboratory mice, Logan *et al.* (2012) found that all mice pups initiated suckling within the first minute of being presented with the nipple of a dam 1 h after birth by caesarean section. However, if the nipple had been washed with clean water, 64% of pups did not begin to suckle within the 15-min test. By applying different odours to the washed nipple, the authors found that only amniotic fluid initiated suckling within a timeframe similar to that observed with unwashed nipples (Fig. 2.3A). When amniotic fluid was used, over 90% of pups began to suckle within the 2-min test, whereas the vast majority of pups did not initiate suckling when nipples were odorized with other odours of varying biological relevance. If the pups were themselves brushed with one of these odours within 5 min of birth, they did initiate suckling similar to that seen with amniotic fluid, except in the (non-odorant) water treatment (Fig. 2.3B). By feeding pregnant mice garlic- or vanillin-flavoured food, Logan *et al.* (2012) could establish that the suckling response initiated by an odour is not innate, but something the pup under normal circumstances learns *in utero*.

Male rats have been observed displaying sexual behaviour (penile erections), not only when exposed to the smell of faeces from a female rat in heat, but also when presented with faeces from female horses (mares) and foxes (vixens) in oestrus (Rampin *et al.*, 2006), pointing towards a certain commonality between oestrus odours. This response is also found in sexually naïve male rats (Nielsen *et al.*, 2016). Within 2 h of birth by caesarean section, rat pups have been found to probe oestrous females more than di-oestrous females (Fillion and Blass, 1986a) indicating that this could be an innate response to oestrus odour. However, given the results by Logan *et al.* (2012) presented earlier, and the finding that male rats reared with lemon-scented dams mate more readily with lemon-scented oestrous females (Fillion and Blass, 1986b), the reaction to oestrus odours may also be a learned response to signature odours acquired perinatally. A female rat usually has a single, post-partum oestrus around 18 h after giving birth (Connor and Davis, 1980), and this may prime the male offspring to recognize (certain components of) oestrus odour.

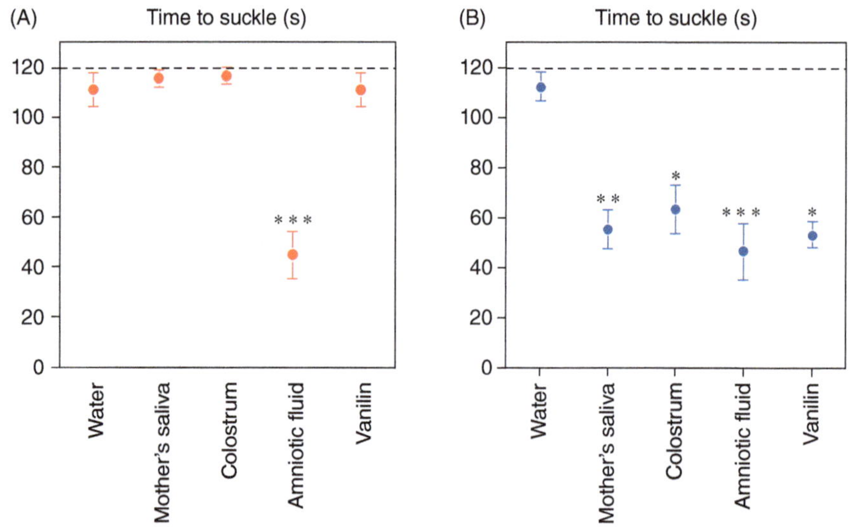

Fig. 2.3. (A) When the nipples of a lactating mouse are washed and subsequently swabbed with either water, her own saliva, the first milk (colostrum), amniotic fluid or a biologically irrelevant odour (vanillin), many of the new-born mouse pups do not begin to suckle within 2 min, and only when amniotic fluid is used to re-odorize the washed nipples do the pups begin to suckle sooner. (B) In contrast, when one of the same odours is used to odorize the pup immediately after birth (by caesarean section), as well as being swabbed on the nipples of the mother, all odours except water lead to faster initiation of suckling. Asterisks indicate statistically significant differences from the water-washed condition (*, $P < 0.05$; **, $P < 0.01$; ***, $P < 0.001$). (Based on data from Logan *et al*., 2012.)

Attraction to certain odours at specific time points is found not only in new-borns. In a variety of mammalian species including dogs and rodents, the smell of amniotic fluids is highly attractive to the dam at the time of parturition (Dunbar *et al.*, 1981; Stern, 1996). In cattle, there is evidence to suggest that the presence of birth fluids as a result of a previous calving influences the calving site selection of the cow (Rørvang *et al.*, 2017). The valence of this odour can change quite dramatically, as, for example, ewes are repelled by the smell of amniotic fluids at all times except immediately before giving birth and in the few hours (< 4h) afterwards (Levy *et al.*, 1983). The aversion observed at all other times appears to be controlled, at least in part, by the physiological state of the female, as ewes in oestrus show a slight reduction in their repulsion towards amniotic fluids (Rubianes, 1992). Indeed, simulating the birth process by cervical stimulation of non-pregnant ewes in oestrus significantly reduces the repulsiveness of amniotic fluids, although never to the point of becoming attractive (Table 2.1). This illustrates how the response to an odour can be influenced by mechanisms other than innateness and learning. The role of odours in connection with maternal behaviour and offspring survival is discussed further in Chapter 8.

Learned odour recognition

Some behavioural responses to odours are clearly learned, as illustrated by the previous example of male rats preferring a female with an odour similar to its mother (Fillion and Blass, 1986b). Such pairings of a neutral odour with an unconditioned stimulus can also happen in adult animals. Kippin and Pfaus (2001) found that male rats preferred to mate with female rats scented with almonds if the males had previously been allowed to copulate with almond-smelling females. The rewarding effect associated with copulation was also seen when comparing the response to oestrus odours of sexually naïve and experienced adult rats, the latter displaying a greater number of penile erections (Nielsen *et al.*, 2013). Conditioning a response to specific odours is commonly used in training animals, e.g. to detect oestrus in cows (dogs; Hawk *et al.*, 1984), to find hidden explosives (elephants; Miller *et al.*, 2015), narcotics (wasps; Bui and Takasu, 2009) and even bumblebee nests (dogs; Waters *et al.*, 2011), as well

Table 2.1. Percentage time spent near feed and amount eaten by non-pregnant ewes given a choice between 70 g of feed sprayed with water or with amniotic fluid. Of the ewes in oestrus, half received 20 min of cervical stimulation prior to the 4-min test. Within rows, different superscripts indicate significant differences. (Adapted from Rubianes, 1992.)

	Non-parturient ewes (n = 7)	Non-parturient ewes in oestrus (n = 9)	Non-parturient ewes in oestrus + cervical stimulation (n = 9)
Time spent (%) at trough with:			
– Feed sprayed with water	67 ± *6*	56 ± *5*	54 ± *3*
– Feed sprayed with amniotic fluid	3 ± *1*[b]	11 ± *5*[b]	24 ± *6*[a]
Consumption (g) of:			
– Feed sprayed with water	70 ± *0*	70 ± *0*	70 ± *0*
– Feed sprayed with amniotic fluid	0 ± *0*[b]	15 ± *11*[ab]	38 ± *15*[a]

as in the diagnosis of diseases, such as tuberculosis (African giant pouched rats; Mgode *et al.*, 2012) and cancer (dogs; Cornu *et al.*, 2011). When sniffer-animals are used for disease detection (see also Chapter 9), it is worth checking that it is indeed a general smell of the illness that is detected. Elliker *et al.* (2014) found that some of the dogs they had trained to detect prostate cancer in human urine samples had memorized the training set of samples and were unable to perform better than random choices when presented with novel urine samples. This, however, emphasizes the ability of macrosmatic animals to learn to recognize very specific odours.

The ability to associate certain odours with an event can also wreak havoc in experimental designs. As a graduate student I worked as a research assistant on a project concerning stereotypic behaviour in tethered sows. We arrived at the sow shed before the break of dawn and observed the behaviour of the sows in the hours before and after feeding. It was baffling how the sows began to show increased activity and vocalization approximately 20 min before they were fed, independent of the fact that feeding was not at a fixed time but differed by up to an hour from one day to the next. It took us embarrassingly long to realize that the sows were reacting to the smell of tobacco smoke from the caretaker, who arrived silently on foot in the farmyard whilst smoking his pipe.

Fig. 2.4. Pigs can learn to associate an arbitrary smell with imminent feeding. (Photo: Liat Romme Thomsen.)

Changes to odour valence through learning

The anecdote above where pigs (Fig. 2.4) learned to associate an otherwise neutral odour with an event such as feeding is an example of how learning can change the meaning of an odour. In the case of sexual preferences for females with a certain odour (such as almond-smelling rats; Kippin and Pfaus, 2001), one inherent experimental design problem is the simultaneous presentation of the odour and the sexually receptive female during both training and testing. This makes it difficult to ascertain if the odour truly has acquired another meaning for the animal tested. Kvitvik *et al.* (2010) tried to separate these effects using male rats and two odours: garlic oil and capelin oil, the latter being an extract of fish, which is aversive to most humans but, like the smell of garlic, neutral to rats. First they ensured the sexual experience of male rats by allowing them to copulate fully with female rats in oestrus on three occasions. They then submitted the rats to a conditioning schedule, where the male rat was put into a square arena with two adjacent cages attached, each holding a female in oestrus (Fig. 2.5A). This was to ensure that the smell of oestrus emanated from both sides of the arena, so that any subsequent choice by the male was not based on oestrus odours. Above each cage, a constant flow of air entered the arena,

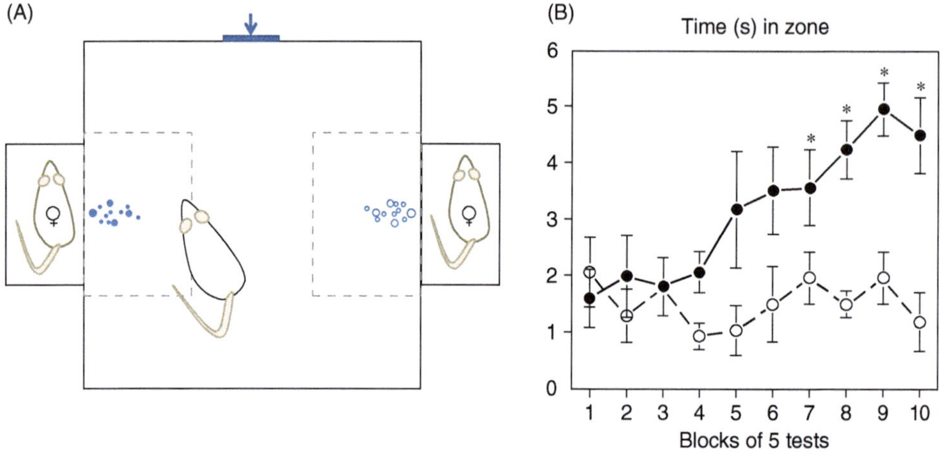

Fig. 2.5. (A) Sexually experienced male rats were tested in an arena, where two oestrous females were present in adjacent cages; the arrow indicates an extractor fan. Air was constantly flowing from two nozzles placed above each cage. After 5 min the air flow above one of the cages was odorized by the conditioning odour (either capelin or garlic oil) and immediately following this, the female from the odour side was released and the male allowed to mount the female until intromission. (B) Time (s; mean ± SE) spent in each of the two rectangular zones in front of the females during the 10 s of odour release: odorized zone (filled circles, solid line) and neutral (air) zone (open circles, stippled line). Asterisks indicate significant differences between zones ($P < 0.05$). (Based on data from Kvitvik *et al*., 2010.)

whilst a wall-mounted extractor fan evacuated the air in the arena. For each test, a male rat was left in the arena for 5 min, at which point the extractor fan stopped and the airflow above one of the adjacent cages was odorized by the conditioning odour (capelin for half the rats, and garlic for the other half) for a period of 10 s. Immediately hereafter, the extractor fan was restarted and the female from the cage on the side of the odour was released. The male was allowed to interact with the female for up to 2 min until intromission occurred, at which point the female was removed and the trial ended. Each rat completed up to 20 tests in any one session, and the side for odour release was randomly assigned. In Fig. 2.5B, the average time spent by the male rats in each of the rectangular zones in front of the females during the 10 s of odour release is presented. Over the course of testing, the male rats developed a clear preference for the odorized zone, demonstrating that they now associated the odour with the possibility for interaction with a sexually receptive female. In another group of rats, where the odours were presented independently of access to the female, no such preference developed (data not shown; Kvitvik *et al.*, 2010).

A similar change in odour valence through conditioning has also been achieved for cadaverine, which smells like putrid flesh and is an odour innately aversive to rats. However, when male rats were allowed to copulate with cadaverine-scented females, they subsequently found cadaverine attractive, whereas for males sexually trained with unscented females, cadaverine retained its repellent properties (Pfaus *et al.*, 2001). Interestingly, when neutral odours through conditioning are associated with sex, they do not evoke the same neural pathways as oestrus odours (Kippin *et al.*, 2003); perhaps indicating that the behavioural response observed with the female's natural odour is of a more fundamental nature. Also, it appears to be difficult, perhaps impossible to change the innate fear-inducing response in rodents to TMT, as Kobayakawa *et al.* (2007) were unable to train mice to associate this odour with a sugar reward (see Fig. 1.3). More recently, it has been found that the aversiveness of TMT appears to be neutralized if it is mixed with an odour the mice find attractive (Saraiva *et al.*, 2016; Fig. 2.6).

Sometimes it may be a question of changing one's own odour away from the norm. Ryon *et al.* (1986) found that wolves rubbed themselves in strong-smelling odours to a different extent depending on the type of odour: the strongest rubbing response was seen with non-natural odours (perfume and motor oil), whereas food smells (salt pork and tuna)

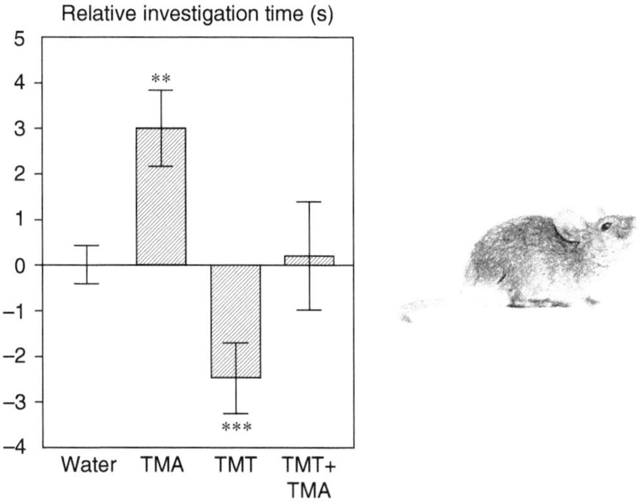

Fig. 2.6. In this test, the time mice (n = 5–11) spend investigating filter paper containing odorant or water is measured during a 3-min period. The average time spent investigating water (7.3 ± 0.4 s) is shown here as $y = 0$. Trimethylamine (TMA) is investigated significantly more than water (**, $P < 0.01$), whereas trimethylthiazoline (TMT) is investigated significantly less (***, $P < 0.001$). However, when the mice are presented with the two odorants mixed, the investigation time is no different from that of water. (Modified from Saraiva *et al*., 2016.)

were rarely used for scent-rubbing. The authors ponder if these results are due to a novelty effect or to the (inherent) attractiveness and aversiveness of certain odours to wolves.

The examples presented in this chapter illustrate the diversity of odour perception, and how olfactory behaviour of a variety of species is dependent not only on olfactory capability and odour sensitivity of the animal in question, but also the context in which an odour is first encountered and the subsequent development of the meaning associated with a given odour. As learning is a major part of odour perception, examples of this can be found throughout this book. Although innate responses to odours exist, these are often associated with pheromones and, as will be discussed in Chapter 3, are much less common than odours that have obtained a valence through learning. Odours can be challenging to work with in an experimental setting, simply due to their volatility and the difficulties associated with knowing when an animal has detected an odorant stimulus. Indeed, the influence of smells may affect animal behaviour experiments even when odours are not part of the set-up, and potentially give rise to spurious results. In Chapter 4, some of the tricks of the trade in behavioural tests of olfaction will be presented.

References

Alaoui-Ismaïli, O., Vernet-Maury, E., Dittmar, A., Delhomme, G. and Chanel, J. (1997) Odor hedonics: connection with emotional response estimated by autonomic parameters. *Chemical Senses* 22, 237–248. DOI: 10.1093/chemse/22.3.237

Apfelbach, R., Parsons, M.H., Soini, H.A. and Novotny, M.V. (2015) Are single odorous components of a predator sufficient to elicit defensive behaviors in prey species? *Frontiers in Neuroscience* 9, 263. DOI: 10.3389/fnins.2015.00263

Badie, A. (1975) Cycle annuel d'activité des fourmis parasitées par les métacercaires de *Dicrocoelium lanceolatum* (Rudolphi, 1819). *Annales de Recherches Vétérinaires* 6, 259–264.

Baldwin, B.A. and Meese, G.B. (1977) The ability of sheep to distinguish between conspecifics by means of olfaction. *Physiology and Behavior* 18, 803–808. DOI: 10.1016/0031-9384(77)90187-1

Bonadonna, F., Villafane, M., Bajzak, C. and Jouventin, P. (2004) Recognition of burrow's olfactory signature in blue petrels, *Halobaena caerulea*: an efficient discrimination mechanism in the dark. *Animal Behaviour* 67, 893–898. DOI:10.1016/j.anbehav.2003.08.013

Bonadonna, F., Miguel, E., Grosbois, V., Jouventin, P. and Bessiere, J.M. (2007) Individual odor recognition in birds: an endogenous olfactory signature on petrels' feathers? *Journal of Chemical Ecology* 33, 1819–1829. DOI: 10.1007/s10886-007-9345-7

Bui, L.H. and Takasu, K. (2009) Learning of narcotic odors by a parasitoid. *Journal of Agricultural Science* 36, 51–56.

Chamaillé-Jammes, S., Malcuit, H., Le Saout, S. and Martin, J.-L. (2014) Innate threat-sensitive foraging: black-tailed deer remain more fearful of wolf than of the less dangerous black bear even after 100 years of wolf absence. *Oecologia* 174, 1151–1158. DOI: 10.1007/s00442-013-2843-0

Christensen, J.W. and Rundgren, M. (2008) Predator odour *per se* does not frighten domestic horses. *Applied Animal Behaviour Science* 112, 136–145. DOI: 10.1016/j.applanim.2007.08.003

Connor, J.R. and Davis, H.N. (1980) Postpartum estrus in Norway rats. I. Behavior. *Biology of Reproduction* 23, 994–999. DOI: 10.1095/biolreprod23.5.994

Cornu, J.N., Cancel-Tassin, G., Ondet, V., Girardet, C. and Cussenot, O. (2011) Olfactory detection of prostate cancer by dogs sniffing urine: a step forward in early diagnosis. *European Urology* 59, 197–201. DOI: 10.1016/j.eururo.2010.10.006

Dunbar, I., Ranson, E. and Buehler, M. (1981) Pup retrieval and maternal attraction to canine amniotic fluids. *Behavioural Processes* 6, 249–260. DOI: 10.1016/0376-6357(81)90004-8

Elliker, K.R., Sommerville, B.A., Broom, D.M., Neal, D.E., Armstrong, S. and Williams, H.C. (2014) Key considerations for the experimental training and evaluation of cancer odour detection dogs: lessons learnt from a double-blind, controlled trial of prostate cancer detection. *BioMed Central Urology* 14, 22. DOI:10.1186/1471-2490-14-22

Fendt, M. and Endres, T. (2008) 2,3,5-Trimethyl-3-thiazoline (TMT), a component of fox odor – just repugnant or really fear-inducing? *Neuroscience and Biobehavioral Reviews* 32, 1259–1266. DOI: 10.1016/j.neubiorev.2008.05.010

Fendt, M., Endres, T. and Apfelbach, R. (2003) Temporary inactivation of the bed nucleus of the stria terminalis but not of the amygdala blocks freezing induced by trimethylthiazoline, a component of fox feces. *Journal of Neuroscience* 23, 23–28.

Fillion, T.J. and Blass, E.M. (1986a) Infantile behavioural reactivity to oestrous chemostimuli in Norway rats. *Animal Behaviour* 34, 123–133. DOI: 10.1016/0003-3472(86)90014-X

Fillion, T.J. and Blass, E.M. (1986b) Infantile experience with suckling odors determines adult sexual behavior in male rats. *Science* 231, 729–731. DOI: 10.1126/science.3945807

Hari Dass, S.A. and Vyas, A. (2014) *Toxoplasma gondii* infection reduces predator aversion in rats through epigenetic modulation in the host medial amygdala. *Molecular Ecology* 23, 6114–6122. DOI: 10.1111/mec.12888

Hawk, H.W., Conley, H.H. and Kiddy, C.A. (1984) Estrus-related odors in milk detected by trained dogs. *Journal of Dairy Science* 67, 392–397.

Kippin, T.E. and Pfaus, J.G. (2001) The development of olfactory conditioned ejaculatory preferences in the male rat. I. Nature of the unconditioned stimulus. *Physiology and Behavior* 73, 457–469. DOI: 10.1016/S0031-9384(01)00484-X

Kippin, T.E., Cain, S.W. and Pfaus, J.G. (2003) Estrous odors and sexually conditioned neutral odors activate separate neural pathways in the male rat. *Neuroscience* 117, 971–979. DOI: 10.1016/S0306-4522(02)00972-7

Kobayakawa, K., Kobayakawa, R., Matsumoto, H., Oka, Y., Imai, T. *et al*. (2007) Innate versus learned odour processing in the mouse olfactory bulb. *Nature* 450, 503–510. DOI: 10.1038/nature06281

Kvitvik, I.L., Berg, K.M. and Ågmo, A. (2010) A neutral odor may become a sexual incentive through classical conditioning in male rats. *Learning and Motivation* 41, 1–21. DOI: 10.1016/j.lmot.2009.06.003

Laska, M., Fendt, M., Wieser, A., Endres, T., Hernandez Salazar, L.T. and Apfelbach, R. (2005) Detecting danger – or just another odorant? Olfactory sensitivity for the fox odor component 2,4,5-trimethylthiazoline in four species of mammals. *Physiology and Behavior* 84, 211–215. DOI: 10.1016/j.physbeh.2004.11.006

Levy, F., Poindron, P. and Le Neindre, P. (1983) Attraction and repulsion by amniotic fluids and their olfactory control in the ewe around parturition. *Physiology and Behavior* 31, 687–692. DOI: 10.1016/S0031-9384(83)80004-3

Logan, D.W., Brunet, L.J., Webb, W.R., Cutforth, T., Ngai, J. and Stowers, L. (2012) Learned recognition of maternal signature odors mediates the first suckling episode in mice. *Current Biology* 22, 1998–2007. DOI: 10.1016/j.cub.2012.08.041

Mgode, G.F., Weetjens, B.J., Nawrath, T., Cox, C., Jubitana, M. *et al*. (2012) Diagnosis of tuberculosis by trained African giant pouched rats and confounding impact of pathogens and microflora of the respiratory tract. *Journal of Clinical Microbiology* 50, 274–280. DOI: 10.1128/JCM.01199-11

Miller, A.K., Hensman, M.C., Hensman, S., Schultz, K., Reid, P., Shore, M., Brown, J., Furton, K.G. and Lee, S. (2015) African elephants (*Loxodonta africana*) can detect TNT using olfaction: implications for biosensor application. *Applied Animal Behaviour Science* 171, 177–183. DOI: 10.1016/j.applanim.2015.08.003

Nielsen, B.L., Jerôme, N., Saint-Albin, A., Rampin, O. and Maurin, Y. (2013) Behavioural response of sexually naïve and experienced male rats to the smell of 6-methyl-5-hepten-2-one and female rat faeces. *Physiology and Behavior* 120, 150–155. DOI: 10.1016/j.physbeh.2013.07.012

Nielsen, B.L., Jerôme, N., Saint-Albin, A., Ouali, C., Rochut, S. *et al*. (2016) Oestrus odours from rats and mares: behavioural responses of sexually naive and experienced rats to natural odours and odorants. *Applied Animal Behaviour Science* 176, 128–135. DOI: 10.1016/j.applanim.2016.01.014

Pelchat, M.L., Bykowski, C., Duke, F.F. and Reed, D.R. (2011) Excretion and perception of a characteristic odor in urine after asparagus ingestion: a psychophysical and genetic study. *Chemical Senses* 36, 9–17. DOI: 10.1093/chemse/bjq081

Pfaus, J.G., Kippin, T.E. and Centeno, S. (2001) Conditioning and sexual behavior: a review. *Hormones and Behavior* 40, 291–321. DOI: 10.1006/hbeh.2001.1686

Poirotte, C., Kappeler, P.M., Ngoubangoye, B., Bourgeois, S., Moussodji, M. and Charpentier, M.J.E. (2016) Morbid attraction to leopard urine in *Toxoplasma*-infected chimpanzees. *Current Biology* 26, R98–R99. DOI: 10.1016/j.cub.2015.12.020

Rampin, O., Jérôme, N., Briant, C., Boué, F. and Maurin, Y. (2006) Are oestrus odours species specific? *Behavioral Brain Research* 72, 169–172. DOI: 10.1016/j.bbr.2006.04.005

Rørvang, M.V., Nielsen, B.L., Herskin, M.S. and Jensen, M.B. (2017) Short communication: calving site selection of multiparous group housed dairy cows is influenced by site of a previous calving. *Journal of Dairy Science* 100, 1467–1471. DOI: 10.3168/jds.2016-11681

Rubianes, E. (1992) Genital stimulation modifies behavior towards amniotic fluid in estrous ewes. *Applied Animal Behaviour Science* 35, 35–40. DOI: 10.1016/0168-1591(92)90014-3

Ryon, J., Fentress, J.C., Harrington, F.H. and Bragdon, S. (1986) Scent rubbing in wolves (*Canis lupus*): the effect of novelty. *Canadian Journal of Zoology* 64, 573–577. DOI: 10.1139/z86-084

Saraiva, L.R., Kondoh, K., Ye, X., Yoon, K., Hernandez, M. and Buck, L.B. (2016) Combinatorial effects of odorants on mouse behavior. *Proceedings of the National Academy of Sciences USA* 113, E3300–E3306. DOI: 10.1073/pnas.1605973113

Schaal, B., Coureaud, G., Langlois, D., Giniès, C., Sémon, E. and Perrier, G. (2003) Chemical and behavioural characterization of the rabbit mammary pheromone. *Nature* 424, 68–72. DOI: 10.1038/nature01739

Sotnikov, S.V., Chekmareva, N.Y., Schmid, B., Harbich, D., Malik, V. *et al.* (2014) Enriched environment impacts trimethylthiazoline-induced anxiety-related behavior and immediate early gene expression: critical role of *Crhr1*. *European Journal of Neuroscience* 40, 2691–2700. DOI: 10.1111/ejn.12624

Soussignan, R., Schaal, B., Marlier, L. and Jiang, T. (1997) Facial and autonomic responses to biological and artificial olfactory stimuli in human neonates: re-examining early hedonic discrimination of odors. *Physiology and Behavior* 62, 745–758. DOI: 10.1016/S0031-9384(97)00187-X

Stern, J.M. (1996) Somatosensation and maternal care in Norway rats. In: Slater, P.J.B., Rosenblatt, J.S., Snowdon, C.T., and Milinski, M. (eds) *Advances in the Study of Behavior*. Academic Press, San Diego, California, pp. 243–294.

Takahashi L.K. (2014) Olfactory systems and neural circuits that modulate predator odor fear. *Frontiers in Behavioral Neuroscience* 8, 72. DOI: 10.3389/fnbeh.2014.00072

Vernet-Maury, E. (1980) Trimethyl-thiazoline in fox feces: a natural alarming substance for the rat. In: van der Starre, H. (ed.) *Olfaction and Taste VII*. IRL Press Ltd, London, p. 407.

Vyas, A., Kim, S.-K., Giacomini, N., Boothroyd, J.C. and Sapolsky, R.M. (2007) Behavioral changes induced by *Toxoplasma* infection of rodents are highly specific to aversion of cat odors. *Proceedings of the National Academy of Sciences USA* 104, 6442–6447. DOI: 10.1073_pnas.0608310104

Waters, J., O'Connor, S., Park, K.J. and Goulson, D. (2011) Testing a detection dog to locate bumblebee colonies and estimate nest density. *Apidologie* 42, 200–205. DOI: 10.1139/z86-084

Wernecke, K.E.A., Vincenz, D., Storsberg, S., D'Hanis, W., Goldschmidt, J. and Fendt, M. (2015) Fox urine exposure induces avoidance behavior in rats and activates the amygdalar olfactory cortex. *Behavioural Brain Research* 279, 76–81. DOI: 10.1016/j.bbr.2014.11.020

Wikelski, M., Arriero, E., Gagliardo, A., Holland, R.A., Huttunen, M.J. *et al.* (2015) True navigation in migrating gulls requires intact olfactory nerves. *Scientific Reports* 5, 17061. DOI: 10.1038/srep17061

Wong, S.T., Trinh, K., Hacker, B., Chan, G.C., Lowe, G., Gaggar, A., Xia, Z., Gold, G.H. and Storm, D.R. (2000) Disruption of the type III adenylyl cyclase gene leads to peripheral and behavioral anosmia in transgenic mice. *Neuron* 27, 487–497. DOI: 10.1016/S0896-6273(00)00060-X

Worth, A.R., Lymbery, A.J. and Thompson, R.C.A. (2013) Adaptive host manipulation by *Toxoplasma gondii*: fact or fiction? *Trends in Parasitology* 29, 150–155. DOI: 10.1016/j.pt.2013.01.004

3 Semiochemicals: Pheromones, Signature Mixtures and Behaviour

Tristram D. Wyatt

Department of Zoology, University of Oxford, UK

Since prehistoric times we have known that smell is important for interactions between animals. For example, a bull sniffing a cow, just as we might observe it now, is featured among the 30,000-year-old prehistoric cave paintings in the Dordogne of southern France. Smell signals between members of the same species, chemical signals that we call pheromones, are especially important in sexual interactions, here between the bull and cow at oestrus. Charles Darwin (1871) wrote about the powerful smells given off by male goats in the breeding season and he speculated that these signals could have evolved by females choosing the smelliest ('most odoriferous') males. These would be the smelly equivalent of the spectacular tail display of the male peacock. Animals also use smell to distinguish different individuals of their own species. For example, a mother sheep learns the individual 'signature mixture' odour of her lamb. She uses this smell memory to recognize her own offspring, rejecting any other lamb that tries to suckle (see Chapter 8).

Pheromones and signature mixtures are examples of semiochemicals (chemical information) used *within* a species (Wyatt, 2010, 2014a). Animals also respond to semiochemicals produced by *other* species, such as predators or prey: these semiochemicals are called allelochemicals (for a discussion of responses to these, see Nielsen *et al.*, 2015; Fig. 3.1). In this chapter, I will focus on chemical information used between members of the same species as many examples of allelochemicals are discussed elsewhere, including Chapters 2 and 10.

One problem with the term 'pheromone' is that sometimes it has been used without care: some molecules have been called 'pheromones' in the scientific literature despite a complete lack of evidence that the molecules really have that function. I will discuss this phenomenon in one of the sections below, using the history of molecules claimed to be 'human pheromones' as a cautionary example.

Pheromones

Since the first pheromone was chemically identified, in the silk moth in 1959, pheromones have been found in every part of the animal kingdom, including mammals. Pheromones can be involved in a wide range of functions including attraction of the sexes, mate choice, territorial behaviour and interactions between parents and offspring, depending on the species (Wyatt, 2014a, 2015a). One way that animals respond to a pheromone signal is with a particular behaviour. This might be a stereotyped behaviour such as rabbit pups searching for a nipple in response to mammary pheromone, or a female pig depressing its back ready for mating (the lordosis response) in response to the male pig's sex pheromones. These immediate behavioural responses are called releaser effects. Another kind of response, called a primer effect, is in the form of longer lasting physiological or developmental changes such as the initiation or delay of puberty (Wyatt, 2010, 2014a). Many pheromones have both releaser and primer effects on the receiving animal.

It is important to note that responses to pheromones can be context dependent. Male moths are among the most reliable responders to female sex pheromones but even they do not respond if they have mated recently. We should not be surprised therefore that mammals' responses are often context dependent, for example, depending on the dominance status of the receiver, its previous experience, including how hungry it is. On a given day, not all female pigs will respond with lordosis when

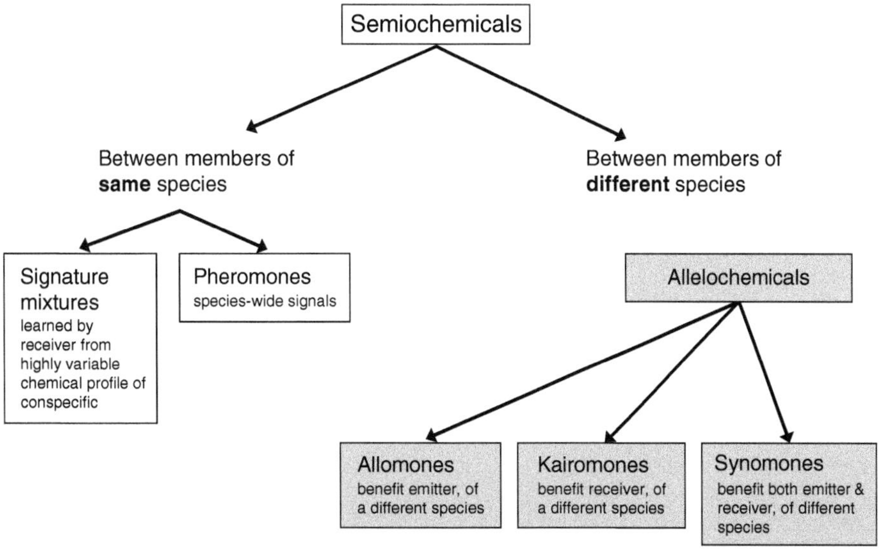

Fig. 3.1. Pheromones are chemical signals used between members of the same species. Pheromones are the same in all sexually mature males, for example, of a species. Individual recognition relies instead on the learning of signature mixtures for individuals (for recognition of siblings, for example), based on differences in chemical profile between individuals. Allelochemical is the term given to chemical information received by members of different species. (Figure from Wyatt, 2014a, reprinted with permission from Cambridge University Press.)

stimulated with male pig pheromone. In addition, the same pheromone can elicit a variety of effects in different kinds of conspecific receivers: for example, male and female mice respond differently to the male pheromone darcin (Roberts *et al.*, 2014).

Generally speaking, pheromones do not require learning: they seem to be 'innate', 'hard-wired', predisposed or 'work out of the box'. However, being innate is not part of the original pheromone definition (Karlson and Lüscher, 1959) or its updated version (Wyatt, 2010). Instead, the defining point for pheromones is that they are species-wide rather than that they are innate (e.g. it would be possible for a pheromone to rely on early learning so long as typically all individuals imprint on the same molecule(s) in normal circumstances). If we were testing only adults which had normal development then this would give the impression of the response being 'innate'.

The idea of 'innate' behaviour is itself a long debated question in animal behaviour (Wyatt, 2017). Seemingly innate behaviours often have developmental and environmental requirements for full expression (Mameli and Bateson, 2011). In practice, trying to separate the contributions of nature (genes) and nurture (environment) to the development of a behaviour is like asking whether the area of a rectangle is due more to its length or width (ascribed to psychologist Donald Hebb, in Meaney, 2001). Just as a mammal's visual cortex does not form correctly if the eyes do not receive visual stimuli during critical periods after birth (Hensch, 2004), olfactory stimuli in amniotic fluid before birth can influence olfactory bulb development (e.g. Todrank *et al.*, 2011).

Almost all pheromones are detected by the sense of smell. Pheromone molecules can be volatile or involatile, soluble or insoluble, large or small, depending on whether they are carried to the receiver in air or water or, for example, deposited on the nose of the receiver (Wyatt, 2014a). While many pheromones in the mouse are detected by the vomeronasal organ (VNO), other mouse pheromones are detected by the main olfactory system (see Chapter 1). This is an important point for mammal pheromones in general as both or either olfactory systems can be involved, depending on the species and signal. For example, the rabbit mammary pheromone is detected by the pups' main olfactory system. Pig pheromones seem to act via the main olfactory system. However, the flehmen responses of horses to sexual odours suggest that their pheromones may act principally via the VNO.

Pheromones can be short or long range, or act on contact. In house mice, volatile pheromone molecules detected at a distance of centimetres include dehydro-*exo*-brevicomin and 2-*sec*-butyl-4,5-dihydrothiazole. By contrast, the non-volatile male peptide pheromone ESP1 (exocrine gland-secreting peptide 1) gets transferred from the male mouse's tear glands to the nose (and thus the VNO) of the female during courtship contact, leading to a lordosis response (Haga *et al.*, 2010). The challenge for scientists even to this day is that chemically identifying the pheromone molecules is hard as pheromones are only released in tiny quantities. In addition, the pheromone molecules are hidden among all the other smells given off by the animals.

So, surprisingly little is known about the chemistry of the pheromones of domesticated animals, from livestock such as cattle and sheep to our companion animals such as cats and dogs. This is despite the many observations related to smell communication in these species, recounted in folklore and by farmers going back generations. For example, the ancient Greeks described how the scent of a female dog in heat attracted males, and how the attractive agent could be transferred to a towel, causing the males to follow that. Even now we are still some distance from full chemical identification (Dzięcioł *et al.*, 2016; Chapter 15). Similarly, while progress has been made, molecules signalling oestrus in cows are still under investigation (Chapter 12). Research on the pheromones of domesticated animals has not made as much progress as research on their productivity and growth. I would be delighted if this book encourages a serious reexamination of the pheromones of domesticated animals, large and small, and continued research on semiochemical-related behaviour in these animals.

Discovering Pheromones

The process needed to identify a pheromone remains much the same as the process established by Butenandt and his team when they published the first chemical identification of a pheromone, the silk moth female sex pheromone, bombykol, in 1959. The essential steps to identifying a pheromone are:

1. A behavioural or physiological response (quantified in a repeatable experiment called a bioassay) which is mediated by a potential chemical stimulus such as a secretion.
2. Isolate, identify and synthesize the bioactive molecule(s).
3. Confirm that the proposed molecule(s) at natural concentrations are necessary and sufficient to recreate the original activity with the original bioassay (Wyatt, 2014a, 2015b).

The bioassay could be any appropriate measure such as the time spent chin-rubbing, sniffing, or longer term primer effects including physiological and behavioural changes such as developing sexual maturity (see later). A key feature of any bioassay is that it should be a reliable measure of the behaviour or physiological response you ultimately want to assess. Designing effective bioassays can be the biggest challenge in pheromone research.

The search for a pheromone is a search for a signal that is not about individual recognition; for example, a male sex pheromone should elicit the response whichever sexually mature male of the species supplied the secretion. Nonetheless, *quantities* of this same pheromone might differ between individuals, for example, between subordinates and dominant males, and females might choose males with more pheromone.

Most pheromones, including the female sex pheromones of most moths and some mammalian pheromones, are *not* single compounds. Instead they tend to be a species-specific multicomponent combination of molecules (Wyatt, 2014a; Logan, 2015). For example, in the house mouse, sulfated oestrogens and a particular fraction of female urine form a multicomponent mouse pheromone, produced by oestrous females, which promotes male mounting (Haga-Yamanaka *et al.*, 2014; Logan, 2015).

Identifying pheromones in mammals presents particular challenges. Pheromones are hidden among the odour background of many hundreds of molecules that any mammal gives off as its complex individual odour profile. This complexity and great variability of odour profiles between individuals was revealed as soon as mammal secretions were analysed with gas chromatography. The complexity encouraged early doubts, later shown to be unjustified, that mammal pheromones would ever be found (Beauchamp *et al.*, 1976). Some of the odour molecules are secreted by the animal itself, while others are related to its diet, bacteria in its gut and scent glands, the environment and even molecules picked up from other individuals during scent marking or grooming.

An indication of the challenge is shown in a study to identify the molecules of a male goat sex pheromone hidden in a forest of other molecules given off by the goats (Fig. 3.2; Murata *et al.*, 2014). This scent of the males causes female goats, anoestrous for the summer, to come into oestrus. Previous bioassays based on behavioural and hormonal responses of females had shown that the active molecules for this 'male effect' were produced from skin on the head of sexually mature male goats. It was also known that the active male secretions were androgen-dependent, so the smells of castrated male goats did not stimulate the females into oestrus. This allowed Murata and co-workers to pick out potential molecules of the pheromone from the hundreds of molecules given off by male goats by comparing the gas chromatogram traces to pick out molecules missing from the secretions of castrated males (Fig. 3.2). The authors then created a synthetic cocktail of 18 of these molecules in the concentrations found in the analysis of the smells given off by intact males.

For their bioassay, Murata *et al.* (2014) used electrophysiological recording of signals in the

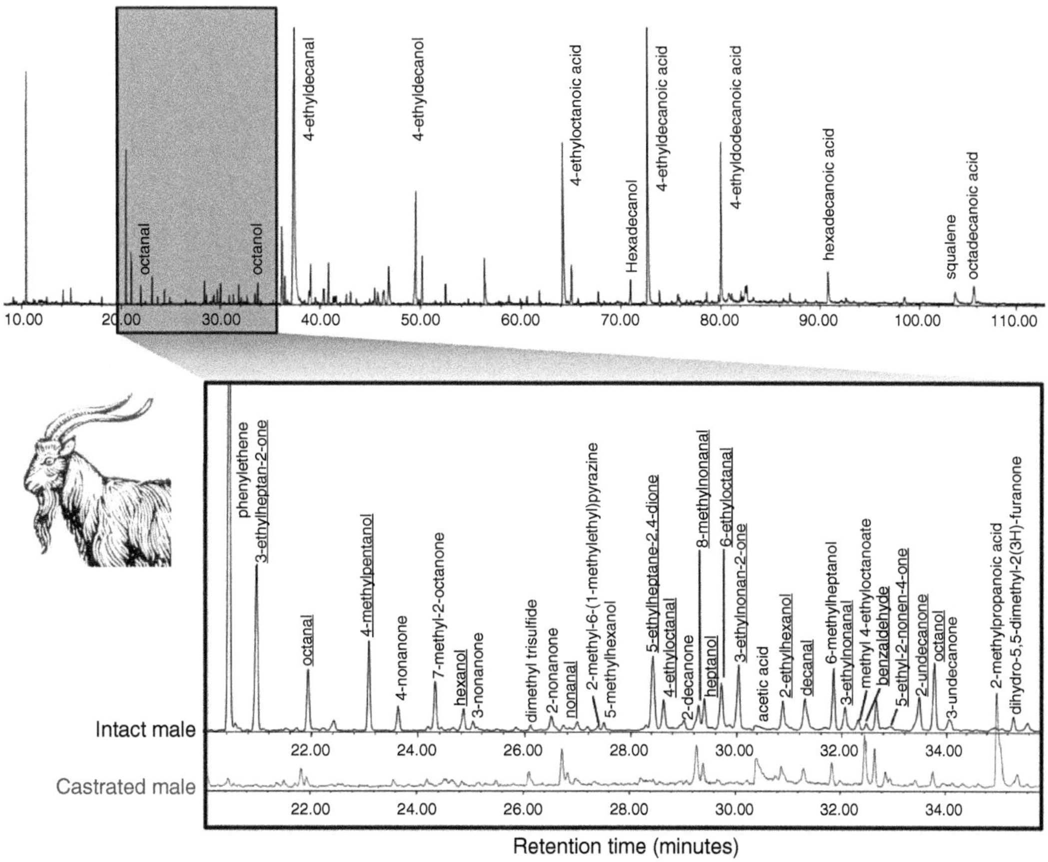

Fig. 3.2. How pheromone molecules were identified against a background of many hundreds of molecules in a sample of volatiles from the heads of male goats (Murata *et al.*, 2014). The shaded section of the gas chromatograph trace, enlarged below, was the region that contained the activity, detected in bioassays of female response. As the pheromone secretion was androgen-dependent, Murata and colleagues looked for molecules missing from the secretions of castrated males (lower chromatogram in enlarged section) and created a synthetic cocktail of 18 molecules (underlined) in the concentrations found in the original headspace analysis of intact males. Note that the candidate molecules were not the highest peaks. (Reprinted from Murata *et al.*, 2014, with permission from Elsevier. Goat image, non-copyright.)

brain. They recorded the signals associated with the pulses of secretion of luteinizing hormone leading to the restarting of oestrus (Fig. 3.3). The synthetic pheromone mixture matched the natural extract's activity. When the candidate molecules were tested individually, the researchers found that most, but not all, of the bioassay activity could be reproduced by one molecule, 4-ethyloctanal. They rightly conclude that the pheromone may include other molecules, from the 18 (or perhaps even others). Interestingly, the key molecule with primer effects, 4-ethyloctanal, is not the 'goaty' smell that we as humans associate with male goats. Instead, the characteristic smell we recognize comes from 4-ethyloctanoic acid, the oxidation product of 4-ethyloctanal (Fig. 3.3). However, 4-ethyloctanoic acid seems to be the pheromone with a releaser effect that attracts the female to come closer to the male.

What Do We Mean by Pheromone? An Operational Definition

When we call something a pheromone, the reader can anticipate that it is a molecule (or a particular combination and ratio of molecules for a multicomponent pheromone) that will be found, for example, in all sexually mature females of a given species. There are a number of other characteristics that we assume but do not often make explicit. I have proposed a five-point pragmatic, operational definition of a pheromone to formalize what most of us are thinking when we call something a pheromone (Wyatt, 2014a):

1. The synthesized molecule/combination of molecules should elicit the same response as the natural stimulus in the bioassay. This is the fundamental basis of the designation of pheromone.
2. It should act in this way at concentrations similar to the natural stimulus. Concentration is important for mammals as for other animals. For example, the response of new-born rabbit pups to the mammary pheromone 2-methylbut-2-enal is limited to a range of concentrations (from 2.5×10^{-9} to 2.5×10^{-5} g/ml in milk, less than a cup of water in an Olympic swimming pool; Schaal *et al.*, 2003; Coureaud *et al.*, 2004). One of the weaknesses of some research on 'putative' human pheromones is that a million times the natural amounts have been used (Wyatt, 2015b).
3. For multicomponent pheromones, experiments should demonstrate that all compounds in the combination are necessary and sufficient. For example, the results suggest that other molecules in addition

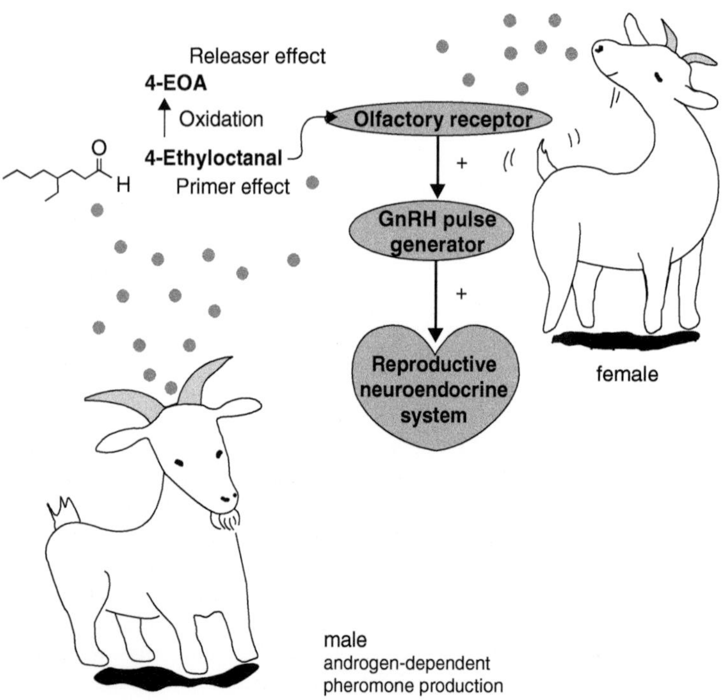

Fig. 3.3. An overview of the primer pheromone 4-ethyloctanal produced by intact male goats and how it affects females by stimulating the gonadotropin-releasing hormone (GnRH) pulse generator. The primer pheromone's oxidation product, 4-ethyloctanoic acid (4-EOA), has a releaser effect on the female, attracting her towards the male. (Reprinted from Murata *et al.*, 2014, with permission from Elsevier.)

to 4-ethyloctanal may be needed to form the full male goat primer pheromone.

4. Only this molecule or the proposed combination of molecules elicits the effect (and other similar molecules or combinations that the animal would encounter do not). In other words, this is not a spurious effect.

5. There should be a credible pathway for the pheromone signal to have evolved by direct or kin selection. In evolutionary terms, to be a signal, it is assumed that both the emission and reception of the pheromone signal have evolved for a particular function. This distinguishes pheromone signals from cues (information used by the receiver but not evolved as a signal by the emitter, as in the case of mosquitoes attracted to the smells we give off just by being alive). However, particularly in mammals, we often have little detailed information about how a pheromone might have evolved. By contrast, for moths, many aspects of pheromone production and release by the female, and reception and processing by the male are well understood.

The work on the male goat pheromone by Murata *et al.* (2014) meets these criteria well as work in progress. Their chemical analysis and bioassay details have been published in detail, including the responses to the original goat odour samples and the proposed pheromone molecule(s). This is a firm basis for future investigation and refinement. The work on rabbit mammary pheromone is similarly very well evidenced (Schaal *et al.*, 2003; Schaal and Al Aïn, 2014). However, there are too many examples of claims that certain molecules are vertebrate 'pheromones' without publication of the needed information. Sometimes only patents are filed, and patents do not provide enough information. If the full details of bioassays and chemical analysis underlying the original discovery of such 'pheromones' have not been published in the peer-reviewed literature, there are real questions to be asked about the claims. Ask for the evidence, however well the story might seem to be established.

To be confident that molecules claimed as 'pheromones' really are pheromones, it must be demonstrated that these molecules are necessary and sufficient for the effect(s) claimed. This is especially important for 'pheromones' that have been commercialized for sale. For example, consider the case of molecules claimed to be 'putative human pheromones' on the basis of patents taken out by a corporation. I discovered there were no published studies demonstrating with robust bioassays that the patented molecules androstadienone and estratetraenol were pheromones meeting the criteria above (Wyatt, 2015b). There was simply no evidence. However, this has not prevented tens of studies by good scientists taking it on faith that the molecules *are* 'human pheromones'. It might strike you as surprising that many studies found positive results with molecules that are unlikely to have any real activity. The experiments were often well designed and carried out blind. However, it is easily possible to get 'false-positive' results if experiments are characterized by small sample sizes and even more so if the molecules have no real effects (Colquhoun, 2014). Such experiments combined with positive publication bias (which leads to only positive results getting published) will give a scientific literature which is self-consistent but wrong (Ioannidis *et al.*, 2014; Wyatt, 2015b). This has led to a situation where there are many 'confirmatory' studies despite the lack of evidence that the molecules had any real function as pheromones in the first place.

The Challenge of Pheromone Primer Effects

We know most about behavioural responses to pheromones. This is largely because we can see the response soon after the pheromone is delivered, making experiments comparatively quick and easy to do. The longer lasting primer effects of pheromones that change the physiology and behaviour of animals are much harder to study as these effects may be delayed and may be hard to measure non-invasively. Primer effects, such as the female goat oestrus response above, work via modulation of hormone signalling and gene expression (Wyatt, 2014a).

Pheromone primer effects are important for coordinating reproduction in many types of organism. They have been particularly well studied in mammals, notably mice and domesticated livestock. The reported primer effects in laboratory studies of rodents include puberty acceleration of females by male pheromones (Vandenbergh effect), oestrus-induction (Whitten effect), oestrus suppression by pheromones from other females in the absence of males (Lee–Boot effect) and luteinizing hormone surges in males in response to female odours (Halpern and Martínez-Marcos, 2003; Doty, 2010). Many primer effects in rodents act in large part via the vomeronasal system. However, in sheep, the stimuli for the primer 'male effects' act via the main olfactory system. A number of the primer effects, including pregnancy block (Bruce effect) in female mice, involve an interaction with the learning of signature mixtures of individual conspecifics (see below).

As mentioned above, primer effects can be harder to study than immediate behavioural responses and many primer effects remain controversial. Studies on the effect of the pheromones of dominant male mice in accelerating puberty in young females (the Vandenbergh effect) illustrate some of the problems. Over decades, different laboratories, using different mouse strains, have proposed different molecules as the active pheromone(s). Among the proposed non-polar molecules in male urine are dehydro-*exo*-brevicomin (abbreviated brevicomin or DHB), 2-(*sec*-butyl)-dihydrothiazole (thiazole or SBT) and 6-hydroxy-6-methyl-3-heptanone (from bladder urine), and α-farnesene and β-farnesene (secreted into urine from the preputial gland). All four are reported to be active in puberty acceleration and oestrus induction, in the absence of protein (Ma *et al.*, 1999; Novotny *et al.*, 1999a,b). By contrast, other researchers have suggested that the primer effect is instead due to the major urinary proteins themselves (Mucignat-Caretta *et al.*, 1995). However, in a thorough set of experiments, Flanagan *et al.* (2011) tested all of the previously proposed molecules (volatiles and proteins), using one mouse strain throughout. Flanagan *et al.* (2011) found that none of the previously proposed molecules, separately or in combination, stimulated the uterus growth (Vandenbergh effect) as much as whole male urine did. Unknown hydrophilic and not particularly volatile molecules appeared to be the best candidates for the effect.

It has been suggested that group-living female mammals, including women, synchronize their oestrus cycles in response to female pheromones. In women, the phenomenon is called the McClintock effect (McClintock, 1971; Stern and McClintock, 1998). However, there is a growing consensus that the apparent effect may be the result of statistical artefacts instead (Yang and Schank, 2006; Doty, 2010; Setchell *et al.*, 2011; Harris and Vitzthum, 2013). No oestrus synchrony was found in a free-ranging population of mandrills *Mandrillus sphinx*, a primate (Setchell *et al.*, 2011).

Normal responses to pheromones may not develop unless species-specific conditions are met. These usually occur as a matter of course in normal development. Experiments that 'dissect' the developmental process can expose the normally hidden mechanism(s) by which a response develops. For example, perhaps surprisingly, early imprinting on species-specific odours can be important in recognition of a mate of the correct species when adult (Owens *et al.*, 1999; Doty, 2010). Normally, as the parents are of its own species, this leads to appropriate courtship choices, but cross-fostering experiments can demonstrate that these olfactory choices are learned in some species such as pygmy mice, house mice, sheep and deer (Doty, 2010). Cross-fostered young are attracted to the species of their foster parents.

In the wild, this learning can be important in sexual selection and speciation in some species (Verzijden *et al.*, 2012). Cross fostering showed that early olfactory imprinting (learning at a sensitive period) by young fish normally contributes to sexual isolation in two stickleback species. The early experience influences adults to choose only the smell of the species they had been reared with (Kozak *et al.*, 2011). When adult, learning may also be involved: male mammals such as rats and mice may need sexual experience before they can distinguish oestrous from di-oestrous female odours (Swaney and Keverne, 2011; see also Nielsen *et al.*, 2016).

Pheromones themselves can prompt learning. While the response of rabbit pups to the mammary pheromone 2-methylbut-2-enal seems hard-wired, the pheromone stimulates learning of other odours, which will then also stimulate suckling (Schaal *et al.*, 2009; Coureaud *et al.*, 2010). Contact with the male mouse protein pheromone, darcin, prompts a female mouse to learn both his volatile individual signature mixture and the location of the scent mark (Roberts *et al.*, 2014).

Individual Recognition: Signature Mixtures

As well as responding to species-wide pheromone signals, animals learn the individual smells of family members. For example, mother sheep learn the smell of their lambs and will only allow their own lambs to suckle. Humans, also, seem to be able to recognize their partner and children by smell, if asked to distinguish between the smell of T-shirts worn by a family member or a stranger. We need a different term, 'signature mixture', for the smells that animals learn and use to distinguish other individuals of their own species (Wyatt, 2010). Animals learn a 'signature mixture' that is characteristic of the individual chemical profile of molecules given off by other animals. Signature mixtures are the subsets of variable molecules from the chemical profile of an individual that are learned as templates by members of the same species and used to recognize an organism as an individual or as a member of a particular social group such as a family, clan or colony. 'Signature' is used as it denotes individuality.

Compared with pheromones, there are two important characteristics of signature mixtures: first, a requirement for learning; and second, the variability of the cues learned, allowing other individuals to be distinguished by their different chemical profiles. Which molecules are being learned by animals, including ourselves, to distinguish individuals is not known, although in mice, major urinary proteins (MUPs), major histocompatibility complex (MHC) peptides or small molecules associated with or related to them, can all be used (Hurst, 2009; Kwak *et al.*, 2010). Mammals are good at detecting differences in smells between individuals and the experiments are designed around this ability. However, because of this experimental design, I wonder if animals may simply be distinguishing between whichever molecules are allowed to vary in that particular study. For example, in some experiments the only difference between individuals being distinguished is in their MHC, in other experiments the MHC is kept constant and only the MUPs differ between individuals.

The way a mother sheep (ewe) learns the odour of her lamb has become a model system for studying the learning of individual odours. In sheep, an enduring bond between a ewe and her lamb is established very rapidly, usually within 2 h of giving birth (Lévy and Keller, 2008, 2009; Sanchez-Andrade and Kendrick, 2009; see Chapter 8). These olfactory cues are perceived by the main olfactory system, not the VNO-accessory olfactory system. Stretching of the vaginocervical area while giving birth sends nerve signals to the brain that trigger a cascade of neurobiological and hormonal mechanisms resulting in three changes in behaviour: first, release of oxytocin, a peptide hormone, mainly in the paraventricular nucleus of the hypothalamus, leads to maternal behaviour towards lambs in general; second, the ewe's response to amniotic fluid changes, making it very attractive (whereas before it was repulsive); and, third, noradrenalin (norepinephrine) and other neurotransmitters released in the main olfactory bulb cause her to learn the individual odour of her lamb while sniffing and licking the amniotic fluid. Afterwards she will only suckle her own lambs, recognized by smell. The sensitive period for learning lasts for between 4 and 12 h after giving birth. In this early period, but not later, an orphan lamb will be accepted and adopted, particularly if coated with amniotic fluid (a method traditionally used by farmers; see Chapter 12).

Working with pheromones and signature mixtures needs different experimental approaches. If a phenomenon, such as a male distinguishing his mate from other females, relies on a learned signature mixture, it would be fruitless to search for a single combination of molecules eliciting individual mate recognition across the species: it is precisely the great variation *between* individuals that makes recognition possible. It is worth being aware when reading the past and current literature that the term 'pheromone' is still used loosely and the term may be used in contexts where 'signature mixture' or 'chemosensory cues' would be more accurate or helpful.

Interactions between Pheromones and Signature Mixtures

In mammals, behavioural responses to signature mixtures can affect exposure or response to pheromones and vice versa. These effects illustrate why it is useful to separate the phenomena (Wyatt, 2014a,b). I briefly discuss two different kinds of interaction here: first, the protein pheromone darcin prompts a female mouse to learn a male signature mixture; and second, a memory for a male mouse's signature mixture blocks the effects of his pheromone on a recently mated female (Bruce effect).

Darcin

A dominant male mouse marks his territory extensively with urine marks containing high concentrations of a variety of MUPs (Roberts *et al.*, 2014). Most MUPs are highly polymorphic but one MUP, Mup20 (named darcin), is the same in all male mice and is a pheromone. A female visiting the territory is attracted by volatiles coming from the urine marks. When she contacts and sniffs a urine mark, darcin reaches her VNO, induces her to spend more time sniffing the urine and stimulates her to learn the location of the mark and the marking male's volatile individual signature mixture (Roberts *et al.*, 2010, 2012). She is then attracted to this male, recognizes him by his airborne signature mixture and mates with him. She is uniquely attracted to him, not other males at this point. Mup20 is probably the MUP that elicited aggression in male mice in earlier experiments by Chamero *et al.* (2007). The explanation for the aggression is presumably that other territorial males will treat Mup20 as an indication of a potential challenger. Darcin does not in itself elicit aggression. Darcin binds and releases the male-specific small molecule mouse pheromone, 2-*sec*-butyl-4,5-dihydrothiazole (thiazole).

Memory and pregnancy block (Bruce effect)

In the laboratory, exposing a recently mated female mouse to the urinary chemosignals of an unfamiliar male mouse initiates a neuroendocrine reflex that results in pregnancy failure and a return to oestrus (the Bruce effect; Brennan and Kendrick, 2006; Brennan, 2009). The physiological explanation for this is as follows: a pheromone in the strange male's urine is sensed by her VNO and activates an excitatory pathway, via the accessory olfactory bulb and medial amygdala, to the hypothalamus. This increases dopamine release, which in turn inhibits prolactin production by the anterior pituitary. Without prolactin, progesterone production by the corpora lutea declines, implantation of the developing embryos fails and the female returns to oestrus.

All male mice produce the pheromone but pregnancy block does not happen when the female meets the stud male she recently mated with. Why not? The reason is that when she mated, vaginocervical signals prompted the individual signature mixture of her male mate to be learnt in her accessory olfactory lobe. This memory selectively prevents his pheromones from eliciting the pregnancy block. Thus, there are two distinct kinds of chemical information: first, a male testosterone-dependent pheromone (the same for all males), as yet unidentified (though it is of low molecular weight; Peele *et al.*, 2003). Second, the male's individual chemical profile including his urinary odour type and peptides related to the MHC. Males that differ only at one MHC locus can cause pregnancy failure (Yamazaki *et al.*, 1983). MHC peptides from an unfamiliar male of a different haplotype can induce pregnancy block and can be detected directly by the V2R receptor-expressing zone of the vomeronasal epithelium (Leinders-Zufall *et al.*, 2004). However, MHC peptide ligands have yet to be identified in mouse urine so there may be other classes of individuality chemosignals (Brennan, 2009). MUPs can carry many different kinds of odour molecules, so they may have a role in carrying the odour molecules of individual difference. It is possible that the female recognizes whatever is different from the signature mixture she has remembered of the first male.

The female's long-term memory of her mate's signature mixture is formed when the mating male's individual odours stimulate a variety of vomerosensory neurons in the VNO to send signals to a subpopulation of mitral cells and associated granule cells in the AOB glomeruli, the first relay in the vomeronasal pathway (Brennan, 2009). The conjunction of the mate's odours and higher levels of noradrenalin (norepinephrine) after mating are thought to prime these inhibitory synapses so that if the same individual odour is smelt again, transmission to the hypothalamus of the pregnancy-blocking signal from male urinary pheromone is stopped. For the memory to be formed, mating and then exposure to the mate's odour during the sensitive period (3–4.5 h after mating) are required.

Contact with the odour source seems to be needed for pregnancy block (Brennan, 2009) but the VNO can also detect volatiles, so the need for contact may reflect the way molecules enter the VNO. The male pheromone responsible for the Whitten effect (oestrus-induction) is different from the one(s) involved in the Bruce effect as some mouse inbred strains induce the Bruce effect but not the Whitten effect (Doty, 2010).

Currently there is no evidence that the Bruce effect occurs in mice or other rodents under natural conditions (Wolff, 2003). One contributory explanation for this might be that in natural conditions the female may control exposure to male odours herself and thus control whether the Bruce effect occurs (Becker and Hurst, 2009). Nonetheless, the Bruce effect has become a very valuable model system for studying olfactory learning and the brain (Brennan, 2009).

Applying a Knowledge of Pheromones and Signature Mixtures

Using a greater understanding of olfactory stimuli including semiochemicals could improve the welfare of farm, zoo, laboratory and companion animals (Chapters 12, 13, 14 and 15). Using the 'male effect' on reproduction of sheep and goats provides a good example of a potential intervention. This could offer a 'clean, green and ethical', hormone-free approach to improve both reproductive efficiency and animal welfare (Jorre de St Jorre *et al.*, 2014). As seen above, the seasonal anoestrus in sheep and goats during the winter and spring can be ended by the introduction of an unfamiliar, sexually active male. This exposure to male pheromones results in synchronized oestrus some days later. In sheep farming, this synchronized oestrus gives a desirable clustering of lambing, which helps labour management and means that the lambs can be marketed as a uniform group. Novelty of the male is crucial to the effect, which suggests the involvement of signature mixtures, for recognition of a male as familiar or unfamiliar, as well as a male pheromone(s) (Jorre de St Jorre *et al.*, 2014). Despite the potential of using the male effect to improve farming practice, Jorre de St Jorre *et al.* explain that there are still many questions to answer and the technique is far from commercial application. This is in part because past experiments have not been designed to separate the effects of male novelty and male pheromones.

A knowledge of semiochemical communication, largely gained from studies of domesticated and laboratory animals, can improve the success of captive breeding programmes for rarer species (Swaisgood and Schulte, 2010; Dehnhard, 2011; Chapter 13). For example, primer effects of pheromones and familiarity with other odour cues can be manipulated to increase the success of captive breeding in the giant panda, *Ailuropoda melanoleuca* (see Swaisgood and Schulte, 2010). The pandas are frequently swapped between pens to expose them to each other's scent and allow them to exchange important messages related to individual identification, sex and reproductive status, reproductive maturity and competitive status. These manipulations appear to increase sexual motivation and reduce aggression prior to the mating introduction.

In a captive breeding programme for the threatened pygmy loris, *Nycticebus pygmaeus*, chemical cues were manipulated to encourage females to mate with particular males chosen for optimal outbreeding (see Swaisgood and Schulte, 2010). The females were presented with artificial marks which indicated that the selected males were territory holders marking a territory and over-marking competitor marks. Females had a tenfold preference for males made familiar in this way and an approximately twofold preference for the over-marking male. Similar techniques were used to influence female choice in captive breeding of harvest mice, *Micromys minutus* (Roberts and Gosling, 2004). A greater knowledge of semiochemicals and their action in the lives of animals could greatly improve our care of animals of all kinds.

Acknowledgements

I warmly thank Birte L. Nielsen and an anonymous reviewer for very helpful comments on drafts of this chapter. I also thank the many colleagues who have helped me by discussing ideas about pheromones and individual odours over many years in person or by correspondence. Some parts of this chapter are based on Wyatt (2014a), with permission from Cambridge University Press.

References

Beauchamp, G.K., Doty, R.L., Moulton, D.G. and Mugford, R.A. (1976) The pheromone concept in mammalian chemical communication: a critique. In: Doty, R.L. (ed.) *Mammalian Olfaction, Reproductive Processes, and Behavior*. Academic Press, New York, pp. 143–160.

Becker, S.D. and Hurst, J.L. (2009). Female behaviour plays a critical role in controlling murine pregnancy block. *Proceedings of the Royal Society B: Biological Sciences* 276, 1723–1729. DOI: 10.1098/rspb.2008.1780

Brennan, P.A. (2009) Outstanding issues surrounding vomeronasal mechanisms of pregnancy block and individual recognition in mice. *Behavioural Brain Research* 200, 287–294. DOI: 10.1016/j.bbr.2008.10.045

Brennan, P.A. and Kendrick, K.M. (2006) Mammalian social odours: attraction and individual recognition. *Philosophical Transactions of the Royal Society B: Biological Sciences* 361, 2061–2078. DOI: 10.1098/rstb.2006.1931

Chamero, P., Marton, T.F., Logan, D.W., Flanagan, K., Cruz, J.R., Saghatelian, A., Cravatt, B.F. and Stowers, L. (2007) Identification of protein pheromones that promote aggressive behaviour. *Nature* 450, 899–902. DOI: 10.1038/nature05997

Colquhoun, D. (2014) An investigation of the false discovery rate and the misinterpretation of *p*-values. *Royal Society Open Science* 1, 140216. DOI: 10.1098/rsos.140216

Coureaud, G., Langlois, D., Sicard, G. and Schaal, B. (2004) Newborn rabbit responsiveness to the mammary pheromone is concentration-dependent. *Chemical Senses* 29, 341–350. DOI: 10.1093/chemse/bjh037

Coureaud, G., Charra, R., Datiche, F., Sinding, C., Thomas-Danguin, T., Languille, S., Hars, B. and Schaal, B. (2010) A pheromone to behave, a pheromone to learn: the rabbit mammary pheromone. *Journal of Comparative Physiology A: Neuroethology, Sensory, Neural, and Behavioral Physiology* 196, 779–790. DOI: 10.1007/s00359-010-0548-y

Darwin, C. (1871) *The Descent of Man, and Selection in Relation to Sex*. John Murray, London.

Dehnhard, M. (2011) Mammal semiochemicals: understanding pheromones and signature mixtures for better zoo animal husbandry and conservation. *International Zoo Yearbook* 45, 1–25.

Doty, R.L. (2010) *The Great Pheromone Myth*. Johns Hopkins University Press, Baltimore, Maryland.

Dzięcioł, M., Szumny, A., Niżański, W., Wrzosek, M., Stańczyk, E. and Polikowska, J. (2016) Semiochemical communication in dogs in the context of reproductive behavior. In: Schulte, A.B., Goodwin, E.T. and Ferkin, H.M. (eds) *Chemical Signals in Vertebrates 13*. Springer, Cham, pp. 381–389.

Flanagan, K.A., Webb, W. and Stowers, L. (2011) Analysis of male pheromones that accelerate female reproductive organ development. *PLoS ONE* 6, e16660. DOI: 10.1371/journal.pone.0016660

Haga, S., Hattori, T., Sato, T., Sato, K., Matsuda, S. *et al.* (2010) The male mouse pheromone ESP1 enhances female sexual receptive behaviour through a specific vomeronasal receptor. *Nature* 466, 118–122. DOI: 10.1038/nature09142

Haga-Yamanaka, S., Ma, L., He, J., Qiu, Q., Lavis, L.D., Looger, L.L. and Yu, C.R. (2014) Integrated action of pheromone signals in promoting courtship behavior in male mice. *eLife* 3, e03025. DOI: 10.7554/eLife.03025

Halpern, M. and Martínez-Marcos, A. (2003) Structure and function of the vomeronasal system: an update. *Progress in Neurobiology* 70, 245–318. DOI: 10.1016/S0301-0082(03)00103-5

Harris, A.L. and Vitzthum, V.J. (2013) Darwin's legacy: an evolutionary view of women's reproductive and sexual functioning. *Journal of Sex Research* 50, 207–246. DOI: 10.1080/00224499.2012.763085

Hensch, T.K. (2004) Critical period regulation. *Annual Review of Neuroscience* 27, 549–579.

Hurst, J.L. (2009) Female recognition and assessment of males through scent. *Behavioural Brain Research* 200, 295–303. DOI: 10.1016/j.bbr.2008.12.020

Ioannidis, J.P.A., Munafò, M.R., Fusar-Poli, P., Nosek, B.A. and David, S.P. (2014) Publication and other reporting biases in cognitive sciences: detection, prevalence, and prevention. *Trends in Cognitive Sciences* 18, 235–241. DOI: 10.1016/j.tics.2014.02.010

Jorre de St Jorre, T., Hawken, P.A.R. and Martin, G.B. (2014) New understanding of an old phenomenon: uncontrolled factors and misconceptions that cast a shadow over studies of the 'male effect' on reproduction in small ruminants. *Turkish Journal of Veterinary and Animal Sciences* 38, 625–636. DOI:10.3906/vet-1404-81

Karlson, P. and Lüscher, M. (1959) 'Pheromones': a new term for a class of biologically active substances. *Nature* 183, 55–56. DOI: 10.1038/183055a0

Kozak, G.M., Head, M.L. and Boughman, J.W. (2011) Sexual imprinting on ecologically divergent traits leads to sexual isolation in sticklebacks. *Proceedings of the Royal Society B: Biological Sciences* 278, 2604–2610. DOI: 10.1098/rspb.2010.2466

Kwak, J., Willse, A., Preti, G., Yamazaki, K. and Beauchamp, G.K. (2010) In search of the chemical basis for MHC odourtypes. *Proceedings of the Royal Society B: Biological Sciences* 277, 2417–2425. DOI: 10.1098/rspb.2010.0162

Leinders-Zufall, T., Brennan, P., Widmayer, P., Chandramani, P., Maul-Pavicic, A. *et al.* (2004) MHC Class I peptides as chemosensory signals in the vomeronasal organ. *Science* 306, 1033–1037. DOI: 10.1126/science.1102818

Lévy, F. and Keller, M. (2008) Neurobiology of maternal behavior in sheep. *Advances in the Study of Behavior* 38, 399–437. DOI: 10.1016/S0065-3454(08)00008-9

Lévy, F. and Keller, M. (2009) Olfactory mediation of maternal behavior in selected mammalian species. *Behavioural Brain Research* 200, 336–345. DOI: 10.1016/j.bbr.2008.12.017

Logan, D.W. (2015) The complexity of pheromone-mediated behaviour in mammals. *Current Opinion in Behavioral Sciences* 2, 96–101. DOI: 10.1016/j.cobeha.2014.10.011

Ma, W.D., Miao, Z.S. and Novotny, M.V. (1999) Induction of estrus in grouped female mice (*Mus domesticus*) by synthetic analogues of preputial gland constituents. *Chemical Senses* 24, 289–293. DOI: 10.1093/chemse/24.3.289

Mameli, M. and Bateson, P. (2011) An evaluation of the concept of innateness. *Philosophical Transactions of the Royal Society B: Biological Sciences* 366, 436–443. DOI: 10.1098/rstb.2010.0174

McClintock, M.K. (1971) Menstrual synchrony and suppression. *Nature* 229, 244–245. DOI: 10.1038/229244a0

Meaney, M.J. (2001) Nature, nurture, and the disunity of knowledge. *Annals of the New York Academy of Sciences* 935, 50–61. DOI: 10.1111/j.1749-6632.2001.tb03470.x

Mucignat-Caretta, C., Caretta, A. and Cavaggioni, A. (1995) Acceleration of puberty onset in female mice by male urinary proteins. *Journal of Physiology* 486, 517–522.

Murata, K., Tamogami, S., Itou, M., Ohkubo, Y., Wakabayashi, Y., Watanabe, H., Okamura, H., Takeuchi, Y. and Mori, Y. (2014) Identification of an olfactory signal molecule that activates the central regulator of reproduction in goats. *Current Biology* 24, 681–686. DOI: 10.1016/j.cub.2014.01.073

Nielsen, B.L., Rampin, O., Meunier, N. and Bombail, V. (2015) Behavioral responses to odors from other species: introducing a complementary model of allelochemics involving vertebrates. *Frontiers in Neuroscience* 9, 226. DOI: 10.3389/fnins.2015.00226

Nielsen, B.L., Jerôme, N., Saint-Albin, A., Ouali, C., Rochut, S. *et al.* (2016) Oestrus odours from rats and mares: behavioural responses of sexually naive and experienced rats to natural odours and odorants. *Applied Animal Behaviour Science* 176, 128–135. DOI: 10.1016/j.applanim.2016.01.014

Novotny, M.V., Ma, W., Zidek, L. and Daev, E. (1999a) Recent biochemical insights into puberty acceleration, estrus induction and puberty delay in the house mouse. In: Johnston, R.E., Müller-Schwarze, D. and Sorensen, P.W. (eds.) *Advances in Chemical Signals in Vertebrates*. Kluwer Academic/Plenum Press, New York, pp. 99–116.

Novotny, M.V., Ma, W.D., Wiesler, D. and Zidek, L. (1999b) Positive identification of the puberty-accelerating pheromone of the house mouse: the volatile ligands associating with the major urinary protein. *Proceedings of the Royal Society of London Series B: Biological Sciences* 266, 2017–2022. DOI: 10.1098/rspb.1999.0880

Owens, I.P.F., Rowe, C. and Thomas, A.L.R. (1999) Sexual selection, speciation and imprinting: separating the sheep from the goats. *Trends in Ecology and Evolution* 14, 131–132. DOI: 10.1016/S0169-5347(98)01578-X

Peele, P., Salazar, I., Mimmack, M., Keverne, E.B. and Brennan, P.A. (2003) Low molecular weight constituents of male mouse urine mediate the pregnancy block effect and convey information about the identity of the mating male. *European Journal of Neuroscience* 18, 622–628. DOI: 10.1046/j.1460-9568.2003.02790.x

Roberts, S.C. and Gosling, L.M. (2004) Manipulation of olfactory signaling and mate choice for conservation breeding: a case study of harvest mice. *Conservation Biology* 18, 548–556. DOI: 10.1111/j.1523-1739.2004.00514.x

Roberts, S.A., Simpson, D.M., Armstrong, S.D., Davidson, A.J., Robertson, D.H., McLean, L., Beynon, R.J. and Hurst, J.L. (2010) Darcin: a male pheromone that stimulates female memory and sexual attraction to an individual male's odour. *BMC Biology* 8, 75. DOI: 10.1186/1741-7007-8-75

Roberts, S.A., Davidson, A.J., McLean, L., Beynon, R.J. and Hurst, J.L. (2012) Pheromonal induction of spatial learning in mice. *Science* 338, 1462–1465. DOI: 10.1126/science.1225638

Roberts, S.A., Davidson, A.J., Beynon, R.J. and Hurst, J.L. (2014) Female attraction to male scent and associative learning: the house mouse as a mammalian model. *Animal Behaviour* 97, 313–321. DOI: 10.1016/j.anbehav.2014.08.010

Sanchez-Andrade, G. and Kendrick, K.M. (2009) The main olfactory system and social learning in mammals: pheromonal communication in higher vertebrates and its implication for reproductive function. *Behavioural Brain Research* 200, 323–335. DOI: 10.1016/j.bbr.2008.12.021

Schaal, B. and Al Aïn, S. (2014) Chemical signals 'selected for' newborns in mammals. *Animal Behaviour* 97, 289–299. DOI: 10.1016/j.anbehav.2014.08.022

Schaal, B., Coureaud, G., Langlois, D., Ginies, C., Semon, E. and Perrier, G. (2003) Chemical and behavioural characterization of the rabbit mammary pheromone. *Nature* 424, 68–72. DOI: 10.1038/nature01739

Schaal, B., Coureaud, G., Doucet, S., Delaunay-El Allam, M., Moncomble, A.-S., Montigny, D., Patris, B. and Holley, A. (2009) Mammary olfactory signalisation in females and odor processing in neonates: ways evolved by rabbits and humans. *Behavioural Brain Research* 200, 346–358. DOI: 10.1016/j.bbr.2009.02.008

Setchell, J.M., Kendal, J. and Tyniec, P. (2011) Do non-human primates synchronise their menstrual cycles? A test in mandrills. *Psychoneuroendocrinology* 36, 51–59. DOI: 10.1016/j.psyneuen.2010.06.004

Stern, K. and McClintock, M.K. (1998) Regulation of ovulation by human pheromones. *Nature* 392, 177–179. DOI: 10.1038/32408

Swaisgood, R. and Schulte, B. (2010) Applying knowledge of mammalian social organization, mating systems, and communication to management. In: Kleiman, D, Thompson, K and Baer, C (eds) *Wild Mammals in Captivity: Principles and Techniques for Zoo Management*, 2nd edn. University of Chicago Press, Chicago, Illinois, pp. 329–343.

Swaney, W.T. and Keverne, E.B. (2011) Genomic imprinting and sexual experience-dependent learning in the mouse. In: Clelland, J.D. (ed.) *Genomics, Proteomics, and the Nervous System*. Springer, New York, pp. 195–225.

Todrank, J., Heth, G. and Restrepo, D. (2011) Effects of *in utero* odorant exposure on neuroanatomical development of the olfactory bulb and odour preferences. *Proceedings of the Royal Society B: Biological Sciences* 278, 1949–1955. DOI: 10.1098/rspb.2010.2314

Verzijden, M.N., ten Cate, C., Servedio, M.R., Kozak, G.M., Boughman, J.W. and Svensson, E.I. (2012) The impact of learning on sexual selection and speciation. *Trends in Ecology and Evolution* 27, 511–519. DOI: 10.1016/j.tree.2012.05.007

Wolff, J.O. (2003) Laboratory studies with rodents: facts or artifacts? *Bioscience* 53, 421–427. DOI: 10.1641/0006-3568(2003)053[0421

Wyatt, T.D. (2010) Pheromones and signature mixtures: defining species-wide signals and variable cues for identity in both invertebrates and vertebrates. *Journal of Comparative Physiology A: Neuroethology, Sensory, Neural, and Behavioral Physiology* 196, 685–700. DOI: 10.1007/s00359-010-0564-y

Wyatt, T.D. (2014a) *Pheromones and Animal Behavior: Chemical Signals and Signatures*, 2nd edn. Cambridge University Press, Cambridge.

Wyatt, T.D. (2014b) Proteins and peptides as pheromone signals and chemical signatures. *Animal Behaviour* 97, 273–280. DOI: 10.1016/j.anbehav.2014.07.025

Wyatt, T.D. (2015a) How animals communicate via pheromones. *American Scientist* 103, 114–121. DOI: 10.1511/2015.113.114.

Wyatt, TD (2015b) The search for human pheromones: the lost decades and the necessity of returning to first principles. *Proceedings of the Royal Society B: Biological Sciences* 282, 20142994. DOI: 10.1098/rspb.2014.2994

Wyatt, T.D. (2017) *Animal Behaviour: A Very Short Introduction*. Oxford University Press, Oxford.

Yamazaki, K., Beauchamp, G.K., Wysocki, C.J., Bard, J., Thomas, L. and Boyse, E.A. (1983) Recognition of H-2 types in relation to the blocking of pregnancy in mice. *Science* 221, 186–188. DOI:10.1126/science.6857281

Yang, Z. and Schank, J. (2006) Women do not synchronize their menstrual cycles. *Human Nature* 17, 434–447. DOI: 10.1007/s12110-006-1005-z

4 Behavioural Tests of Olfaction

Markus Fendt[1], Raimund Apfelbach[2] and Burton Slotnick[3]

[1]*Otto-von-Guericke-Universität Magdeburg, Magdeburg, Germany;* [2]*University of Tübingen, Tübingen, Germany;* [3]*American University, Washington, DC, USA*

Introduction

Interest in the olfactory ability of animals from both the general public and research scientists stems from the seemingly amazing abilities many animals have for detecting odours and following odour-trails. Their keen sense of smell also has practical advantages such as detecting drugs or explosive devices, buried bodies, tracking criminals and, as shown more recently, detecting disease states. While the species best known for such exceptional olfactory abilities is the dog, almost all other species including insects, fish, reptiles, most mammals and, yes, even many species of birds have very well-developed olfactory systems with smell playing an important, and in many cases, critical role in their everyday life.

Studies of the sense of smell are not as advanced as are those of other sensory modalities. In large part this is due to unique problems in generating, controlling and measuring odours, ones that are not encountered in studies using visual, auditory, tactile or gustatory stimuli. Whether odours exist as gaseous vapour or, for aquatic animals, as solutions in water, odour molecules either diffuse freely such that concentration decreases with distance from their source or, with turbulence, follow a path or paths difficult, at best, to define. To further complicate matters there are literally 10,000 or more different molecules that can be detected, at least by most mammals, and their vapour pressure, rate of diffusion and solubility varies considerably. In short, controlling and measuring odours is a major problem in studies of olfaction.

In this chapter we will describe some of the behavioural methods (in field studies as well as laboratory research) used to investigate the sense of smell in fish, birds and rodents. Insects, perhaps the class of animals most dependent upon the sense of smell, will not be covered because the entomological literature is vast and varied and would require a separate chapter. In addition, many of the experimental procedures used in insect studies are not applicable to or relevant for studies using vertebrates, which are the focus of this book.

Not all studies require precise control of either the animal or the stimulus. Thus, if the question at hand is only whether an animal can detect an odour or can discriminate between two odours then quite simple observational methods may be used, such as determining if a subject's behaviour is altered by an odour that is allowed to freely diffuse from a point source. Quantitative measures of olfactory sensitivity and discrimination require more stringent tests and more precise control of odours. In the following account we recognize two classes of tests commonly used with most species: non-associative and associative. Generally, in non-associative tests the experimenter presents an animal with an odour and notes its response. These relatively simple observational studies might take place in the animal's natural environment or in a simulated natural setting in a laboratory. The outcome of such tests can provide evidence not only for odour detection but, potentially, also give a rough estimate of odour sensitivity, odour preference (or aversion) and odour discrimination.

In associative or conditioning tests animals are trained to make a specific response to one type of odour and another response to a different odour or the absence of an odour. Typically, these involve laboratory studies using a special apparatus for generating and delivering odours to test subjects.

Naturally, when experimenting with animals, investigators should always respect each animal as an individual and follow approved institutional protocols. Only animals treated with appropriate concern for their welfare and cooperation will bring us reliable results.

Olfaction in Fish

Fishermen have long known that chumming, i.e. spreading food extracts in the water, attracts fish. Early studies (Nagel, 1894; Parker, 1910) demonstrated that, unlike controls, catfish with transection of the olfactory nerves failed to respond to food wrapped in cheesecloth. Major advances in our knowledge of fish olfaction were made through field studies on homing in the Coho salmon (*Oncorhynchus kisutch*). Sheer (1939, cited in Hasler and Wisby, 1951), found that 'salmon hatched and reared in a particular freshwater region will, upon returning to fresh water, return in the great majority of cases to the same region, even from considerable distances'. Normally a saltwater inhabitant, sexually mature salmon spawn upstream in freshwater streams. The larvae hatch about 2 months later, remain in fresh water for another 18 months and then migrate to the ocean. When sexually mature 1–3 years later, they return to spawn most often to the same freshwater regions where they developed. As detailed below, experimental studies have shown clearly that olfaction plays a major role in identifying these freshwater streams (Cooper *et al.*, 1976).

In large-scale studies, Scholz *et al.* (1976) and Hasler *et al.* (1978) raised 45,000 salmon for 1.5 years under identical conditions. Fish were then individually marked with fin clips and divided into three groups of 15,000 animals each and raised for another 6 weeks (during their pre-smolt and smolt stages) in large water tanks. A low concentration (5×10^{-5} mg/l) of the synthetic chemical morpholine (C_4H_9NO) was added to the water of one group and 1×10^{-3} mg/l of phenethyl alcohol ($C_8H_{10}O$, PEA) was added to the water of a second group. These artificial odours can be detected by salmon but are not normally found in lakes or rivers. The third (control) group was raised without being exposed to an artificial odour. After 6 weeks all animals were released into Lake Michigan between the rivers Little Manitowoc and Two Rivers whose estuaries are about 15 km apart. About 18 months later, during spawning migration, morpholine (5×10^{-3} mg/l) and PEA (5×10^{-5} mg/l) were added to the Little Manitowoc and Two Rivers, respectively. Then these two scented streams and an additional 17 streams along Lake Michigan were surveyed. More than 90% of the two experimental groups returned to the streams that contained their 'imprinted' odour, while control fish, raised in non-scented water, showed no preference and were found at numerous locations. The results clearly supported the olfactory hypothesis for homing salmon first presented by Hasler and Wisby in 1951. While the potential role of other factors such as water temperature, salinity or currents that might guide the fish from the ocean to coastal areas has not been studied in detail, most researchers agree that 'the final freshwater phase of the salmon's homing migration is governed primarily by the olfactory discrimination of home stream water' (Dittman, 1994).

Odour sensitivity has also been studied in some detail in the eel (*Anguilla anguilla*) by Teichmann (1957, 1959). Both salmon and eels migrate between fresh water and the ocean for their reproduction. While salmon as an anadromous fish, spends most of its life in the ocean and returns to fresh water only to spawn there, the catadromous eel migrates from fresh water to the Sargasso Sea in the ocean for spawning. Teichmann (1959) maintained young eels (about 12–15 cm long) in freshwater tanks equipped with hiding places (tubes) and trained them to respond to food (*tubifex*, a genus of sludge worms). To train the eels, Teichmann modified an experimental set-up described by Karl von Frisch[1] (1941) to study odour detection in fish (Fig. 4.1). When *tubifex* were offered, the eels left their hiding tubes, approached the odour source and fed. Next, in separate trials, the odorants PEA or eugenol were added to the water but food was given only in the PEA trials.

Eels eventually learned to search for food only in PEA trials but they appeared more motivated to seek hiding places than to feed. Thus, in further trials eels were housed individually in 40 × 25 × 9 cm tanks in which three separate 130 cm/min streams of fluid could be released into hiding tubes extending into the tank. Two streams consisted only of water and one contained a dilute solution of PEA. In separate 20-min trials the eel could enter any of the three tubes but was only allowed to stay in the tube carrying the odour.

At high concentrations (0.0025%) eels approached and hid in the tube releasing PEA. Then, in separate

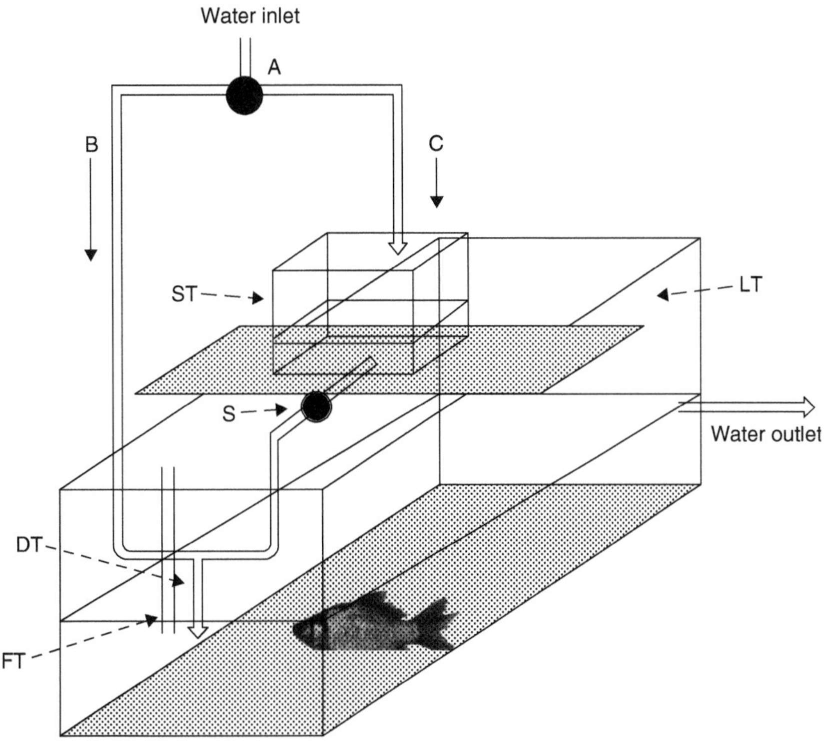

Fig. 4.1. Experimental set-up to examine the ability of fishes to detect conspecific odours, heterospecific odours or water soluble odorants. Using a switch (A), water was directed through line B during control and training sessions to the large tank (LT) where the experimental fish were trained to approach the delivery tube (DT). Whenever the fish approached the DT it was rewarded with food from a feeding tube (FT) extending into the water. S indicates a stopcock. During odour trials, water was directed via line C through a small tank (ST) containing the odour source. Fish were rewarded only in odour trials. (Modified from von Frisch, 1941.)

trials, the alcohol concentration was gradually reduced until the eel showed no tube preference. Fifteen out of 16 eels responded to concentrations as low as 10^{-15} parts of PEA and the lowest concentration detected by four eels was a 10^{-18} dilution of the alcohol. At that dilution there are fewer than two odour molecules in each microlitre of water.

In more recent laboratory studies a variety of conditioning methods have been used to demonstrate odour detection and discrimination in goldfish. For example, in a series of well-controlled laboratory studies, Zippel examined olfactory learning in goldfish (Zippel *et al.*, 1993a,b). Typically, fish were first habituated to being fed *tubifex* from a funnel tip inserted into the water. Next, an odorant was released through the funnel prior to feeding. Test trials without food were used to assess the anticipatory feeding responses (nosing and biting at the funnel tip) to odorant release alone. Goldfish were initially not attracted to relatively low concentrations of novel odorants (10^{-6} – 10^{-7} M concentrations of amyl acetate, α-ionone and β-phenylethanol) but, after extensive training, learned to associate these odours with food. They were attracted to the release of amino acids at concentrations of 10^{-5} M and higher and in threshold tests quickly learned to respond to these acids and discriminate between them at concentrations of 10^{-6} or 10^{-7} M. In all cases the ability to detect odours was lost after removing the olfactory bulbs (bilateral olfactory bulbectomy). In contrast, in fish with crushed olfactory nerves or transection of the forebrain olfactory tract, odour detection was lost but, after sufficient time for regeneration as documented in anatomical studies, performance returned to or near preoperative levels.

Nowadays, the zebrafish (*Danio rerio*) has become a model animal for genetic, sensory, behavioural and other studies (e.g. Suboski *et al.*, 1990; Fetcho and Liu, 1998; Braubach *et al.*, 2009, 2011; Blaser and Vira, 2014). In the Braubach *et al.* (2011) study, similar conditioning methods were used. A flow system was used to produce 30-s exposures of zebrafish to odorized water. An observer counted the number of sharp turns the fish made during odour exposure. Naïve zebrafish responded to the compounds l-alanine and l-valine but not to PEA. But anticipatory feeding responses to PEA as well as increased responding to amino acids could be conditioned by pairing these odours with food flakes. This allowed an estimate of odour sensitivity to be obtained in subsequent trials with lower odour concentrations. A simple odour conditioning method such as this one has proven useful in subsequent experimental studies of olfaction in zebrafish (e.g. Paskin and Byrd-Jacobs, 2012; White *et al.*, 2015). However, because these methods are time-consuming they are ill-suited for psychophysical studies in which many trials and precise control of stimulus concentration are generally needed. A variant of the Braubach *et al.* flow system combined with operant conditioning was used by Manton *et al.* (1972) to study olfaction in young sea turtles and those methods could be adaptable for fish. More recently, automated or semi-automated and computer-controlled operant conditioning methods have been described for studying fish vision and audition (Chase and Hill, 1999; Mueller and Neuhaus, 2012), and automated video tracking systems are being developed to measure experimentally induced changes in zebrafish behaviour (Stewart *et al.*, 2014). Such systems are expensive and somewhat complex but their use, combined with pattern analysis, could provide efficient automated monitoring for use in the study of olfaction in a species whose behaviour is difficult to characterize using more traditional methods.

Olfaction in Birds

For many decades zoologists did not consider olfaction an important sensory modality for birds. Early experiments to elicit physiological or behavioural responses to odours in birds failed (Walter, 1942) although later anatomical studies (e.g. Bang, 1971; Rieke and Wenzel, 1978) revealed that many species of birds have well-developed olfactory systems. Further, contemporary molecular biological studies have found that the olfactory receptor gene family in birds is quite large and diverse, making it highly likely that birds have a well-developed sense of smell (Steiger *et al.*, 2010). Indeed, initial studies by Wenzel (Sieck and Wenzel, 1969; Wenzel, 1971) demonstrated changes in heart rate and olfactory bulb electrical responses of pigeons exposed to a stream of odorized air. This pioneering work provided the impetus for further field and laboratory behavioural studies of bird olfaction. Although, for the most part, quite simple behaviour methods have been used, every species tested has been shown to respond to odours and, in many cases, to relatively low concentrations of odours (Clark *et al.*, 2014).

A Y-maze test has proven useful in a number of these studies: the subject bird is placed in the stem of the maze and is then allowed to enter the right or left arm. Different odours (or odorous objects) are placed at the end of these arms and the bird, entering the arm of its choice, provides a measure of its odour preference. Thus, for example, adult zebra finches (*Taeniopygia guttata*) tested in a Y-maze could discriminate between conspecifics based on their odour (Krause *et al.*, 2014). Zebra finch fledglings given a choice of nest odours from a strange and natal nest moved to the side of a test box containing natal nest odour (Caspers and Krause, 2010). Budgerigars, tested in a Y-maze could use the odour from secretions of the uropygial preen gland to discriminate between male and female conspecifics (Zhang *et al.*, 2010), and sea birds (Antarctic prions) tested in a Y-maze entered the arm containing odour from dimethyl sulfide (DMS). Oceanic DMS is produced by phytoplankton and constitutes the principal source of natural sulfur in the atmosphere, which is a known foraging cue for this bird species at sea (Nevitt and Bonadonna, 2005).

A wide variety of field studies have also implicated olfaction in food selection and mating; thus, courtship behaviours including the grunt whistle and head-up-tail-up display of male ducks during courting were significantly disrupted and even blocked after transection of the olfactory nerves (Balthazart and Schoffeniels, 1979). In terms of food selection, feral Oriental honey buzzards (*Pernis ptilorhynchus*) – specialist feeders of bees and wasps – given a choice of dough balls almost always chose those containing hidden pollen (Yang *et al.*, 2015).

Of particular interest is the kiwi, a flightless and nocturnal New Zealand bird that has small eyes but prominent nostrils at the end of its long beak. The beak tip falls outside of the bird's visual field and it is therefore unlikely that kiwis use vision in probing the substrate for insects and other invertebrates. But they do have a particularly well-developed olfactory epithelium and olfactory bulb (Corfield, 2014). In a set of simple tests, Wenzel (1971) demonstrated that given a choice of containers, kiwis were able to use olfactory cues in choosing to forage in those that contained food and to avoid those that contained presumably aversive odours of chemical solvents. Using similar but more sophisticated choice tests in captive kiwis, these outcomes were confirmed by Cunningham *et al.* (2009). By comparing success with active vs still mealworms they also found that, in the absence of olfactory cues, the birds could use bill-tip touch receptors (the 'scolopacid type bill-tip organ') to detect movements of active prey but were even more successful when olfactory cues were available.

Among the more dramatic examples of animal navigation are those of pelagic sea birds like petrels, albatrosses and shearwaters. These birds forage and find widely distributed patches of food over thousands of kilometres of ocean, even in overcast skies. In a series of observational and laboratory studies, Nevitt and her co-workers (2004) have shown that many sea birds use olfactory cues to find food. Sea birds can respond to odours from prey species or those associated with prey. For example, a number of petrels feed on Antarctic krill, which produce the fishy odour of pyrazine. In one study they used commercially available 3-methyl pyrazine dissolved in vegetable oil and deployed pyrazine-scented vegetable oil slicks and control vegetable oil slicks on the ocean surface at seven different locations near the sub-Antarctic South Georgia Island. Observers, blind with respect to the type of slick, identified and counted birds attracted to the slicks. Cape petrels, giant petrels, white-chinned albatrosses and black-browed albatrosses were preferentially attracted to the scented slicks, while other species including shearwaters, prions, wandering albatrosses, black-bellied storm-petrels and Wilson's storm-petrels showed no preference. Most species were attracted to another slick containing herring oil. Other such field studies provide further evidence that odours may be used in foraging and that different sea bird species are attracted by different odours (review in Nevitt, 2008). Two different models provide further support for odour foraging. According to a simulation by Clark and Shah (1992), a patch of krill 0.5 m in diameter is predicted to be detectable by storm-petrels several kilometres away, a far greater distance than the visual range of these birds when foraging close to the surface of the ocean. Models of odour dispersion indicate that vapours disperse laterally and downwind of the odour source where wind turbulence breaks the odour plumes into irregular patchy concentrations. If so, then an optimal foraging strategy for sea birds would be to fly against the wind to encounter a plume of odour and then fly in a zigzag pattern to better localize the prey (Nevitt *et al.*, 2004). In contrast, if the prey were localized visually then the bird should fly directly to it regardless of wind direction. This hypothesis was tested using 19 wandering albatrosses from a nesting colony in the Indian Ocean. The birds were fitted with GPS transmitters, their flight paths were recorded, and wind direction and wind speed data were obtained from the National Aeronautics and Space Administration/Jet Propulsion Laboratory. In brief, about 50% of the flight paths were in line with the model, the remaining paths supported a direct visually guided approach to prey. The results are consistent with at least some searches being guided by airborne vapours, and, the authors speculate, that even the 'visual' search patterns may be multimodal, with the final approach requiring both visual and olfactory input. It is assumed that the olfactory input will become more important if visual input is disturbed by bad weather or sea conditions.

A quite unexpected role of olfaction in birds relates to cues used in homing pigeons. About 45 years ago, Papi and colleagues (1972) discovered that after producing anosmia by sectioning the olfactory nerves, pigeons released from familiar locations still succeeded in navigating home but, unlike controls, failed to find their way home when released at unfamiliar locations. They hypothesized that visual cues were sufficient for navigation from familiar surroundings but that some, as yet unknown, odour cue was used when birds were released from an unfamiliar site. While this suggestion met with considerable scepticism (Schmidt-König, 1987) it received strong support from the studies of Benvenuti and Gagliardo (1996) and Gagliardo (2013). In the later study the olfactory epithelium of one nostril was destroyed using zinc sulfate and either the ipsilateral or contralateral nostril was plugged with a rubbery paste. Upon release, normal pigeons and those with plugs ipsilateral to the treated nostril oriented towards home, where most arrived on the same or next day. Experimental birds, those with nasal plugs contralateral to the treated nostril, oriented in a variety of directions upon release and most birds did not return. Recently, Pollonara *et al.* (2015) obtained similar results with pelagic shearwaters: untreated control birds, those with magnets glued to the head (to disrupt magnetic navigation) and those made anosmic or hyposmic with intranasal zinc sulfate were released over the Mediterranean Sea about 400 km from the colony home and 100 km from any coastline. Birds wore a GPS data logger allowing experimenters to track their flight paths. From the release point the paths of the control and magnetic birds did not differ; they flew in a north westerly direction towards home but anosmic birds flew north and, at least for those that returned, were, upon encountering a coastline, presumably able to use visual cues. The role of odour cues in bird navigation remains an active, fascinating and still controversial area of research (Nevitt and Bonadonna, 2005; Wallraff, 2014).

These field and simple laboratory studies made clear that birds are attentive to and make use of odours. But these methods are not well suited for examining the range of odours that can be detected, determining odour sensitivity and making cross-species comparisons under controlled conditions. Such aims would be best served by a standardized test that provides good control of the odour concentration and discrete delivery of the stimulus. These objectives were largely met in the 1959 study of Michelsen, whose work stands as a major advance in laboratory studies of olfaction with birds. Pigeons were trained using operant conditioning to respond by pecking a key for a food reward when they detected the presence of an odour. As shown in Fig. 4.2, response keys were located in a tube connected to an odour

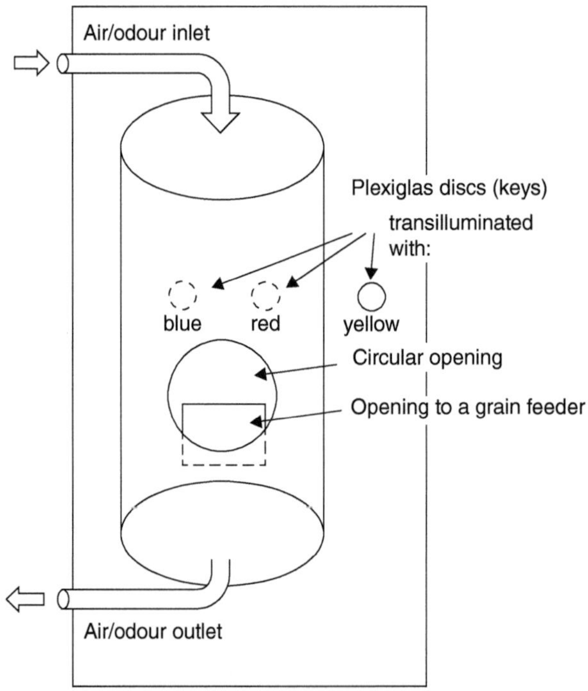

Fig. 4.2. Diagram of odour sampling tube and response keys of Michelsen (1959) for assessing odour detection in pigeons. The end wall of the bird's chamber contained an aluminium cylinder with a round opening that gave the bird access to two illuminated Plexiglas discs (response keys blue and red). An illuminated yellow key is mounted outside the cylinder. A rectangular opening to a grain feeder was located below the response keys. Odour or clean air from an odour generator could be directed through the aluminium tube to an exhaust line. The chamber and cylinder floors were covered with activated charcoal to trap lingering odours. All structures with broken lines are inside the cylinder. A trial began with the illumination of the yellow key. A peck on this key initiated the flow of clean air or odorized air through the cylinder. Nine seconds later another peck dimmed the yellow key light and illuminated the red and blue keys. Food reward was provided if the bird pecked on the blue key on odour trials or on the red key in clean air trials. (Modified from Michelsen, 1959.)

generator whose flows were controlled by electrically operated valves (solenoids). This allowed the experimenter to provide a timed stream of clean or odorized air to the subject bird. In each trial, pigeons pecked on one key until either an S+ (odour) or S− (no odour) stimulus was produced. Pecking on a second key in an S− trial or a third key in an S+ trial was rewarded with food; pecking the wrong key terminated the trial without reward. After initial training, birds achieved greater than 80% accuracy in a single 50-trial session. Michelsen's report was followed by further developments in operant conditioning methods combined with well-engineered precision odour generators for detailed studies of odour detection and discrimination in pigeons. Thus, Henton *et al.* (1966) reported on sensitivity to amyl acetate, butyl acetate and butyric acid, and later demonstrated that acquisition of an amyl acetate detection task in pigeons was abolished after sectioning the olfactory nerves (Henton *et al.*, 1969). Shumake *et al.* (1969) provided the first quantitative measure of the ability of birds to discriminate between two different intensities of the same odour (odour intensity difference threshold) and Walker (1983) determined the absolute detection threshold for amyl acetate in pigeons. This line of research reached its zenith in the study by Stattelman *et al.* (1975) in which absolute detection thresholds for pentane, hexane and heptane were assessed in pigeons, quail and chickens. In that study, charcoal-filtered air flowed continuously into a Teflon-lined test chamber and stimuli were generated by injecting an odour into the air stream. Air samples taken from the chamber were quantified using gas–liquid chromatography. Birds were reinforced with food for pecking a key when, in initial sessions, a relatively high concentration of odour was used. In subsequent sessions odour concentration was lowered in steps and training was continued until performance fell to chance levels. The results demonstrated significant differences in sensitivity among species (chickens proved to be particularly sensitive to pentane and hexane) but, more importantly, provided estimates of odour thresholds in units of physical concentration. Except for a single study comparing odour and visual learning in pigeons (Duncan and Slotnick, 1990), this promising line of bird research combining conditioning with precise control of odour stimuli appears not to have been continued.

Olfaction in Rodents

Non-associative methods

By far, the most widely used non-associative measures of olfaction in rats and mice are habituation and dishabituation (Fig. 4.3). For habituation, an odorant with which the animal cannot make physical contact is placed in the subject's cage and its odour is allowed to diffuse over a short (e.g. 2-min) test period. During this time an observer cumulates the amount of time the animal is attending to the odour source. Generally, attending is defined as the animal's snout being within 1 cm of the object containing the odorant. The odorant is removed and some minutes later the procedure is repeated and then repeated again one or two times more. In these repeated trials the subject habituates to the odour; that is, it shows less and less interest in the odour over trials, and this habituation is taken as evidence that the odour can be detected. If, on a further trial, a different odour is used and the animal then shows a renewed interest in the odour (dishabituation) this is taken as evidence that the animal discriminates between the habituated odour and the new odour (Fig. 4.3).

There are many advantages of these simple methods: they can generate evidence for odour detection and discrimination in a series of brief tests, they involve no special apparatus or training, and rats and mice can be tested in their home cage. When preceded by no odour trials (e.g. Fig. 4.3A and C) as a control to ensure that initial responses are not simply to the insertion of a new object in the cage, an increased response on the first odour trial provides convincing evidence for odour detection. In the same way, an increase in responding to a new odour after habituation to the prior odour provides evidence for odour discrimination. The importance of the initial no-odour control trials is manifest from a number of studies showing that rats and mice will investigate and show habituation to even non-odorous objects inserted into their cage (Slotnick and Coppola, 2015).

The habituation method does have a number of disadvantages: it provides no control over the odour other than its concentration at the point of entry, and subject responses depend entirely on the animal's voluntary inspection of the object containing the odorant. Furthermore, response measures depend on the judgement of an observer and/or are difficult to automatically register (e.g. by infrared beams or video tracking software). Because animals can probably detect the odour

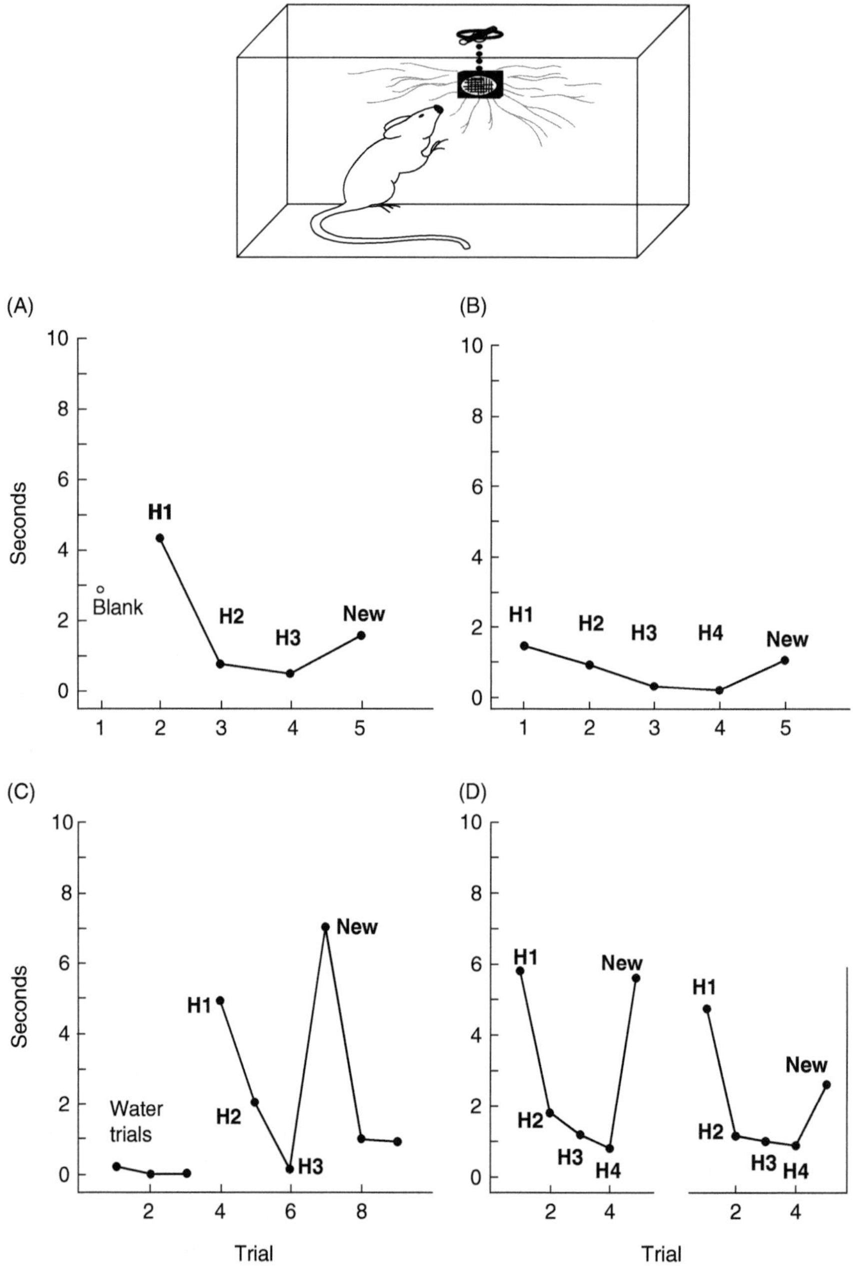

Fig. 4.3. Some typical results obtained using habituation–dishabituation tests. (A) Following a blank (no odour) stimulus, mice were habituated to odours from mice of the B6 strain and then tested with odours from bm1 strain mice (Carroll *et al.*, 2002). In this and the remaining figures H indicates the habituation trials and NEW indicates a trial with a different odour. (B) After four habituation trials to one odour (e.g. ethyl acetate) mice were tested with a homologous odour (e.g. amyl acetate; Freedman *et al.*, 2013). (C) After three trials with water, male mice (C57BL strain) were habituated to the odour of urine from male mice of the same strain, then tested three times with the odour of urine from female C57BL mice (Pankevich *et al.*, 2004). (D) Left graph: B6 strain mice were habituated to the odour of limonene (+) and then tested with the enantiomer limonene (–); right graph: the identical test with mice lacking ERK5 MAP kinase (Zou *et al.*, 2012).

even when not sniffing directly at its point source, sniff scores underestimate odour sampling time. Indeed, in most 2-min tests the average 'sniff' score on the first introduction of an odour is only about 5 s or about 4% of the test period, and the animal's contact with the odour in the remaining test period is unknown. Habituation is also highly parameter dependent; for example, it occurs more slowly, if at all, with high intensity stimuli and, because it recovers with time, it is ill-suited for studies of odour memory (McNamera *et al.*, 2008). Partly as a consequence of the brief exploration of novel odours, the effect size in most studies tends to be relatively small; evidence for habituation and dishabituation may involve changes of only a few seconds in attending to the stimulus (Fig. 4.3) and, given that observers are seldom blind to which trial is being run or which odour is being used, unintended observer bias in judging when the subject meets the arbitrary criterion of odour sampling could be a concern. To avoid or minimize such an observer bias, automated measurement procedures should always be used if possible.

An interesting constraint is that the habituation method does not lend itself to multiple within-group tests. That is because there are strong carry-over effects and after one or two exposures to a habituation–dishabituation routine, animals may show little interest in exploring new novel odours (habituation of dishabituation; Thompson and Spencer, 1966).

Finally, null results, e.g. the failure to find a difference between experimental groups or failure to provide evidence for odour detection or discrimination as a result of some treatment, must be evaluated with caution. This is because habituation provides a relatively weak measure of olfaction: positive results in well-controlled tests certainly provide evidence of olfactory function but there may be multiple reasons for a failure to attend to a novel odour or to detect differences among treatments or groups. As discussed in greater detail below, a number of studies using conditioning combined with precision olfactometry have failed to confirm several claims based on null outcomes of the habituation–dishabituation method.

Simple non-automated associative conditioning methods

Conditioning methods are a departure from simple habituation or free choice studies because they employ motivated, often hungry or thirsty subjects that are trained using associative conditioning to respond to an odour for a reward. They generally use a special albeit simple apparatus and a training procedure aimed at having a defined (and measured) behaviour displayed depending on the stimulus provided (e.g. the stimulus robustly induces a behavioural response). While these methods are more time-consuming than non-associative tests, they have marked advantages: they use objective, easily scored measures of response accuracy and allow multiple, even hundreds of trials to be run on individual subjects to assess odour sensitivity as well as a variety of simple and complex odour discrimination tasks.

A number of somewhat idiosyncratic methods such as fear conditioning, odour aversion, special types of mazes or searching for a buried food object have been employed but have not gained wide acceptance. Other non-automated associative methods have proven quite useful. Maze learning is probably the most widely used instrumental conditioning method and some laboratories have generated a large body of data using a maze (e.g. Y-maze training, Yamaguchi *et al.*, 1981). Although effective, maze training is labour intensive; several sessions are generally required to adapt animals to the maze and many, often hundreds, of trials may be required to train mice on detection and discrimination tasks. More recently, two simpler and more efficient conditioning methods for mice have been described: using odours to indicate which of two (or more) containers to dig in to obtain a food reward (the 'dig' test), and training mice to use odours to avoid ingesting a bitter tastant, the OCTA (odour cued taste avoïdance) test.

In the dig test, rodents are first trained to dig in a cup to obtain a food reward that is covered with some odorized material (e.g. sand or woodchip bedding). After this initial training the subject is presented with two cups that are identical except that one contains the training odour and the other no odour (for simple odour detection tasks) or a different odour (odour discrimination). Identical food items are present in both but are accessible only in the cup with the 'positive' odour previously associated with food.

Rats and mice learn quickly to dig to obtain a food reward and acquire detection and discrimination tasks in relatively few trials (Fig. 4.4A and B). The test provides two easily obtained and objective measures of response accuracy: time spent digging in each container and initial choice of container. The latter measure is perhaps the more useful as, once trained, rats and mice often sniff at one container before deciding which to dig into. Provided the usual precautions are taken (ensuring that the two containers differ only with regard to the test odours contained and that the right/left positions of the stimuli are randomized) the dig method provides a robust and objective measure of odour detection and discrimination. Carry-over effects are not an issue and, as shown in Fig. 4.4C and D, the same subjects can be tested over a large number of problems to assess, for example, odour detection threshold (Fig. 4.4C), stimulus generalization (Fig. 4.4D) and a variety of more complex instrumental and cognitive tasks (e.g. Dusek and Eichenbaum, 1997; Dreumont-Boudreau *et al.*, 2006). The disadvantages to the method are that, like the habituation test, the odours of interest are allowed to diffuse freely and, because animals dig vigorously, the minor but troublesome inconvenience of cleaning the arena and using fresh stimuli for the next trial.

A quite different non-automated associative conditioning procedure is the OCTA test (Fig. 4.5) (Kimura *et al.*, 1991; Darling and Slotnick, 1994; Nathan *et al.*, 2004; Slotnick and Coppola, 2015). For this, a thirsty subject is given a series of five to ten trials to drink water or drink an aqueous solution of an odorant (the S+ stimulus). Next, it is presented with one of two liquids in separate trials. In some trials the S+ stimulus is presented, while in other trials a solution containing a different odour dissolved in a bitter tasting substance (e.g. quinine) is presented (the S− stimulus). In each trial, the subject is allowed a fixed amount of time (e.g. 30 s) to make contact with the solution. If contact is made it is allowed a few seconds (e.g. 5 s) to drink. At the end of the trial, the subject is briefly confined in the start compartment while the investigator sets up the next stimulus. The subject is then released and the next trial is initiated. A touch-sensitive circuit is used to detect contact with the liquid stimulus. A computer is used to record inputs and time events. The outcome of each trial is measured by response latency and, if contact with fluid is made, the percentage of time spent drinking. Computer control over time intervals and recording of fluid contact is not necessary but is useful in maintaining a record of outcomes and freeing the experimenter from timing intervals on each trial. If a computer is not used then a touch circuit can be wired to illuminate an LED to signal contact with the fluid stimulus (Slotnick, 2009). As currently practised, the OCTA has one important disadvantage: the fact that many water-soluble bitter substances themselves may have a detectable odour. The odour of quinine is easily masked by commonly used odorants but post-learning S− trials without the bitterant should be run to ensure performance is based on the odour of interest and not that of the bitterant (e.g. Fig. 4.5B).

One clear advantage of the OCTA method is that acquisition of odour detection and odour discrimination is both remarkably rapid and uniform among subjects: after only one or two contacts with the bitter tasting solution, rats and mice will quickly and consistently drink the S+ solution but avoid or hesitate before sampling the S− solution (Fig. 4.5). Subjects can be clearly observed to carefully sniff at the fluid before making or refusing to make fluid contact. The same mice can be trained in a variety of discrimination tasks and tested for memory of prior learning. Finally, OCTA is one of the few simple tests of olfaction that has been validated using anosmic animals: olfactory bulbectomized rats and mice cannot recall prior odour learning and are unable to acquire a new odour detection task (Darling and Slotnick, 1994; Slotnick and Coppola, 2015).

Olfactometry – combining precision control of odours with operant conditioning

Strictly speaking, an olfactometer is an instrument used to measure odours. In common parlance, however, it refers to a device with multiple functions: the dilution and generation of a quantifiable odour from an odorant source, the channelling of the odour to the test subject, and control of the onset and offset of the odour vapour. Olfactometers not only allow reasonably precise control of odours but they can also be automated and computer-controlled (review in Slotnick and Schellinck, 2002). When combined with operant conditioning, olfactometers provide the most precise, versatile, and powerful method for studying olfactory psychophysics, learning and memory in normal animals, and can be used to evaluate the effects of drugs, toxins, surgical interventions and genetic manipulations on the sense of smell.

An important function of an olfactometer is dilution of the odorant vapour. This is accomplished either by air dilution of the saturated vapour (air dilution olfactometry) or by sampling the head space of a liquid dilution of the odorant (liquid dilution

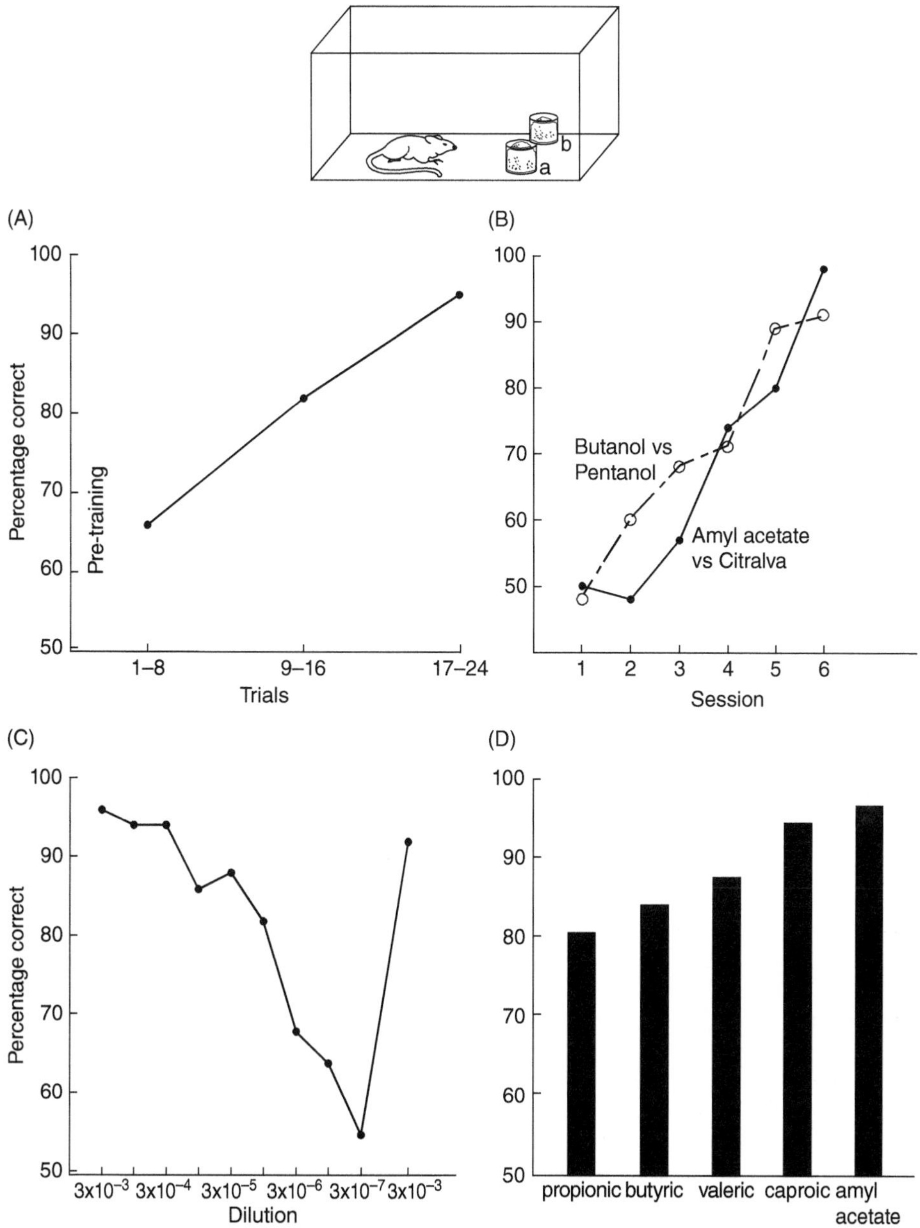

Fig. 4.4. Representative outcomes of studies in which mice were trained to dig into a (S+) scented substrate for a food reward. (A) Mean acquisition for six mice trained in three daily eight-trial sessions to choose nutmeg (S+) over cinnamon (non-rewarded; S–) scented sand. Mice were initially trained to dig in nutmeg-scented sand for a food reward (Mihalick *et al.*, 2000). (B) Using similar methods, mice were trained to discriminate between structurally similar pairs of odours in daily eight-trial sessions (Zou *et al.*, 2012). (C) Male mice were trained to detect sequentially lower concentrations of male mouse urine. The last session served to ensure that, when detectable, the mice were still attending to the stimulus (Sorwell *et al.*, 2008). (D) Odour similarity across a homologous series of odorants. Mice were trained to discriminate between acetic acid odour (S+) from those varying only in carbon chain length and, as a control, from an unrelated odour amyl acetate (Cleland *et al.*, 2002). Note that discrimination difficulty is related to the difference in carbon chain length between acetic acid (C2) and the homologous odorants (C3 to C6).

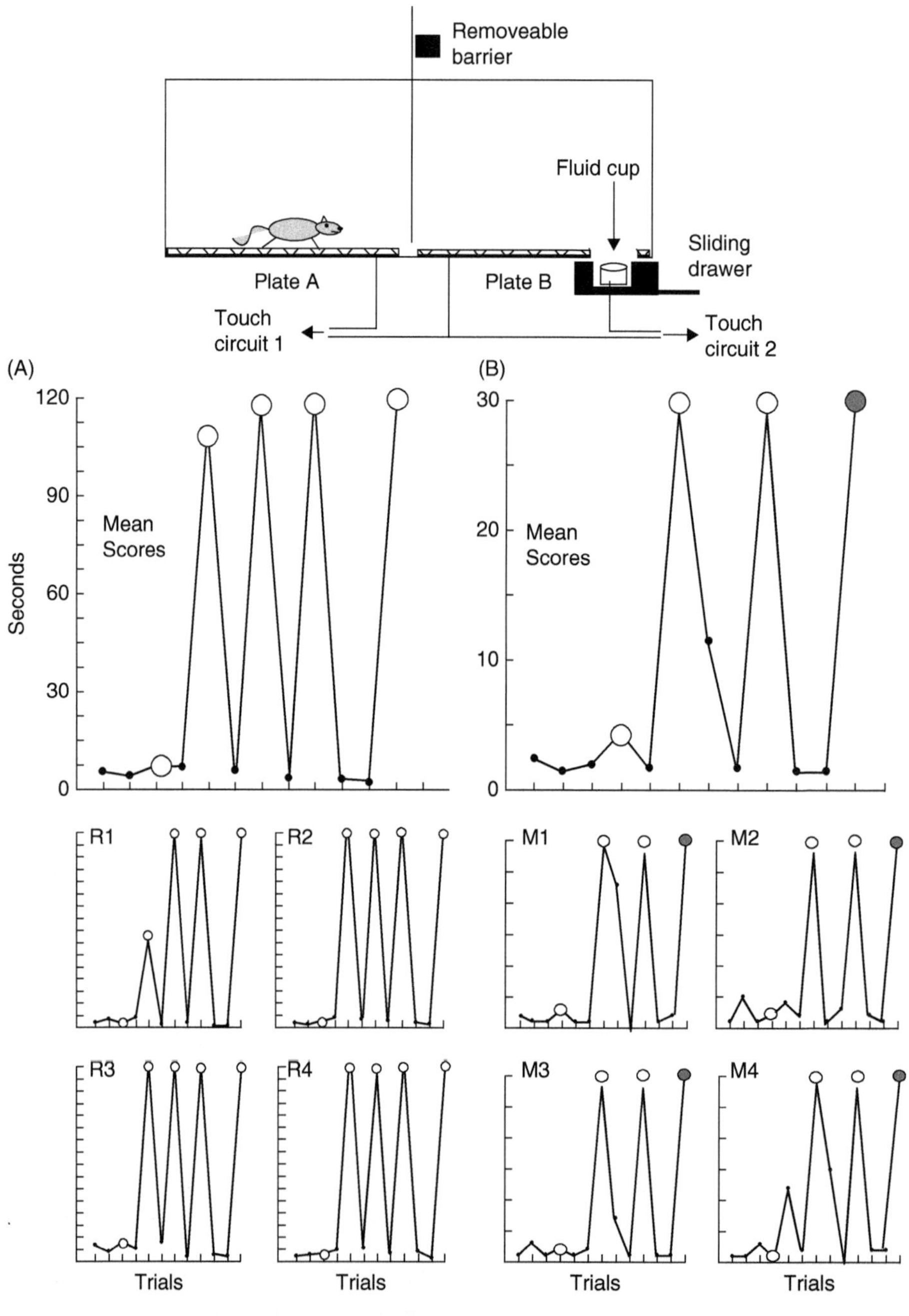

Fig. 4.5. Chamber for conducting odour cued taste avoidance (OCTA) tests. The animal is confined in the rear of the chamber during the inter-trial interval. Crossing from Plate A to Plate B signals approach to the stimulus and contact with fluid is recorded when the animal completes the circuit between Plate B and the fluid cup by touching the latter. (A) Mean response latencies for four rats (R1–R4) trained to discriminate between the S+ of water (small solid circles) and the S– aqueous solution of 0.1% amyl acetate and 0.5% quinine (open circles; Darling and Slotnick, 1994). (B) Mean response latencies for four mice (M1–M4) trained to discriminate between the S+ solution of 5% McCormick almond flavouring (small solid circles) and the S– solution of McCormick vanilla flavouring and 5 mM quinine (open circles; Slotnick and Coppola, 2015). Quinine was not used in the last S– trial (shaded circles).

olfactometry). In practice, air dilution systems have the advantage of generating odour vapours of known molecular concentration. Unfortunately, air dilution systems tend to be somewhat complex, have multiple components, are difficult to maintain, require large volumes of odour-free air and are capable of generating only one or two odours. In contrast, liquid dilution systems are much simpler, have relatively few components, require relatively small volumes of air and are capable of generating multiple different odours or different concentrations of the same odour. The chief disadvantage of a liquid dilution system is that few odorants form ideal mixtures with non-odorous solvents and, therefore, it is difficult to predict the molecular concentration of the odorant in the head space above the diluted odorant. Although approximations of head space concentration can be calculated and, in some cases, have been measured using gas chromatography (see Slotnick and Restrepo, 2005) essentially all studies using a liquid dilution olfactometer report odour concentration as percentage dilution of the odorant in a solvent.

A second and equally important function of an olfactometer is precise control of odour onset and offset. This is generally accomplished using electrically operated solenoid valves. Generally, the odour container for each channel is isolated by two normally closed valves, one upstream and one downstream of the container; operation of these valves introduces a stream of odorized air into the carrier flow (Fig. 4.6). A commonly used small animal olfactometer provides for both liquid and air dilution of the odour. As shown in Fig. 4.6, the odour is generated by adding a small volume of air from the odour saturator to a much faster and continuous clean air carrier flow. Thus, the original odorant is diluted twice: the major dilution is with a liquid solvent and the vapour sample from that dilution is manifolded with clean air carrier streams. A three-way valve (final valve, FV, in Fig. 4.6), when operated, diverts all air flow from the animal sampling port into an exhaust line. This valve operates for 1 s in each trial; operation of this valve serves to alert the subject of the imminent arrival of the odour stimulus and diverts flow from the subject sampling tube during the initial mixing of the odour and clean air streams. The animal operant chamber, constructed of Plexiglas, has a stainless steel floor and a small intake fan in the wall. Less than 3 l/min of air are required by the odour generator. The opposite end contains a glass odour sampling port and water delivery spout. Snout insertions into the odour sampling port are monitored by an infrared photocell unit, and tongue contacts with the water spout are detected by a low current (~0.5 μA) touch circuit.

Thirsty rats or mice are trained using traditional operant conditioning methods. Initially subjects are trained to insert their snout into the odour sampling port and, upon detecting an odour (the S+ stimulus), lick at the water spout. Successful responses are rewarded by water delivery from a gravity fed and solenoid controlled reservoir. In subsequent sessions this procedure is continued except that in each trial either the S+ stimulus or the S– stimulus (either no odour or a different odour) is presented. Each type of stimulus is presented an equal number of times in each 20-trial block and 200–400 trials can be given in daily sessions. Responding correctly in an S+ trial (licking the water delivery tube in the presence of the odour) is scored as a *Hit*, failing to respond in an S+ trial is scored as a *Miss*. Not responding in an S– trial is scored as a *Correct Rejection* and responding in an S– trial is scored as a *False Alarm*. A brief (5-s) inter-trial interval separates trials and, on average, rats and mice complete 100 trial sessions in about 20–30 min.

Early studies using automated olfactometers in conjunction with operant conditioning demonstrated that rats and mice learned to respond on the water delivery tube on S+ trials and to inhibit responding on S– trials generally within a single 100-trial session. Simple odour detection and discrimination tasks were acquired rapidly, often in 20–60 trials (Fig. 4.7) and estimates of odour detection threshold could be obtained in a single session (Fig. 4.7B), although additional training would be required for better precision. The olfactometer has proved useful for other psychophysical measures including determining intensity difference thresholds, assessing odour masking and testing a variety of odour mixture tasks. Capacity to learn odours is probably quite large; rats and mice quickly learned to discriminate among eight different odours (Fig. 4.7C) and, after learning a number of such odour sets, showed excellent memory for ones learned earlier (Slotnick *et al.*, 1991). Rodents also performed well in more cognitive tasks including acquisition of a learning set (Fig. 4.7D) and matching and non-matching to sample. The olfactometer method is widely used to assess odour memory, the effects of neurotoxins, diet, discrete

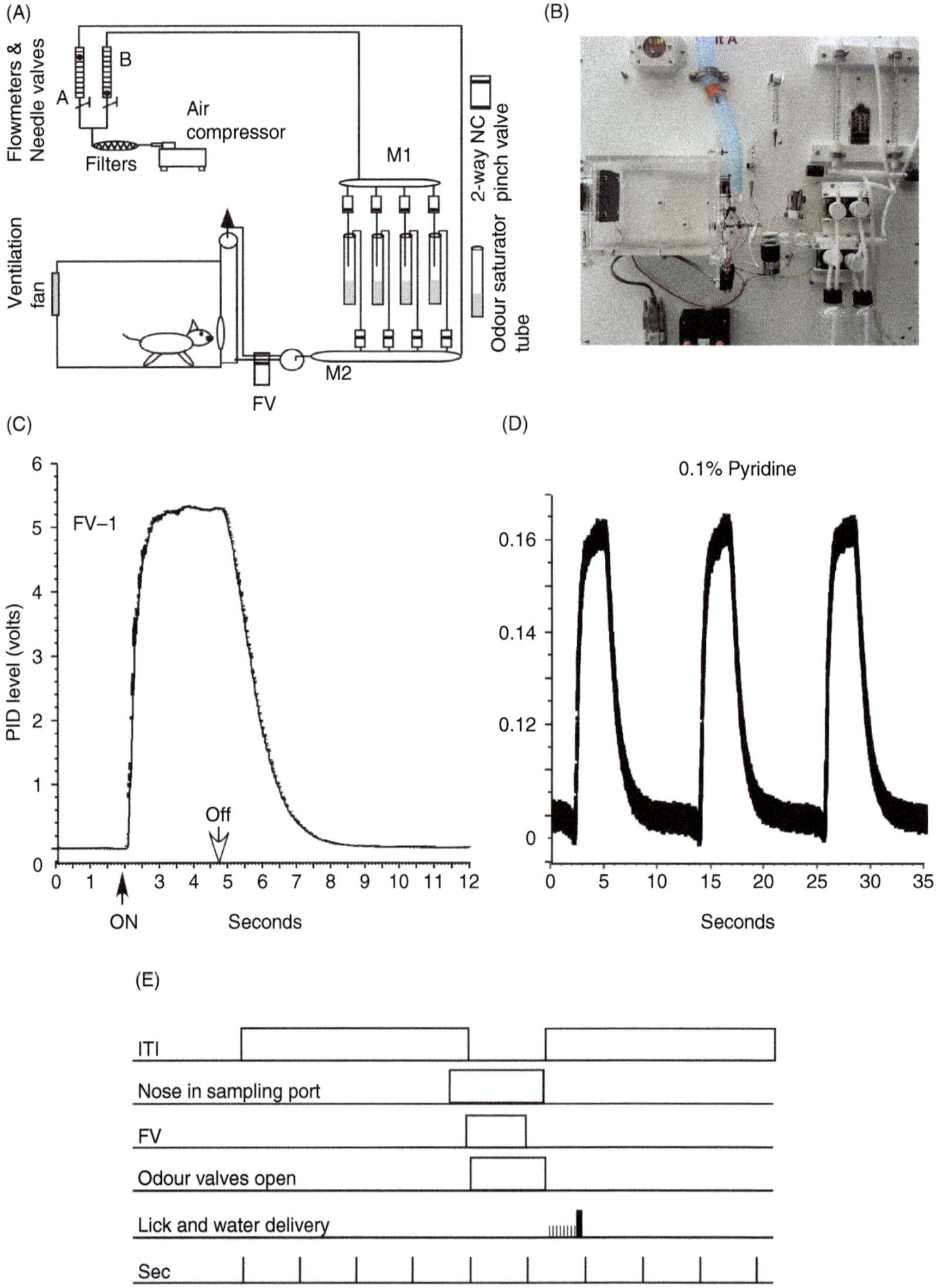

Fig. 4.6. (A) Diagrammatic representation of olfactometer components. M1 and M2, manifolds; FV, final valve. (B) Simple two-channel olfactometer for mice. (C) Photo ionization recording from the animal odour sampling tube showing the changes in odour level at the onset and offset of a 2.7-s trial. (D) Photo ionization recording showing reliability in odour production in three consecutive trials separated by 8-s inter-trial intervals. (E) Events during a typical S+ trial. ITI, inter trial interval; FV, final valve operation.

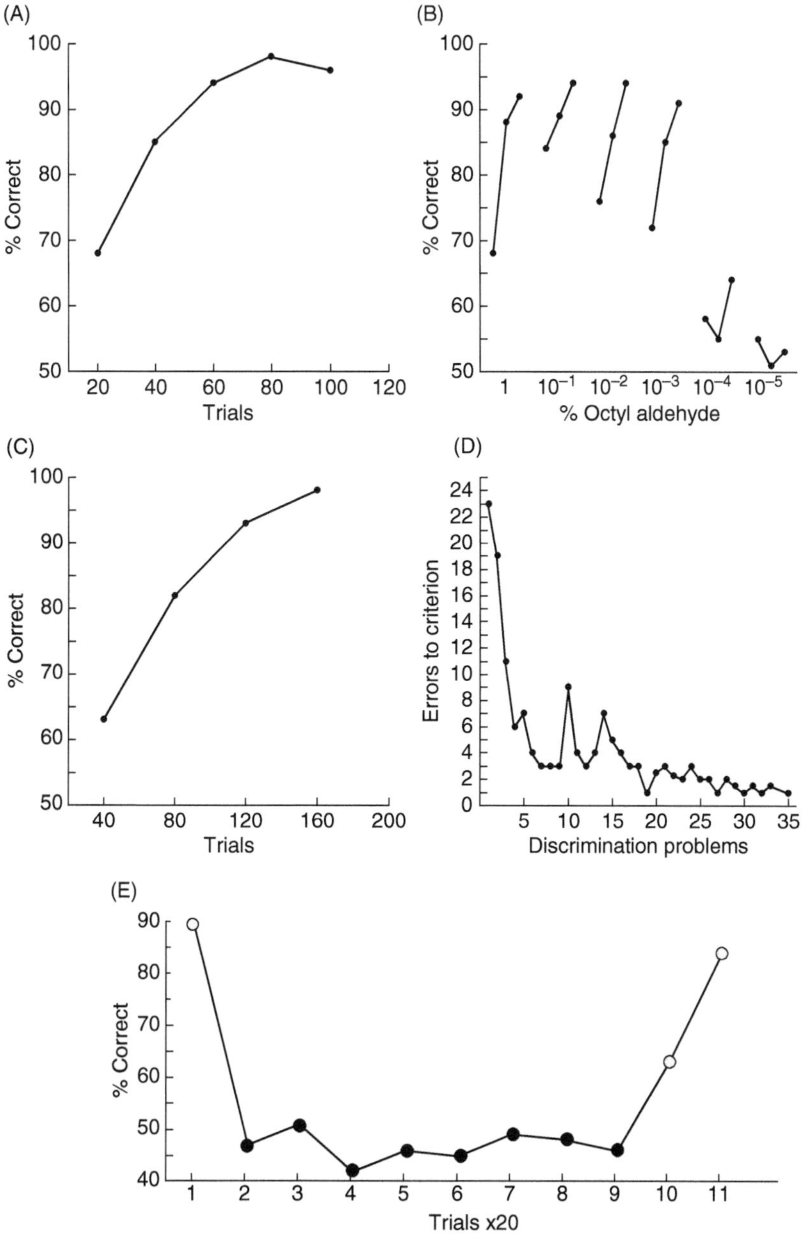

Fig. 4.7. (A) Rapid acquisition of a simple odour discrimination task: mean performance of ten mice trained to discriminate 1% ethyl acetate (S+) from 0.1% amyl acetate (S−). (B) Performance accuracy for the same ten mice trained to detect descending concentrations of octyl aldehyde. With additional training (not shown), mice were able to detect much lower concentrations of the odour. (C) Mean accuracy of four mice trained to discriminate among eight different odours in a 200-trial session. Four odours served as S+ stimuli and four as S− stimuli; each odour was presented five times in each block of 40 trials. (D) Acquisition of an olfactory learning set: mean number of errors for eight rats trained on 35 consecutive two-odour discrimination tasks. After problem 25 most rats had scores of 90–100% correct responding on the first 20 trials of a new task. (E) Mean performance for two mice in a no-odour control test (% correct out of 20 trials). Trial block one shows terminal performance in a simple odour detection task (1% ethyl acetate S+ and mineral oil S−). Both channels contained only the mineral oil solvent in trial blocks 2–9; the S+ stimulus was back in place in trial blocks 10 and 11. (All data are from Slotnick *et al.*, 2001, and Slotnick and Restrepo, 2005.)

lesions of the olfactory system, and changes in genetically modified mice and neuroplasticity.

An interesting recent development in rodent olfactometry is the fixed head preparation. For some studies, including those using single unit or field recording or imaging of bulbar activity in the behaving animal, it is necessary to have the head of the animal fixed in place. Surgical procedures are described in detail by Guo *et al.* (2014) but, briefly, involve cementing a titanium bar to the skull. This allows the head to be held rigidly in place during behaviour training. After preliminary adaptation to the apparatus, mice can be trained using traditional discrimination methods to lick at a water delivery port in the presence of one odour and to inhibit licking in the presence of another odour. Head fixed mice rapidly learn a simple odour discrimination task (Guo *et al.*, 2014).

Using head fixation is particularly useful in optogenetic studies of genetically modified mice whose olfactory sensory neurons contain light sensitive channel-rhodopsins. Small clusters of olfactory sensory neurons in these mice can be activated by a high power LED, and mice can be trained to respond to both odorized air and light onset (e.g. Baranauskas, 2015; Gschwend *et al.*, 2015; Lehmann *et al.*, 2016). Because light can be controlled with much greater precision than can odours, optogenetics provides a unique opportunity to stimulate small discrete areas of the olfactory epithelium or olfactory bulb and examine the spatiotemporal characteristics of olfactory physiology.

Testing the valence of odours

Odours can have a large impact on the life of an animal and often provide the first cues to biologically important stimuli such as food, approach of group members, mating partners, predators or diseases. These stimuli have a high valence for the survival and very often are characterized by inducing both innate behavioural responses and/or by very rapid (one-trial) learning. Especially for small animals, predators are the most dangerous environmental threats and it is of pivotal importance for them to recognize a potentially dangerous animal as early as possible to avoid an encounter or to prepare for an encounter with the predator.

Responses to potentially positive (appetitive) or negative (avoidance) stimuli are easily assessed in simple enclosures by noting approach or avoidance to an odorant source (e.g. Amiri *et al.*, 1998). Basically, odours with a positive impact for the survival of an animal induce appetitive behavioural responses, while odours with a negative impact trigger aversive behavioural responses. However, the behavioural responses of an animal are not always that simple and are modulated by a number of variables such as previous experience, environmental context or distance to the odour source. Changes in the valence of novel and familiar odours (including nest and food odours) produced by associative conditioning can also be assessed in open field choice tests (Terry and Johanson, 1996).

The effects of predator odours on animal behaviour are well investigated (for a review, see Apfelbach *et al.*, 2005; Takahashi *et al.*, 2005). Of particular interest is the response of laboratory rats to the odours of predators. Laboratory rats have been bred under controlled conditions for about 100 years without any contact with predators. Such rats uniformly show defensive reactions to predator odours (e.g. urine samples of domesticated cats, foxes or bobcats) or synthesized components of predator odours such as trimethylthiazoline (TMT). Therefore, their defensive response to the odours of predators is considered to be innate (see Chapter 2).

When tested in a simple enclosure, rats may cautiously approach the odour source in a very typical manner, the so-called flat-back approach, then withdraw to the furthest corner and even stay very still (e.g. Blanchard and Blanchard, 1989; Apfelbach *et al.*, 2005; Takahashi *et al.*, 2005). Other behavioural modulations observed in rats during exposure to a predator odour include the emission of ultrasonic alarm calls, an enhanced startle response or – when a shelter is available – the animals will spend more time within this shelter and express so-called 'head out behaviour' to assess potential risks (Dielenberg and McGregor, 2001). Often, animals also reduce their locomotor activity but are still highly alert. In the most extreme case, they will 'freeze' – no movements are visible except those necessary for breathing. These defensive changes are accompanied by stress-like physiological changes, like an increase in heart rate, blood pressure, breathing rate and the release of stress hormones (Takahashi, 2014).

As with most studies, the control conditions are of particular importance. Depending on the goal of the study, different control conditions have to be used. A negative control would be an 'odour' sample without any smell (e.g. purified water). If a solvent is used to dissolve an odour, the solvent should also serve as a further control stimulus. A positive control is meaningful, for instance an odour for which the expected valence of the to-be-tested odour was already demonstrated. In the best case, an established odour with an opposite valence should be used. Such an odour would demonstrate that the used behavioural paradigm is sensitive to both negative and positive valence.

For a very basic experiment, naïve rats, a rectangular experimental arena and an odour (e.g. fox urine) are needed. It is important that the rats show no preference for a specific corner under control conditions. Then, a small urine sample (1–2 ml) can be placed into a glass vial or dropped onto a filter paper and deposited in one corner of the arena. If many animals are tested, the 'odorized' corner should be changed in a balanced way and clean test arenas should be used each time. Another approach might be to put the rat into an experimental box (a shelter) located in the arena. It is very likely that the rat will leave the box and start to explore the arena (due to a motivation to investigate new surroundings); however, the predator odour will modulate this exploratory behaviour and induce behavioural defence responses. If the perceived danger is moderate for the rat it may display a flat-back approach, and may even have nose contact with the odour sample. Both behaviours are part of defensive risk assessment behaviour in rodents. If the perceived danger is high, the corner will be less often visited or even avoided compared with the other corners or compared with a control odour.

Figure 4.8A and B summarizes a typical experiment, using a square test arena. The question was whether urine samples of the fishing cat (*Prionailurus viverrinus*) are aversive, neutral or appetitive for rats. Since fishing cats are carnivores, the working hypothesis of this study was that the odour of these urine samples would have negative valence. Ten male rats were exposed daily to one of four different odours that were presented in a balanced order and the corner used for odour presentation was randomized. The percentage of time the rats spent in the quadrant of odour presentation was used as a main measure. Since a quadrant is one-fourth of the test arena, the chance level for being in the odour quadrant is 25%. When the neutral control (water) was presented, the time spent in the water quadrant did not differ significantly from chance level. In contrast, the positive control bobcat urine (several publications have already demonstrated an aversive valence of bobcat urine, e.g. Fendt, 2006) led to a presence of only 2.9% in the odour quadrant, which is significantly less than chance level. This was also the case for the fishing cat odour (12.9%), although to a lesser extent. These findings demonstrate that the set-up and the protocol are able to detect the negative valence of an odour. As an additional odour, urine samples of female rats were tested as a negative control. According to the literature, female urine should have appetitive effects and, indeed, male rats spent significantly more time than chance level in the quadrant with female urine – indicating that also a positive valence can be detected.

Using a similar experimental set-up, the hypothesis that 2-phenylethylamine (a component of carnivore urine) is responsible for the negative valence of carnivore urine was tested on rats (data from Ferrero *et al.*, 2011). This hypothesis was based on the observation that 2-phenylethylamine is highly concentrated in carnivore urine but not in herbivore urine and that the trace amine-associated receptor 4 (TAAR4) is highly responsive to both carnivore urine and 2-phenylethylamine. Most probably, 2-phenylethylamine is a metabolic product of the carnivore diet. In a first experiment, the researchers tested whether this compound has a negative valence similar to carnivore urine. As shown in Fig. 4.8, 2-phenylethylamine was avoided like the positive controls (urine samples from lions and coyotes). Benzylamine, which is chemically closely related to 2-phenylethylamine but does not activate TAAR4, served as an additional control. Benzylamine as well as the neutral control water did not induce any avoidance behaviour. Furthermore, 2-phenylethylamine induced an increase in the corticosterone level of exposed rats. In a second experiment, 2-phenylethylamine was added to urine samples from giraffe, an odour source that is not avoided by rats. With increasing 2-phenylethylamine concentrations in the urine the avoidance reaction increased, confirming that this type of test can give an indication of the relative strength of odour valence. In a last experiment, lion urine was depleted from 2-phenylethylamine by adding the enzyme monoaminoxidase B, and this 2-phenylethylamine-depleted lion urine did not induce an avoidance response. However, after re-spiking the depleted lion samples by adding 2-phenylethylamine, the avoidance response returned. In summary, David Ferrero and

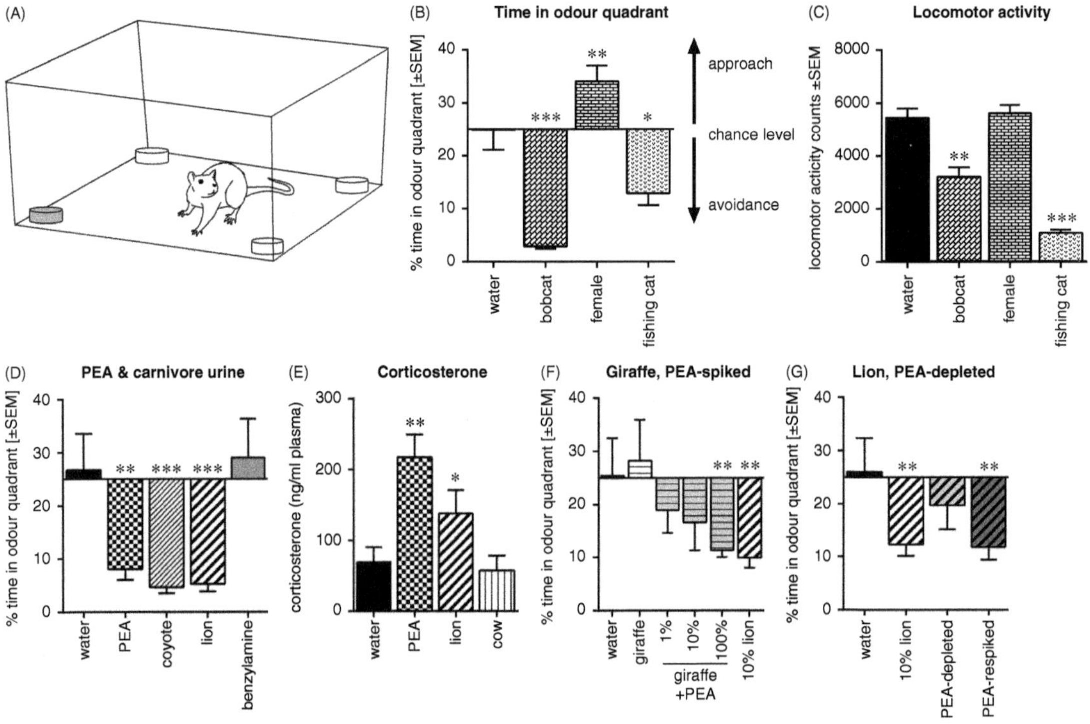

Fig. 4.8. Testing odour valence in an open field arena. (A) Experimental arena with a rat and one glass vial in each corner. One vial contains the odour sample. (B) The percentage time rats spent in the odour quadrant of the arena is depicted. Urine samples from bobcat and fishing cat are avoided, whereas urine from female rats is approached. (C) This avoidance behaviour is typically accompanied by a reduction in general locomotor behaviour. (D) 2-phenylethylamine (a ubiquitous component of carnivore urine; PEA) is similarly avoided as are coyote and lion urine, whereas benzylamine (a derivate of 2-phenylethylamine) has no effect. (E) 2-phenylethylamine and lion urine induce an increased release of the stress hormone corticosterone. (F) Urine samples of the herbivore giraffe do not induce avoidance behaviour. However, 2-phenylethylamine-spiked giraffe urine is avoided as is lion urine. (G) 2-phenylethylamine-depleted lion urine is avoided less but adding 2-phenylethylamine to the depleted samples restores avoidance. (All data are from Ferrero *et al.*, 2011, and Wernecke, 2016.)

his co-workers nicely demonstrated that (i) 2-phenylethylamine is highly concentrated in carnivore urine; (ii) is avoided by rodents; (iii) herbivore urine is avoided if 2-phenylethylamine is added; and (iv) 2-phenylethylamine-depleted carnivore urine is not avoided.

Note

[1] Karl von Frisch, a pioneer in studying animal behaviour and sensory physiology was awarded the Nobel prize in Physiology and Medicine (1973) jointly with Konrad Lorenz and Nikolaas Tinbergen 'for their discoveries concerning organization and elicitation of individual and social behaviour patterns'.

References

Amiri, L., Dark, T., Noce, K.M. and Kirstein, C.L. (1998) Odor preferences in neonatal and weanling rats. *Developmental Psychobiology* 33, 157–162. DOI: 10.1002/(SICI)1098-2302(199809)33:2<157::AID-DEV6>3.0.CO;2-I

Apfelbach, R., Blanchard, C.D., Blanchard, R.J., Hayes, R.A. and McGregor, I.S. (2005) The effects of predator odors in mammalian prey species: a review of field and laboratory studies. *Neuroscience and Biobehavioral Reviews* 29, 1123–1144. DOI: 10.1016/j.neubiorev.2005.05.005

Balthazart, J. and Schoffeniels, E. (1979) Pheromones are involved in the control of sexual behaviour in birds. *Naturwissenschaften* 66, 55–56.

Bang, B.G. (1971) Functional anatomy of the olfactory system in 23 orders of birds. *Acta Anatomica* (Basel) 79, 1–76.

Baranauskas, G. (2015) Can optogenetic tools determine the importance of temporal codes to sensory information processing in the brain? *Frontiers in Systems Neuroscience* 9, 174. DOI:org/10.3389/fnsys.2015.00174

Benvenuti, S. and Gagliardo, A. (1996) Homing behaviour of pigeons subjected to unilateral zinc sulfate treatment of their olfactory mucosa. *Journal of Experimental Biology* 199, 2531–2535.

Blanchard, R.L. and Blanchard, D.C. (1989) Antipredator defensive behaviour in a visible burrow system. *Journal of Comparative Psychology* 103, 70–82. DOI: 10.1037/0735-7036.103.1.70

Blaser, R.E. and Vira, D.G. (2014) Experiments on learning in zebrafish (*Danio rerio*): a promising model of neurocognitive function. *Neuroscience and Biobehavioral Reviews* 42, 224–231.

Braubach, O.R., Wood, H.D., Gadbois, S., Fine, A. and Croll, R.B. (2009) Olfactory conditioning in the zebrafish (*Danio rerio*). *Behavioural Brain Research* 198, 190–198. DOI: 10.1016/j.bbr.2008.10.044

Braubach, O.R., Wyeth, R.C., Murray, A., Fine, A. and Croll, R.B. (2011) A simple and effective method to condition olfactory behaviors in groups of zebrafish. In: Kalueff, A.V. and Cachat, J.M. (eds) *Zebrafish Neurobehavioral Protocols*, Neuromethods Vol 51, pp. 85–97. DOI: 10.1007/978-1-60761-953-6_7

Carroll, L.S., Penn, D.J. and Potts, W.K. (2002) Discrimination of MHC-derived odors by untrained mice is consistent with divergence in peptide-binding region residues. *Proceedings of the National Academy of Sciences USA* 99, 2187–2192. DOI: 10.1073/pnas.0422899

Caspers, B.A. and Krause, E.T. (2010) Odour-based natal nest recognition in the zebra finch (*Taeniopygia guttata*), a colony breeding song bird. *Biological Letters* 7, 184–186.

Chase, A.R. and Hill, W. (1999) Reliable operant apparatus for fish: audio stimulus generator, response button, and pellet-dispensing nipple. *Behavior Research Methods Instrumentation and Computing* 31, 470–478. DOI: 10.3758/BF03200728

Clark, L. and Shah, P.S. (1992) Information content of door plumes: what do Leach's storm-petrels know? In: Doty, R.L. and Müller-Schwarze, D. (eds) *Chemical Signals in Vertebrates VI*. Plenum Press, New York, pp. 421–427.

Clark, L., Hagelin, J. and Werner, S. (2014) The chemical senses in birds. USDA National Wildlife Research Center – Staff Publications, paper 1618.

Cleland, T.A., Morse, A., Yue, E.L. and Linster, C. (2002) Behavioral models of odor similarity. *Behavioral Neuroscience* 116, 222–231. DOI: 10.1037//0735-7044.116.2.222

Cooper, J.C., Scholz, A.T., , R.M., Hasler, A.D. and Madison, D.M. (1976) Experimental confirmation of the olfactory hypothesis with homing, artificially imprinted Coho salmon (*Oncorhynchus kisutch*). *Journal of the Fisheries Research Board of Canada* 33, 703–710. DOI: 10.1139/f76-087

Corfield, J.R., Eisthen, H.L., Iwaniuk, A.N. and Parsons, S. (2014) Anatomical specializations for enhanced olfactory sensitivity in kiwi, *Apteryx mantelli*. *Brain Behavior and Evolution* 84, 214–226. DOI: 10.1159/000365564

Cunningham, S.J., Castro, I. and Potter, M.A. (2009) The relative importance of olfaction and remote touch in prey detection by North Island brown kiwis. *Animal Behavior* 78, 899–905. DOI: 10.1016/j.anbehav.2009.07.015

Darling, F.M.C. and Slotnick, B. (1994) Odor-cued taste avoidance: a simple and efficient method for assessing olfaction in rats. *Physiology and Behavior* 55, 817–822.

Dielenberg, R.A. and McGregor, I.S. (2001) Defensive behaviour in rats towards predatory odors: a review. *Neuroscience and Biobehavioral Reviews* 25, 597–609. DOI: 10.1016/S0149-7634(01)00044-6

Dittman, A.H. (1994) Behavioral and biochemical mechanisms of olfactory imprinting and homing by Coho salmon. Digital.lib.washington.edu. Available at: http://hdl.handle.net/1773/10321 (accessed 1 November 2016).

Dreumont-Boudreau, S.E., Dingle, R.N., Alcolado, G.M. and LoLordo, V.M. (2006) An olfactory biconditional discrimination in the mouse. *Physiology and Behavior* 87, 634–640. DOI: 10.1016/j.physbeh.2006.01.001

Duncan, H. and Slotnick, B. (1990) Olfactory and visual reversal learning in pigeons. *Chemical Senses* 15, 59–73. DOI: 10.1093/chemse/15.1.59

Dusek, J.A. and Eichenbaum, H. (1997) The hippocampus and memory for orderly stimulus relations. *Proceedings of the National Academy of Sciences USA* 94, 7109–7114. DOI: 10.1073/pnas.94.13.7109

Fendt, M. (2006) Exposure to urine of canids and felids, but not of herbivores, induces defensive behavior in laboratory rats. *Journal of Chemical Ecology* 32, 2617–2627. DOI: 10.1007/s10886-006-9186-9

Ferrero, D.M., Lemon, J.K., Fluegge, D., Pashkovski, S.L., Korzan, W.J., Datta, S.R., Spehr, M., Fendt, M. and Liberless, S.D. (2011) Detection and avoidance of a carnivore odor by prey. *Proceedings of the National Academy of Sciences USA* 108, 11235–11240. DOI: 10.1073/pnas.1103317108

Fetcho, J.R. and Liu, K.S. (1998) Zebrafish as a model system for studying neuronal circuits and behavior. *Annals of the New York Academy of Sciences* 860, 333–345. DOI: 10.1111/j.1749-6632.1998.tb09060.x

Freedman, K.G., Radhakrishna, S., Escanilla, O. and Linster, C. (2013) Duration and specificity of olfactory

nonassociative memory. *Chemical Senses* 38, 369–375. DOI: 10.1093/chemse/bjt010
Gagliardo, A. (2013) Forty years of olfactory navigation in birds. *Journal of Experimental Biology* 216, 2165–2171. DOI: 10.1242/jeb.070250
Gschwend, O., Abraham, N.M., Lagier, S., Begnaud, F., Rodriguez, I. and Carleton, A. (2015) Neuronal pattern separation in the olfactory bulb improves odor discrimination learning. *Nature Neuroscience* 18, 1474–1482. DOI: 10.1038/nn.4089
Guo, Z.V., Hires, S.A., Li, N., O'Connor, D.H., Komiyama, T. *et al*. (2014) Procedures for behavioral experiments in head-fixed mice. *PLoS ONE* 9, e88678. DOI: 10.1371/journal.pone.0088678
Hasler, A.D. and Wisby, W.J. (1951) Discrimination of steam odors by fishes and its relation to parent stream behavior. *The American Naturalist* 85, 223–238.
Hasler, A.D., Scholz, A.T. and Horrall, R.M. (1978) Olfactory imprinting and homing in salmon. *American Scientist* 66, 347–355.
Henton, W.W., Smith, J.C. and Tucker, D. (1966) Odor discrimination in pigeons. *Science* 153, 1138–1139.
Henton, W.W., Smith J.C. and Tucker, D. (1969) Odor discrimination in pigeons following section of the olfactory nerves. *Journal of Comparative and Physiological Psychology* 69, 317–323.
Kimura, Y., Miwa, T., Furukawa, M. and Umeda, R. (1991) Effects of topical application of steroids on olfactory disturbance in mice. *Chemical Senses* 16, 297–302. DOI: 10.1093/chemse/16.4.297
Krause, E.T., Brummel, C., Kohlwey, S., Baier, M.C., Müller, C., Bonadonna, F. and Caspers, B.A. (2014) Differences in olfactory species recognition in the females of two Australian songbird species. *Behavioral Ecology and Sociobiology* 68, 1819–1827. DOI: 10.1007/s00265-014-1791-y
Lehmann, A., D'Errico, A., Vogel, M. and Spors, H. (2016) Spatio-temporal characteristics of inhibition mapped by optical stimulation in mouse olfactory bulb. *Frontiers in Neural Circuits* 10, 15. DOI: 10.3339/fncir.2016.00015
Manton, M.I., Karr, A. and Ehrenfeld, D.W. (1972) An operant method for the study of chemoreception in the green turtle, *Chelonia mydas*. *Brain, Behavior and Evolution* 5, 188–201.
McNamara, A.M., Magidson, P.D., Linster, C., Wilson, D.A. and Cleland, T.A. (2008) Distinct neural mechanisms mediate olfactory memory formation at different timescales. *Learning and Memory* 15, 117–125. DOI: 10.1101/lm.785608
Michelsen, W.J. (1959) Procedure for studying olfactory discrimination in pigeons. *Science* 130, 630–631.
Mihalick, S.M., Langlois, J.C., Krienke, J.D. and Dube, W.V. (2000) An olfactory discrimination procedure for mice. *Journal of the Experimental Analysis of Behavior* 73, 305–318. DOI: 10.1901/jeab.2000.73-305
Mueller, K.P. and Neuhaus, S.C. (2012) Automated visual choice discrimination learning in zebrafish (*Danio rerio*). *Journal of Integrative Neuroscience* 11, 73–85. DOI: 10.1142/S0219635212500057
Nagel, W.A. (1894) Vergleichend physiologische und anatomische Untersuchungen über den Geruchs-und Geschmackssinn und ihre Organe mit einleitenden Betrachtungen aus der allgemeinen Vergleichenden Sinnesphysiologie. No. 18. E. Nägele, 1894.
Nathan, B.P., Yost, J., Litherland, M.T., Struble, R.G. and Switzer, P.V. (2004) Olfactory function in apoE knockout mice. *Behavioral Brain Research* 150, 1–7. DOI: 10.1016/S0166-4328(03)00219-5
Nevitt, G.A. (2008) Sensory ecology on the high seas: the odor world of the procellariiform seabirds. *Journal of Experimental Biology* 211, 1706–1713. DOI: 10.1242/jeb.015412
Nevitt, G.A. and Bonadonna, F. (2005) Sensitivity to dimethyl sulphide suggests a mechanism for olfactory navigation by seabirds. *Biology Letters* 1, 303–305. DOI: 10.1098/rsbl.2005.0350
Nevitt, G., Reid, K. and Trathan, P. (2004) Testing olfactory foraging strategies in an Antarctic seabird assemblage. *Journal of Experimental Biology* 207, 3537–3544. DOI: 10.1242/jeb.01198
Pankevich, D.E., Baum, M.J. and Cherry, J.A. (2004) Olfactory sex discrimination persists, whereas the preference for urinary odorants from estrous females disappears in male mice after vomeronasal organ removal. *Journal of Neuroscience* 24, 9451–9457. DOI: 10.1523/JNEUROSCI.2376-04.2004
Papi, F., Fiore, L., Fiaschi, V. and Benvenuti, S. (1972) Olfaction and homing in pigeons. *Monitore Zoologico Italiano* 6, 85–95.
Parker, G.H. (1910) Olfactory reactions in fishes. *Journal of Experimental Zoology* 8, 535–542.
Paskin, T.R. and Byrd-Jacobs, C.A. (2012) Reversible deafferentation of the adult zebrafish olfactory bulb affects glomerular distribution and olfactory-mediated behavior. *Behavioral Brain Research* 235, 293–301. DOI: 10.1016/j.bbr.2012.08.018
Pollonara, E., Luschi, P., Guilford, T., Wikelski, M., Bonadonna, F. and Gagliardo A. (2015) Olfaction and topography, but not magnetic cues, control navigation in a pelagic seabird: displacements with shearwaters in the Mediterranean Sea. *Science Reports* 5, 16486. DOI: 10.1038/srep16486
Rieke, G.K. and Wenzel, B.M. (1978) The forebrain projections of the pigeon olfactory bulb. *Journal of Morphology* 158, 41–56.
Schmidt-König, K. (1987) Bird navigation: has olfactory navigation solved the problem? *Quarterly Review of Biology* 62, 31–47. DOI: 10.1086/415266
Scholz, A.T., Horrall, R.M., Cooper, J.C. and Hasler, A.D. (1976) Imprinting to chemical cues: the basis for home stream selection in salmon. *Science* 192, 1247–1249.

Shumake, S.A., Smith, J.C. and Tucker, D. (1969) Olfactory intensity difference thresholds in the pigeon. *Journal of Comparative Physiological Psychology* 67, 64–69.

Sieck, M.H. and Wenzel, B.M. (1969) Electrical activity of the olfactory bulb of the pigeon. *Electroencephalography and Clinical Neurophysiology* 26, 62–69.

Slotnick, B. (2009) A simple 2-transistor lick or contact detector circuit. *Journal of the Experimental Analysis of Behavior* 91, 253–255. DOI: 10.1901/jeab.2009.91-253

Slotnick, B., and Coppola, D.M. (2015) Odor-cued taste avoidance: a simple and robust test of mouse olfaction. *Chemical Senses* 4, 269–278. DOI: 10.1093/chemse/bjv005

Slotnick, B.M. and Restrepo, D. (2005) Olfactometry with mice. In: Crawley, J.N., Gerefen, C.R., Rogawski, M.A., Sibley, D.R., Skolnick, P. and Wray, S. (eds) *Current Protocols in Neuroscience*. Wiley, New York, pp 1–24.

Slotnick, B. and Schellinck, H. (2002) Methods in olfactory research with rodents. In: Simon, S.A. and Nicolelis, M. (eds) *Frontiers and Methods in Chemosenses*. CRC Press, Boca Raton, Florida, pp. 21–61.

Slotnick, B., Kufera, A. and Silberberg, A.M. (1991) Olfactory learning and odor memory in the rat. *Physiology and Behavior* 50, 555–561. DOI: 10.1016/0031-9384(91)90545-Y

Slotnick, B., Bodyak, N. and Davis, B.J. (2001) Olfactory marker protein immunohistochemistry and the anterograde transport of horseradish peroxidase as indices of damage to the olfactory epithelium. *Chemical Senses* 26, 605–610. DOI: 10.1093/chemse/26.6.605

Sorwell, K.G., Wesson, D.W. and Baum, M.J. (2008) Sexually dimorphic enhancement by estradiol of male urinary odor detection thresholds in mice. *Behavioral Neuroscience* 122, 788–793. DOI: 10.1037/0735-7044.122.4.788

Stattelman, A.J., Talbot, R.B. and Coulter, D.B. (1975) Olfactory thresholds of pigeons (*Columba livia*), quail (*Colinus virginianus*) and chickens (*Gallus gallus*). *Comparative and Biochemical Physiology A* 50, 807–809.

Steiger, S.S., Fidler, A.E., Mueller, J.C. and Kempenaers, B. (2010) Evidence for adaptive evolution of olfactory receptor genes in 9 bird species. *Journal of Heredity* 101, 325–333. DOI: 10.1093/jhered/esp105

Stewart, A.M., Braubach, O., Spitsbergen, J., Gerlai, R. and Kalueff, A.V. (2014) Zebrafish models for translational neuroscience research: from tank to bedside. *Trends in Neuroscience* 37, 264–278. DOI: 10.1016/j.tins.2014.02.011

Suboski, M.D., Bain, S., Carty, A.E., McQuoid, L.M., Seelen, M.I. and Seifert, M. (1990) Alarm reaction in acquisition and social transmission of simulated-predator recognition by zebra danio fish (*Brachydanio rerio*). *Journal of Comparative Psychology* 104, 101–112. DOI: 10.1037/0735-7036.104.1.101

Takahashi, L.K. (2014) Olfactory systems and neural circuits that modulate predator odor fear. *Frontiers in Behavioral Neuroscience* 8, 72. DOI: 10.3389/fnbeh.2014.00072

Takahashi, L.K., Nakashima, B.R., Hong, H. and Watanabe, K. (2005) The smell of danger: a behavioral and neural analysis of predator odor-induced fear. *Neuroscience and Biobehavioral Reviews* 29, 1157–1167. DOI: 10.1016/j.neubiorev.2005.04.008

Teichmann, H. (1957) Das Riechvermögen des Aals (*Anguilla anguilla* L.). *Die Naturwissenschaften* 44, 242.

Teichmann, H. (1959) Über die Leistung des Geruchsinnes beim Aal (*Anguilla anguilla* L.). *Zeitschrift für vergleichende Physiologie* 42, 206–254.

Terry, L.M. and Johanson, I.B. (1996) Effects of altered olfactory experiences on the development of infant rats' responses to odors. *Developmental Psychobiology* 29, 353–377. DOI: 10.1002/(SICI)1098-2302(199605)29:4<353::AID-DEV4>3.0.CO;2-P

Thompson, R.F. and Spencer, W.A. (1966) Habituation: a model phenomenon for the study of neuronal substrates of behavior. *Psychological Review* 73, 16–43.

von Frisch, K. (1941) Die Bedeutung des Geruchssinnes im Leben der Fische. *Die Naturwissenschaften* 29, 321–333.

Walker, J.C. (1983) An operant procedure for testing olfactory capacities in restrained pigeons. *Physiology and Behavior* 30, 165–168. DOI: 10.1016/0031-9384(83)90054-9

Wallraff, H.G. (2014) Pigeon homing from unfamiliar areas: an alternative to olfactory navigation is not in sight. *Communicative and Integrative Biology* 7, e28565. DOI: 10.4161/cib.28565

Walter, W.G. (1942) Some experiments on the sense of smell in birds: studied by the method of conditioned reflexes. Doctoral dissertation, University of Amsterdam, 92 pp. Library.wur.nl.

Wenzel, B.M. (1971) Olfactory sensation in the kiwi and other birds. *Annuals of the New York Academy of Science* 188, 183–193.

Wernecke, K.E.A. (2016) Predator odor-induced fear in rats: a behavioral characterization and neural substrate analysis. PhD thesis, University of Magdeburg, Germany.

White, E.J., Kounelis, S.K. and Byrd-Jacobs, C.A. (2015) Plasticity of glomeruli and olfactory-mediated behavior in zebrafish following detergent lesioning of the olfactory epithelium. *Neuroscience* 284, 622–631. DOI: 10.1016/j.neuroscience.2014.10.036

Yamaguchi, M., Yamazaki, K., Beauchamp, G.K., Bard, J., Thomas, L. and Boyse, E.A. (1981) Distinctive urinary odors governed by the major histocompatibility locus of the mouse. *Proceedings of the National Academy of Sciences USA* 78, 5817–5820. DOI: 10.1073/pnas.78.9.5817

Yang, S.Y., Walther, B.A. and Weng, G.J. (2015) Stop and smell the pollen: the role of olfaction and vision of the oriental honey buzzard in identifying food. *PLoS ONE* 10, e0130191. DOI: 10.1371/journal.pone.0130191

Zhang, J., Wei, W., Zhang, J. and Yang, W. (2010) Uropygial gland-secreted alkanols contribute to olfactory sex signals in budgerigars. *Chemical Senses* 34, 375–382. DOI: 10.1093/chemse/bjq025

Zippel, H.P., Meyer, H.M. and Zeman, D.L. (1993a) Functional and morphological regeneration of the olfactory tracts and subtracts in goldish. *Journal of Comparative Physiology A* 172, 91–99. DOI: 10.1007/BF00214718

Zippel, H.P., Voigt, R., Knaust, M. and Luan, Y. (1993b) Spontaneous behaviour, training and discrimination training in goldfish using chemosensory stimuli. *Journal of Comparative Physiology A*, 172, 81–90. DOI: 10.1007/BF00214717

Zou, J., Pan, Y., Wang, Z., Chang, S., Wang, W., Wang, X., Tournier, C., Storm, D.R. and Xia, Z. (2012) Targeted deletion of ERK5 MAP kinase in the developing nervous system impairs development of GABAergic interneurons in the main olfactory bulb and behavioral discrimination between structurally similar odorants. *Journal of Neuroscience* 32, 4118–4132. DOI: 10.1523/jneurosci.6260-11.2012

5 Is There Such a Thing as a Bad Smell?

ROLAND SALESSE[1] AND LAURENT DORMONT[2]

[1]*Neurobiology of Olfaction, INRA, Université Paris-Saclay, Jouy-en-Josas, France;* [2]*CNRS Centre d'Ecologie Fonctionnelle et Evolutive, Montpellier, France*

Qualifying odours as good or bad implies that a classification of odours should exist. It seems that, at least in modern times and in Western countries, the most salient dimension of odours has been hedonism. As recalled by Dubois (2006), classifications in the 18th century rated odours from pleasant (*Odores suaveolentes* or *Odores fragrantes*) to unpleasant (*Odores foetores* or *Odores nausei*). This classification may seem straightforward, but still, in the 21st century after the discovery of olfactory receptors, there is no agreement on an absolute classification of odorants, suggesting that there is no biological ground on which to base such a nomenclature.

Many of us can agree that decaying materials, faeces, pungent or irritant products, rotten food and gas leaks all have bad smells. Yet, when looking in more detail, a chemical like isovaleric acid may be disgusting when smelled in an old sock, while it is appreciated positively in the exhalations of a ripe cheese. Or the chemical skatole, a signature of faeces, also gives its heady facet to flower fragrances such as jasmine or jacinth. Neighbours to a foul-smelling industrial plant will strongly dislike its stink while workers earning their life in the same plant often express a softer opinion (Pourtier, 2012). On the other hand, olfactory receptor gene polymorphism is so huge that it may give rise to vastly different appraisals of an odorant: some people may display hypersensitivity to the above-mentioned isovaleric acid (Menashe *et al.*, 2007), while others, lacking the receptor for androstenone, won't be perturbed by this chemical that otherwise evokes a potent pig urine reek for those endowed with the functional receptor (Keller *et al.*, 2007). Thus, it seems that, in the human species, defining a smell as bad needs also to define the historical, psychological, social, occupational, environmental and biological conditions of this judgement.

The same holds true for animals. We can infer they 'like' or 'dislike' a given odorant from their behaviour: they would approach attractive (good) smells and be repelled by aversive (bad) ones. However, as a function of species, sex, physiological and psychological status, or learning, preferences or aversions may change. In this chapter, we look at some of the historic accounts of good and bad odours, and raise a few examples from humans and animals to elucidate the various issues associated with odour valence.

Recent Olfactory History: Tackling Foul Smells in the 19th and 20th Centuries

In his seminal book, *Le miasme et la jonquille*, Alain Corbin (1982) describes the 'olfactory revolution' occurring throughout the 19th century in France. Eighteenth-century cities are described, both by historians and novelists, as foetid and reeking (Barnes, 2006); personal as well as social hygiene was neglected because people believed water to be potentially unhealthy. Thus, not only houses, but public places were likely to be ill-smelling, and this covered a large range of locations, including streets, hospitals, barracks, palaces and slums. Seen from the 21st century, the situation was paradoxical: on the one hand, pungent body odours were supposed to indicate virile vigour, but, on the other hand, people still believed that foul smells would bring about diseases and general unhealthiness. Yet, the 19th century witnessed both the rise of the bourgeoisie, who wanted to keep away from the stench of the common people, and the development of modern science, noticeably bacteriology, personified by Louis Pasteur in France. The rising awareness that bacteria were, most of the time, the origin of bad smells and diseases, progressively led to the collection of garbage and the digging of sewage pipe networks, at least in the largest cities. Eventually, this revolution benefited all social classes, and resulted in an increased deodorization of individuals and of public and private places, and the consideration that public health was associated with 'olfactory silence'.

Nowadays, eighteenth-century perfumes would be described as unpleasant, as they relied largely on heavy, animal-scented products like ambergris, civet or musk. Indeed, by the end of the nineteenth century, floral notes, especially for young women, came into fashion and opened up to novel, synthetic chemicals like coumarin or vanillin, which are still regarded as good-smelling.

In the 20th century, we began to master the struggle against filth and germs. Indeed, we gradually replaced the foul-smelling odours with other scents, such as those emitted by morning coffee, toiletries, deodorants (often odorized!), added food flavours, washing powder, home fragrances, scented cars and even olfactory brand logos. It is almost impossible to find places or products without the presence of an added odour. This has now got to the stage where many people find it intrusive; so much so that a few consumer associations have begun to fight to ban perfumes from toiletries and public places. This brief historical summary shows that the social context of the use of odours and odorants and their perceived pleasantness is always changing according to politics, economics, science and fashion.

Predicting the Perceived Valence of Odours

In the previous section we highlighted the historical and sociological influences on odour valence, but are there ways to predict how a compound will smell based on its inherent properties: chemical structure, molecular weight or the presence or absence of certain elements? Consistent with this objective, Khan *et al.* (2007) tried to link statistically the physico-chemical properties of odorants and their hedonic rating, before trying to match them. They first reduced odorant chemical characteristics (no less than 1514 word descriptors!) to four principal components, the first two of which corresponded roughly to molecular size (molecular weight) and to denseness (or compactness) of the molecules. Concerning perception, starting from the 146 verbal descriptors proposed in Dravnieks' *Atlas of Odour Character Profiles* (1985), they statistically reduced them to a couple of perceptual dimensions. Subsequently, they experimentally assimilated the first – and major – perceptual dimension with hedonism thanks to a human panel, who rated a set of odorants from 'fragrant' to 'sickening'. However, when the authors tried to generalize their findings to 27 new odorants (not present in the initial set), they weren't able to draw such a firm conclusion of a straightforward relationship between chemical structure and odour perception. Thus, depending on the reader's tendency towards optimism or pessimism, this research would seem either encouraging or disappointing.

Another way to reduce the multidimensional odour space would be to start from human actual feelings, i.e. to use a psychological approach. Such an approach should take into account cross-cultural differences because 'tastes' vary around the world. This motivated Zarzo's study (2008), where the author was able to propose only two dimensions – hedonism and edibility – as the two principal components emerging from his statistical analysis of a previously collected dataset (Chrea *et al.*, 2005), which compared the rating of daily odorants (odorant blends mimicking daily fragrances) between three groups of 30 students: American (Dallas), French (Dijon) and Vietnamese (Danang). Each panel tested 40 odorants, among which 17 were expected to be equally familiar for all three cultures, while three other sets were each supposed to be more familiar to a specific culture. Participants were invited to rank each odorant on a numeric scale in every one of 11 categories, which had been defined in a preliminary study: *fruit*, *flower*, *candy*, *cleaner*, *animal*, *musty*, *bakery*, *cosmetic*, *spice*, *medicine* and *nature*. We should note that none of these descriptors is explicitly understood as good or bad, except *musty*, which was categorized as 'bad' in a previous experiment (Chrea *et al.*, 2004). Nevertheless, the first dimension (hedonism) accounted for one-third of descriptors, and the second (edibility) for one-fifth, which means that approximately half of the odorant characteristics could be assigned to these two categories. These results were in fair agreement with previous ones (Moncrieff, 1966; Dravnieks *et al.*, 1984), provided that some correlations were made between the descriptors of both studies.

Figure 5.1 displays a schematic representation of these results. We modified Zarzo's original figure in order to highlight the pleasant/unpleasant dimension on the x-axis and the edible/inedible dimension on the y-axis. What becomes apparent is that two graph quarters are occupied by clusters of odours: the pleasant-edible and the unpleasant-inedible. Implicitly, this means that the two other categories, pleasant-inedible and unpleasant-edible would not exist, which makes some sense at least at first: perfumes and flowers are not classified as edible and who would like to eat stinking food? However, exceptions do exist: as noted in the introduction, cheese odorants, although not appreciated worldwide, are nevertheless prized by gastronomes; the durian tropical fruit of *Durio zibethinus*, albeit foul-smelling, has a pleasant, sweet taste when eaten. Interestingly, the 40 odorants in the study by Zarzo (2008) were selected on the basis of familiarity and not of edibility. Yet, if neurogastronomy – as championed by Gordon Shepherd (2012) – did play a great part in the evolution of human olfaction, it is no wonder that edibility would appear among the very first odour qualities.

Olfactory World Tour in a Nutshell

Another way of relativizing scent appreciation is to travel around the world. Ethnologists often report local, national customs, which may appear surprising to foreigners. Olfaction does not escape the rule. Far from being merely ornamental, odours bear powerful and symbolic meanings concerning social and sexual relationships, as well as human history and our place in nature. They amply demonstrate the prepotency of culture in the classification of odours. As the matter has been extensively studied, we shall limit this section to a couple of issues related to certain perceived bad smells.

Constance Classen *et al.* (1994) described a few ethnic groups. For the farming and cattle-herding Dassaneth in Ethiopia, the year is marked by the alternation of two dry seasons and two wet seasons. Dry seasons are characterized by rather unpleasant odours: plant decay, burning grass.

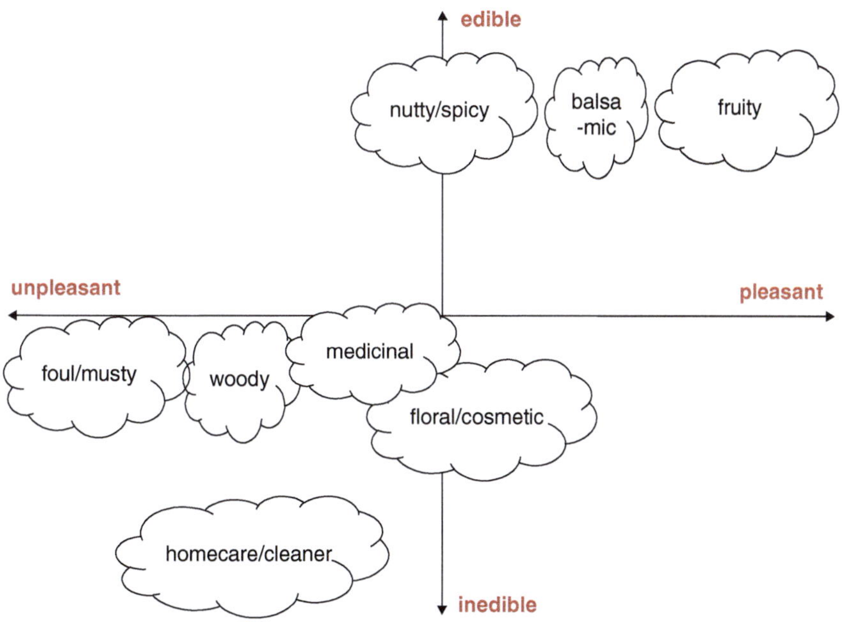

Fig. 5.1. A simplified representation of Zarzo's (2008) results based on data from Chrea *et al.* (2005) concerning the psychological evaluation of odours. Location of odours in the pleasant/unpleasant vs edible/non-edible space is represented by labelled clouds. Axes are not graduated because of adaptation to data: in Zarzo's original figure, the dividing line between edible and inedible products was not horizontal but tilted approximately 20 degrees clockwise from it.

However, enduring this stench is deemed necessary to comply with the yearly rhythm of nature. The only scent Dassaneth really hate is that of fish, believed to escape the – on earth unescapable – dry/wet natural cycle by living all the time in water. For them, time and odours match together, setting the scene of their daily life.

In the Andaman Islands, an archipelago in the Bay of Bengal, the Ongee people (a small aboriginal population) believe that one's odour emanates from one's bones, where scents are concentrated and touching one's nose means 'myself'. Dying is equivalent to being robbed of one's odour by odourless spirits called *tomya*, which in some way reincarnate a human odour when they are absorbed by a fertile woman. Ongee are one of the rare examples for whom menstrual blood is no stench; rather, it is believed to have a purifying function and men, who lack this ability to regulate their odour, have to go through a 2-day initiation upon passage into manhood, in which women play a great role by breathing men's odour and by spreading their female scent onto males.

In a study in Senegal, Dupire (1987) reported that the Serer N'dut considered their own body odour, as well as that of the close Bambara ethnic group, as pleasant (termed *hen*). On the contrary, young children and Europeans were found to smell like urine (*sun*) in the same category as monkeys, dogs and diuretic plants. Finally, corpses and farts were assigned the term *hot* (rotten). Here too, odour categorization is a mix of perceptual as well as metaphoric and ritual evaluation, which again underscores the importance of olfactory culture.

During his stay in Pondicherry (India), geographer Robert Dulau (2004) observed the Tamoul's religious rites in the homes of notables, and particularly the *kolam* Vishnu rite on Fridays, comprising many odorous steps, some of which would seem malodorous to a westerner. Indeed, the first woman to get up in the morning spreads diluted cow dung in front of the house threshold (reflecting the importance of cow and cow products for these people). Subsequently, several deity celebrations occur, some of them fragrant (flowers, camphor burning). One particular instance is the lavish incense rite: devout people burn not just a few incense sticks, but bunches of them, with the smoke creating a heady atmosphere.

Western cities are not devoid of malodours. Lucile Grésillon (2010) investigated how inhabitants perceived the smellscape of five Paris districts. Despite many rules concerning odorant pollution (Pourtier, 2012), city dwellers mainly complain of four kinds of malodours: car exhausts, restaurant kitchen exhausts, garbage dumps and urine stench. Even 'beaux quartiers' are not free from such nuisances, since Rue Lagrange, just a few steps away from Notre-Dame, in the Latin Quarter of Paris, is used for parking by tourist buses, whose drivers let the motors run, spreading burnt diesel stench. Nearby, in Rue de la Huchette, local residents are sick of smelling the lingering stench of burnt fat from restaurants, of rotting food bags and of urine from alleyways and street corners. These questions may come up for consideration in town planning. In an approach similar to Grésillon's, Victoria Henshaw (2014) performed 'smellwalks' to map urban smellscapes and establish a user's olfactory representation. Roughly speaking, she finds the same nuisances in English towns as did Grésillon in Paris. However, as an urbanist, she also reflects on how to lay out cities in order to prevent or at least blunt malodours. First, sources (garbage, traffic, food leftovers) must be reduced. Second, slowing down the traffic, or taking it away from pedestrian areas, increasing air replacement by exploiting the wind and laying out green spaces would bring about an objective improvement, which moreover makes people feel better. This may seem to be simple common sense, but necessitates long-lasting consultations with inhabitants, elected representatives and designers.

Yet town odours, foul as they can be, may be a source of inspiration for an artist-perfumer such as Sissel Tolaas, who 'olfactorily' investigated the scents of Paris, Berlin, Cape Town, Mexico and Kansas City. In every city, she sniffed out the olfactory characteristics, identifying year-long lasting smells (more than once those of pollution, as stated above) and recomposing them in her laboratory. Finally, dwellers were invited to smell their own city in an exhibition or in an olfactory treasure hunt through the streets (Barbery-Coulon, 2012). Among her many original fulfilments, Sissel Tolaas also composed the First World War Scent (a rather nauseating odour for the Military History Museum in Dresden), and showed wallpapers impregnated with human fear odour in the exhibition *Belle Haleine – Der Duft der Kunst* (Art Smell) in Tinguely Museum in Basel (February–May 2015). This is another example of how art, and culture in general, can relativize our seemingly well-grounded perception of odour hedonism according to the context.

Occupational Malodour

Some occupations may be practised in an ill-smelling environment. Candau and Jeanjean (2006) investigated the strategies by which workers such as sewage workers, firemen, forensic scientists, nurses, morgue attendants and embalmers cope with malodours. The authors reported interviews performed at the place of work, focusing on the specific words used in different occupations.

First of all, these professionals had acquired their olfactory competence on the job without prior training, all the more as each job has its own characteristic odours. Moreover, this practical shaping appeared to be largely devoid of words: many employees were able to smell 'something' or 'something different' that urged them to move in, for instance to clean a sewage pipe, but without naming the odour. The authors mention a nurse who, throughout a 45-min interview, used the adjectives 'special' and 'characteristic' for the hospital odours 23 times, without qualifying them more specifically. Others turn to childhood memories: in a water-treatment plant, someone compared the prevailing odour to the cow dung smell she remembered from when she was young; another recalled a rotting jellyfish on the beach when describing the smell of a drowned body.

Often, respondents explained that they had become accustomed to bad odours. In particular, the initial experience seemed a determining factor: those unable to stand the first disgusting impression gave up the job, while the others hung on, either because they relied on a deliberate professional expertise, or because they had to earn a living, in both cases putting aside the hedonic–emotional dimension. Another way interviewees coped with malodour was by consolation, citing worst odour conditions: a water-treatment employee stated he wouldn't be able to work in a morgue. Surprisingly, some people described coping with and even appreciating the odour from their own water-treatment plant, while they couldn't endure that of another plant which used a different treatment method. Last, social considerations were not absent from malodour evaluation: according to interviewees, sewage smelled worse in areas whose population they disliked. Such sensory bias has also been noted by anthropologists, who have studied how the 'smell of others' is one of the ways to discriminate human groups or individuals, according to ethnicity, social status or age (Le Breton, 1998).

Independent of profession, the worst odours appeared to be those of corpses. All of us feel an apparently innate repulsion towards the smell of decaying flesh. Accordingly, these odours are qualified as 'horrible', 'unbearable' and even 'dreadful'. Moreover, they are also reported as 'penetrating' or 'pervasive', suggesting that even the smeller's body integrity is at stake. As exposed by Candau and Jeanjean (2006), this is reminiscent of Emmanuel Kant's opinion, who judged odours to be intrusive because, as we are obliged to respire, we are also obliged to incorporate the smell of others. And, indeed, professional morgue attendants complained of not being able to get rid of these odours, which stuck to their hair, clothes, skin and even mouth. Aside from wearing a bio-hazard suit with a gas mask – which firemen sometimes do – protections are illusory, and most workers declared that odours had to be accepted whilst performing their tasks. In the case of children's corpses, some morgue attendants even denied that they had an odour at all, although the authors themselves did smell the stink upon visiting the place. According to Candau and Jeanjean, this was characteristic of an olfactory denial, the idea of a dead child being even more unbearable than that of a dead adult.

More than in other professions, people working in morgues are directly faced with death, and psychological reports indicate that, in these conditions, managing disgust is an important reaction in these people to protect themselves against both physical and psychological threats (Kelley *et al.*, 2015). Also, these occupations may trigger the so-called cognitive dissonance reaction, which is the discomfort experienced by individuals who are faced with contradictory beliefs. People having made a (supposedly free) choice (being a morgue attendant instead of an office worker, for example) display an ad hoc rationalization bias in that they re-evaluate their chosen item more positively, and their rejected item more negatively (Chen and Risen, 2010). Though none of these theories has been tested with persons working in ill-smelling atmospheres, it could be that these psychological strategies help them to cope with extreme occupational environments.

Chemistry of Human Body Odours

Numerous studies have investigated the chemical composition of odours from different parts of the human body (Bernier *et al.*, 2000; Ara *et al.*, 2006;

Natsch *et al.*, 2006; Gallagher *et al.*, 2008). These studies mainly focused on volatiles emitted by human skin (in particular, armpits, hands and feet) as well as breath. Others have been interested in the odours from various body fluids suspected to produce volatile odoriferous compounds, such as urine, faeces, blood and saliva (Garner *et al.*, 2007; Amann *et al.*, 2014). One of the main goals of these studies was to identify what kinds of olfactory cues help blood-sucking insects to locate and choose their vertebrate hosts for blood meals, with particular attention to anthropophilic mosquitoes that may transmit pathogens to humans (Logan *et al.*, 2008; Syed and Leal, 2009). However, chemical analyses of human odours also met other objectives, such as searching for putative human pheromones, providing available tools for applications in forensic studies (e.g. by using dog abilities to detect human body emanations; Palmieri *et al.*, 2016), understanding the skin-mediated chemical transformation of perfumes and evaluating means to reduce malodorous releases.

Recently, the study of chemical compounds emanating from the human body has proven to be promising for the diagnosis of human diseases and infections (Prugnolle *et al.*, 2009; Kim *et al.*, 2012). For example, the composition of exhaled breath can be used for the diagnosis of lung cancer, and many key volatiles in breath samples are now available as potential biomarkers for diverse human diseases (Buchbauer and Buljubasic, 2015). Volatile scents from human faeces have also shown their potential for diagnosis of gastrointestinal disease, following observations by both patients and healthcare professionals that faeces often smell abnormal during gastrointestinal disease (Garner *et al.*, 2007). In Chapter 9 of this book, a more detailed account of the role of odours in disease detection is given.

The total number of volatile organic compounds isolated from the human body has reached 1850 molecules, following the review of Costello *et al.* (2014). Most of these compounds were recorded from breath analyses (870) and from skin emanations (530). However, even when the same part of the body is sampled, the number of volatile compounds varies greatly among studies. Some authors isolated and identified more than 300 compounds from human skin extracts (Bernier *et al.*, 2000), whereas other studies reported only 20 to 30 volatile compounds from roughly similar skin samples (Zhang *et al.*, 2005). A crucial step for the identification of human odours is the choice of the sampling method. The chemical profile of human volatiles varies greatly in the scientific literature, and the use of diverse sampling procedures (solvent extraction, dynamic headspace adsorption onto porous polymers, solid-phase micro-extraction) is likely to be responsible for many inconsistencies between studies in the compounds detected (Dormont *et al.*, 2013).

Many compounds found in exhaled breath come from the external environment and are inhaled regularly by human individuals. The chemical composition of breath emissions thus reflects the pollution level of the environment, as well as the smoking habits of the person exhaling. Nonetheless, breath analyses are considered a non-invasive means of diagnosing several human health diseases, such as lung disease, oxidative stress, gastrointestinal disease, metabolic disorders and *Helicobacter pylori* infection. Quantitative analysis of breath volatile samples of patients suffering from lung cancer showed higher levels of particular compounds, such as 2-propanol, dimethyl sulfide, 2-butanone and 2-pentanone when compared to healthy controls (Schmidt and Podmore, 2015). Breath samples can even be used as a diagnostic method without the aid of gas chromatography: the presence of acetone in the exhaled breath of diabetic patients makes breath smell like a 'rotten apple', and a urine-like smell in breath emissions may characterize failing kidneys (Kim *et al.*, 2012).

Numerous studies have examined the volatiles emitted by human skin. The production of volatile organic compounds by skin is mainly governed by the secretion of three types of glands. Eccrine glands (producing odourless sweat) are the most abundant and widely distributed on the skin surface, and are particularly concentrated on the hands and feet. Apocrine glands (which secrete lipids, proteins and steroids) characterize the axillae (armpits) and genital regions, while sebaceous glands (which secrete lipids and an oily substance called sebum) are distributed all over the body. Populations of commensal cutaneous microorganisms, i.e. skin microbiota, are also key contributors to the formation of human body odour. Moreover, the use of various fragranced products by humans typically results in the detection of numerous exogenous compounds in skin odour analyses. Before odour sampling, volunteer subjects are often asked to avoid fragranced soap/shampoo, deodorants and perfumes, and must also follow particular instructions related to diet, such as avoiding spicy food and garlic several days before odour collection. However, some compounds not known to occur in nature are regularly cited in the literature as found in skin volatile extracts, such as the industrial chemical (R,S)-2-ethyl-1-hexanol, which is sometimes even considered a biomarker compound of human body odour.

The chemical composition of human skin volatiles is highly diversified, but only a few families of compounds are represented, such as carboxylic acids (both short-chain acids and fatty acids) and derivative esters, aldehydes, alkanes, short chain alcohols and some ketones. In particular, four compounds have been often reported to be largely predominant in the volatile profile of human skin, regardless of the subject's ethnic background: 6-methyl-5-hepten-2-one, nonanal, decanal and geranylacetone (Syed and Leal, 2009; Dormont *et al.*, 2013). The axillary region has been proven to be a particularly important source of diverse volatile compounds. The three types of skin glands commonly co-occur at high densities in the axillae, together with a large diversity of microbial flora, such as lipophilic corynebacteria, which are likely involved in the emanation of malodorous volatile compounds. Odours from axillae mostly consist of alkanes and short-chained carboxylic acids, and are dominated by two key odoriferous compounds, 3-methyl-2-hexenoic acid and 3-hydroxy-3-methylhexanoic acid (Natsch *et al.*, 2006). Several axillary steroids and other compounds have often been suggested as potential human pheromones, but no active compound has yet been conclusively identified. In contrast to axillae and hands, the chemical composition of foot odours has received limited attention until recently. As for the axillary region, populations of skin-inhabiting bacteria are especially dense on feet, resulting in the production of many malodorous volatiles. Carboxylic acids are the main components of foot volatiles, and a few short-chain fatty acids, such as isovaleric acid, are likely responsible for strong foot odour (Ara *et al.*, 2006).

As well as in breath and from axillae and feet, malodorous volatile compounds from the human body may obviously be found in rectal gas. Odours from flatus are chemically very different from those isolated in other sources of bad-smelling human odours. The gases responsible for unpleasant odours in flatus have been identified to be mainly sulfur-containing compounds (hydrogen sulfide, methanethiol, dimethyl disulfide; Suarez *et al.*, 1998).

Recent studies investigated the volatiles emitted by the whole human body by placing human volunteers inside individual aluminized plastic bags, with only their heads outside (Logan *et al.*, 2008). Six main 'classic' compounds were isolated from whole body odours, four short-chained aldehydes (heptanal, octanal, nonanal, decanal), together with 6-methyl-5-hepten-2-one and geranylacetone. A number of upcoming studies will focus on the analysis of volatiles released by the human body as potential diagnostic biomarkers for human diseases and disorders: this new field of scientific research of growing importance is now referred to as 'the scent of disease' (Buchbauer and Buljubasic, 2015).

Epigenetics Push the Boundaries between Nature and Nurture: The Example of Olfactory Aversive Conditioning

In all animals, including man, neutral and even good odours can become aversive after a bad experience. This is called classical (or Pavlovian) conditioning, where for instance pairing a harmful event (illness) with food intake results in further aversion towards this food. As food is always endowed with odours, the sole presentation of this scent is enough to trigger aversion. This has been largely documented in rodents and in some farm animals (see also Chapter 6), and relies on the activation of some brain structures including the infralimbic and insular cortices and the basolateral amygdala, which are the sites where olfactory, gustatory and visceral information converge (Lévy and Ferreira, 2012). Interestingly, in humans, the same regions are activated upon inhalation of chemicals producing a strong feeling of disgust. Moreover, sniffing the odorants, or seeing others making a facial grimace of disgust while sniffing them, both activated the same sites in the anterior insula and to a lesser extent in the anterior cingulate cortex (Wicker *et al.*, 2003). Thus, in the same way as observing movements in others activates mirror neurons in the observer's motor cortex (Di Pellegrino *et al.*, 1992), observing an (olfactory) emotion would activate the neural support of that emotion.

Interesting as they may be, these works describe only standard classical conditioning that is restricted to individuals. What Dias and Ressler reported in 2013 was more astonishing. In a first experiment with mice, they induced a conditioned aversion to acetophenone (a chemical reminiscent of almond or cherry, not aversive, at least for humans) by associating it with an electric shock. Subsequently, the animals displayed fear behaviour in the presence of acetophenone, even without a shock. The important novelty was that this reaction was passed on to their pups, without reconditioning, and that mating a conditioned parent with a naïve one was sufficient to obtain reactive pups. Despite never having encountered acetophenone in their lives, the offspring exhibited odorant-induced startling. In addition, *in vitro* fertilization with the gametes of either of the conditioned parents resulted in pups being scared by acetophenone. The third generation of mice also inherited this behavioural sensitivity and the neuroanatomical changes, even after *in vitro* fertilization. Finally, cross-fostering experiments ruled out the possibility of a social transmission of these characteristics.

How could acetophenone aversion be passed on to descendants? Dias and Ressler found that there were no gene alterations in the offspring, but rather epigenetic modifications. Epigenetic modifications do not change the sequence of DNA but rather alter its capacity to drive the expression of some genes, here related to odorant-triggered fear behaviour, and these properties were passed on to the progeny. This kind of modification (previously named 'imprinting', but not to be mistaken for the sensitive period phenomenon of the same name) had been observed in obesity-prone infants whose birth weight was lower than normal (Yang and Huffman, 2013). Gametes are involved in this transmission since epigenetic plasticity of sperm DNA has been demonstrated. In humans, in the case of obesity, which is a complex, heritable condition involving the nervous system, hormones and metabolism, a comprehensive profiling of sperm epigenome from lean and obese men showed markedly different small non-coding RNA expression and DNA methylation patterns (Donkin *et al.*, 2016). Moreover, after bariatric surgery (i.e. a bypass of the stomach), loss of weight was rapidly (1 week) accompanied by a dramatic remodelling of sperm DNA methylation, notably at genetic locations implicated in the central control of appetite.

Epigenetic changes were no surprise for those working with invertebrates. In its critical first larval stage (L1) *Caenorhabditis elegans* (a nematode worm commonly used in laboratories) undergoes a lifelong olfactory imprinting that increases attraction to imprinted odorants and enhances odorant-induced egg-laying rates. Remy (2010) demonstrated that the preference for an imprinting odorant (benzaldehyde or citronellol) – provided it was administered at the L1 stage during five successive generations – could be passed on to the progeny without further learning, and became innate-like. This epigenetic transmission appears to be supported by small RNAs that are transferred by the gametes (Rankin, 2015). Potentially, these mechanisms offer a way to manipulate olfactory preference and aversion, not only for individuals, but for their progeny, thus blurring the boundaries between innate and acquired odour appraisal (see also Chapter 2).

How to Change an Aversive Odour to a Preferred One through Evolution

Many insects feed and breed on substrates humans would find foul-smelling and disgusting. As most invertebrates are devoid of teeth, it is important for them to find decaying material to feed on. An intriguing case is that of *Drosophila sechellia*, for which the rubiaceous shrub *Morinda citrifolia* is the exclusive host plant for feeding and oviposition, whereas the closest species, *Drosophila simulans*, is repelled by this plant (Fig. 5.2). The *D. sechellia/simulans* branch diverged from *Drosophila melanogaster* some

Fig. 5.2. *Drosophila sechellia* on the *Morinda citrifolia* fruit (or noni fruit, or Tahiti noni). The fruit contains poisonous organic acids that are a deterrent for all *Drosophila* species, except for *D. sechellia*, which feeds and breeds on the plant. (Photograph courtesy of Corbin Jones and Betty Wanjiru, University of North Carolina at Chapel Hill and Carolina Center for Genome Sciences, North Carolina, USA.)

3 million years ago, while *D. sechellia* and *D. simulans* separated only 0.5 million years ago (McBride, 2007). Although these two latter species retain cross-fertility, hybrids are not viable because *M. citrifolia* is a poison, except for pure-bred *D. sechellia*. Olfactory cues of *M. citrifolia* are detected by the two *Drosophila* species, the latter by attraction, the former by aversion. As *M. citrifolia* is also aversive to *D. melanogaster*, it can be considered a deterrent for all *Drosophila* species, except for *D. sechellia*.

Progressively, the chemicals in *M. citrifolia* responsible for this species difference have been disclosed. Most of them are 6-carbon aliphatic chemicals: hexanoic (or caproic) acid and hexanoate esters, as well as the 7-carbon ketone heptanone (Ibba *et al.*, 2010). Interestingly, olfactory preferences are accompanied by anatomical changes on the drosophila antennae in the number of sensilla, which house the olfactory neurons sensitive to odorants, as well as changes in the antennal lobe anatomy (the equivalent of the vertebrate olfactory bulb) with increased glomerulus size (Stensmyr *et al.*, 2003; Dekker *et al.*, 2006; Ibba *et al.*, 2010).

These anatomical changes could explain the higher sensitivity to *Morinda* volatiles, but not the inversion in preference. A first genetic study reported that three or four groups of genes could be responsible for this switch (R'Kha *et al.*, 1991). Two of these genes could be the odorant binding proteins (OBPs) Obp57e and Obp57d, the mutations of which makes *D. sechellia* tolerant to the noni fruit (Matsuo *et al.*, 2007). This is consistent with the importance of OBPs in insect olfaction, as exemplified by LUSH (Obp76a): initially identified for supporting ethanol avoidance by *D. melanogaster*, LUSH was eventually recognized for its key role in the complex molecular recognition of the aggregation pheromone *cis*-vaccenyl acetate (Xu *et al.*, 2005). More recently, in a combined genetic approach and tolerance assay to octanoic acid, other OBP genes have been implicated in *D. sechellia* specificities (Hungate *et al.*, 2013).

Lately, one of the possible clues for *D. sechellia* adaptation to *M. citrifolia* has been reported by Lavista-Llanos *et al.* (2014). *D. sechellia* suffers from a lack of dopamine, which causes an arrest in oogenesis. However, when fed with noni, females resume oogenesis because the fruit provides them with L-DOPA (or 3,4-dihydroxyphenylalanine), a precursor of dopamine. The authors conclude that the need for dopamine has driven *D. sechellia* evolution to become a *M. citrifolia* obligate specialist. Thus, the picture of *D. sechellia* adaptation to an otherwise deterrent, ill-smelling plant is gradually completed and suggests a complex convergence of sensory, reproductive and perhaps detoxification genes during the course of evolution.

Conclusion: In Praise of Bad Smells

Some odours may seem foul or disgusting, but according to context, place, time, species, learning and evolution, their meaning could change, and consequently their appraisal, as well as the behaviours they may elicit. Although it may seem easy to say 'all is relative', it is important when working on odour valence not only to control the experimental set-up, but also to take into account genetics, the environmental and social context and previous experience before even attempting to evaluate the hedonic valence of odours.

Most of all, we ought to place olfaction into its ethological context. What does smell mean for animal behaviour and welfare? Olfaction is implicated in such vital functions as feeding, social life, reproduction and environmental investigation. Thus, because hedonic value is generally considered a dominant aspect of olfactory perception, bad smells could represent spoiled food, ill congeners, non-reproductive status or environmental threats, be this knowledge either learned or innate. Recent results are in favour of an innate hedonic mapping of odours in mice olfactory bulbs (Kermen *et al.*, 2016): attractive (deemed pleasant) odorants stimulated the anterior–ventral part of the bulb, while aversive ones (e.g. sulfur compounds) activated its posterior–dorsal moiety.

In humans, such an anterior–posterior topography between the olfactory epithelium and the bulb is not known. However, Lapid *et al.* (2011) found that a mucosa location that responded maximally to a pleasant odorant was likely to respond strongly to other pleasant odorants, and a location that responded maximally to an unpleasant odorant was likely to respond strongly to other unpleasant odorants, here too suggesting some innate wiring.

One important function of olfaction is detecting diseases, many of them producing a characteristic, often foul, odour. In medicine, after having been forgotten for years, olfactory diagnostics seem to have come back into favour, thanks to new powerful analytical instruments (Prugnolle *et al.*, 2009; Kim *et al.*, 2012; Buchbauer and Buljubasic, 2015), and to realizing what dogs (Palmieri *et al.*, 2016) and rats can do in this context, such as detecting cancers or infectious diseases (Pajot-Augy, 2012). In animals, staying clear of ill congeners is safer, while avoiding diseased sexual partners on the basis of olfactory cues appears to be one of the means employed to prevent the propagation of potentially unfavourable genes (Blaustein, 1981). There is even a strong indication that olfaction has evolved in parallel to the immune system since the major histocompatibility complex (MHC) is recognized as a possible source of individual specific body odours (Penn and Potts, 1998). Because MHC polymorphism is correlated with a high degree of gene diversity in natural populations (thus multiplying the chances of adaptation), it should be maintained by some selective behavioural mechanisms (Eggert *et al.*, 1998). Thus, it may be that bad odours, besides keeping individuals away from danger, have played a decisive role during evolution in contributing to the selection of beneficial (good-smelling?) genes.

Acknowledgements

We thank Corbin Jones and Betty Wanjiru (Department of Biology, University of North Carolina at Chapel Hill and Carolina Center for Genome Sciences, Chapel Hill, North Carolina, USA) for allowing us to use their *D. sechellia* photograph.

References

Amann, A., Costello, B.d.L., Miekisch, W., Schubert, J., Buszewski, B., Pleil, J., Ratcliffe, N. and Risby, T. (2014) The human volatilome: volatile organic compounds (VOCs) in exhaled breath, skin emanations, urine, feces and saliva. *Journal of Breath Research* 8, 034001. DOI: 10.1088/1752-7155/8/3/034001

Ara, K., Hama, M., Akiba, S., Koike, K., Okisaka, K., Hagura, T., Kamiya, T. and Tomita, F. (2006) Foot odour due to microbial metabolism and its control. *Canadian Journal of Microbiology* 52, 357–364. DOI: 10.1139/W05-130

Barbery-Coulon, L. (2012) Sissel Tolaas: odeurs de trouble. In *Le Monde magazine.* Available at: http://www.lemonde.fr/m-styles/article/2012/11/23/sissel-tolaas-odeurs-de-trouble_1794326_4497319.html (accessed 16 January 2017).

Barnes, D.S. (2006) *The Great Stink of Paris and the Nineteenth-Century Struggle against Filth and Germs.* The Johns Hopkins University Press, Baltimore, Maryland.

Bernier, U.R., Kline, D.L., Barnard, D.R., Schreck, C.E. and Yost, R.A. (2000) Analysis of human skin emanations by gas chromatography / mass spectrometry. 2. Identification of volatile compounds that are candidate attractants for the yellow fever mosquito (*Aedes aegypti*). *Analytical Chemistry* 72, 747–756. DOI: 10.1021/ac990963k

Blaustein, A. (1981) Sexual selection and mammalian olfaction. *The American Naturalist* 117, 1006–1010. DOI: 10.1086/283786

Buchbauer, G. and Buljubasic, F. (2015) The scent of human diseases: a review on specific volatile organic compounds as diagnostic biomarkers. *Flavour and Fragrance Journal* 30, 5–25. DOI: 10.1002/ffj.3219

Candau, J. and Jeanjean, A. (2006) Des odeurs à ne pas regarder. *Terrain* 47, 51–68.

Chen, M.K. and Risen, J.L. (2010) How choice affects and reflects preferences: revisiting the free-choice paradigm. *Journal of Personality and Social Psychology* 99, 573–594.

Chrea, C., Valentin, D., Sulmont-Rossé, C., Ly Mai, H., Hoang Nguyen, D. and Abdi, H. (2004) Culture and odour categorization: agreement between cultures depends upon the odours. *Food Quality and Preference* 15, 669–679. DOI: 10.1016/j.foodqual.2003.10.005

Chrea, C., Valentin, D., Sulmont-Rossé, C., Hoang Nguyen, D. and Abdi, H. (2005) Semantic, typicality and odour representation: a cross-cultural study. *Chemical Senses* 30, 37–49. DOI: 10.1093/chemse/bjh255

Classen, C., Howes, D. and Synnott, A. (1994) *Aroma: The Cultural History of Smell.* Routledge, London and New York, pp. 95–158.

Corbin, A. (1982) *Le miasme et la jonquille*. Flammarion, Paris.

Costello, B.d.L., Amann, A., Al-Kateb, H., Flynn, C., Filipiak, W., Khalid, T., Osborne, D. and Ratcliffe, N.M. (2014) A review of the volatiles from the healthy human body. *Journal of Breath Research* 8, 014001. DOI: 10.1088/1752-7155/8/1/014001

Dekker, T., Ibba, I., Siju, K.P., Stensmyr, M.C. and Hansson, B.S. (2006) Olfactory shifts parallel superspecialism for toxic fruit in *Drosophila melanogaster* sibling, *D. sechellia*. *Current Biology* 16, 101–109. DOI: 10.1016/j.cub.2005.11.075

Di Pellegrino, G., Fadiga, L., Fogassi, L., Gallese, V. and Rizzolatti, G. (1992) Understanding motor events: a neurophysiological study. *Experimental Brain Research* 91, 176–180.

Dias, B.G. and Ressler, K.J. (2013) Parental olfactory experience influences behaviour and neural structure in subsequent generations. *Nature Neuroscience* 17, 89–96. DOI: 10.1038/nn.3594

Donkin, I., Versteyhe, S., Ingerslev, L.R., Qian, K., Mechta, M., Nordkap, L., Mortensen, B., Appel, E.V., Jorgensen, N., Kristiansen, V.B., Hansen, T., Workman, C.T., Zierath, J.R. and Barres, R. (2016) Obesity and bariatric surgery drive epigenetic variation of spermatozoa in humans. *Cell Metabolism* 23, 369–378. DOI: 10.1016/j.cmet.2015.11.004

Dormont, L., Bessière, J.-M. and Cohuet, A. (2013) Human skin volatiles: a review. *Journal of Chemical Ecology* 39, 569–578. DOI: 10.1007/s10886-013-0286-z

Dravnieks, A. (1985) *Atlas of Odour Character Profiles*. American Society for Testing and Materials, Philadelphia, Pennsylvania.

Dravnieks, A., Masurat, T. and Lamm, R.A. (1984) Hedonics of odours and odour descriptors. *Journal of Air Pollution Control Association* 34, 752–755.

Dubois, D. (2006) Des catégories d'odorants à la sémantique des odeurs. *Terrain* 47, 89–106.

Dulau, R. (2004) Deux lieux, deux approches du champ odorant: Pondichéry, Paris. In: J. Cobbi and R. Dulau, *Sentir. Pour une anthropologie des odeurs*. L'Harmattan, Paris, pp. 129–148.

Dupire, M. (1987) Du goût et des odeurs: classifications et universaux. *L'Homme* 27, 5–25.

Eggert, F., Müller-Ruchholz, W. and Ferstl, R. (1998) Olfactory cues associated with the major histocompatibility complex. *Genetica* 104, 191–198. DOI: 10.1023/A:1026402531196

Gallagher, M., Wysocki, C.J., Leyden, J.J., Spielman, A.I., Sun, X. and Preti, G. (2008) Analyses of volatile organic compounds from human skin. *British Journal of Dermatology* 159, 780–791. DOI: 10.1111/j.1365-2133.2008.08748.x

Garner, C.E., Smith, S., Costello, B.d.L., White, P., Spencer, R., Probert, C.S.J. and Ratcliffe, N.M. (2007) Volatile organic compounds from feces and their potential for diagnosis of gastrointestinal disease. *FASEB Journal* 21, 1675–1688. DOI: 10.1096/fj.06-6927com

Grésillon, L. (2010) *Sentir Paris. Bien-être et matérialité des lieux*. Quae, Versailles, Paris, coll. 'Indisciplines'.

Henshaw, V. (2014) *Urban Smellscapes: Understanding and Designing City Smell Environments*. Routledge, New York, London.

Hungate, E.A., Earley, E.J., Boussy, I.A., Turissini, D.A., Ting, C.T., Moran, J.R., Wu, M.L., Wu, C.I. and Jones, C.D. (2013) A locus in *Drosophila sechellia* affecting tolerance of a host plant toxin. *Genetics* 195, 1063–1075. DOI: 10.1534/genetics.113.154773

Ibba, I., Angioy, A.M., Hansson, B.S. and Dekker, T. (2010) Macroglomeruli for fruit odours change blend preference in *Drosophila*. *Naturwissenschaften* 97, 1059–1066. DOI: 10.1007/s00114-010-0727-2

Keller, A., Zhuang, H., Chi, Q., Vosshall, L.B. and Matsunami, H. (2007) Genetic variation in a human odorant receptor alters odour perception. *Nature* 449, 468–473. DOI: 10.1038/nature06162

Kelley, N.J., Crowell, A.L., Tang, D., Harmon-Jones, E. and Schmeichel, B.J. (2015) Disgust sensitivity predicts defensive responding to mortality salience. *Emotion* 15, 590–602. DOI: 10.1037/a0038915

Kermen, F., Midroit, M., Kuczewski, N., Forest, J., Thevenet, M., Sacquet, J., Benetollo, C., Richard, M., Didier, A. and Mandairon, N. (2016) Topographical representation of odour hedonics in the olfactory bulb. *Nature Neuroscience* 19, 876–878. DOI: 10.1038/nn.4317

Khan, R.M., Luk, C.H., Flinker, A., Aggarwal, A., Lapid, H., Haddad, R. and Sobel, N. (2007) Predicting odour pleasantness from odorant structure: pleasantness as a reflection of the physical world. *Journal of Neuroscience* 27, 10015–10203. DOI: 10.1523/jneurosci.1158-07.2007

Kim, K.-H., Jahan, S.A. and Kabir, E. (2012) A review of breath analysis for diagnosis of human health. *Trends in Analytical Chemistry* 33, 1–8. DOI: 10.1016/j.trac.2011.09.013

Lapid, H., Shushan, S., Plotkin, A., Voet, H., Roth, Y., Hummel, T., Schneidman, E. and Sobel, N. (2011) Neural activity at the human olfactory epithelium reflects olfactory perception. *Nature Neuroscience* 14, 1455–1461. DOI: 10.1038/nn.2926

Lavista-Llanos, S., Svatos, A., Kai, M., Riemensperger, T., Birman, S., Stensmyr, M.C. and Hansson, B.S. (2014) Dopamine drives *Drosophila sechellia* adaptation to its toxic host. *eLife* 3, e03785. DOI: 10.7554/eLife.03785

Le Breton, D. (1998) Imaginaire sensoriel du racisme. Odeur de l'autre. In: Méchin C., Bianquis I. and Le Breton D. (eds) *Anthropologie du sensoriel. Les sens dans tous les sens*. L'Harmattan, Paris/Montréal, coll. 'Nouvelles études anthropologiques', pp. 7–21.

Lévy, F. and Ferreira, G. (2012) Les apprentissages alimentaires et sociaux régulés par l'olfaction. In: Salesse, R. and Gervais, R. (eds) *Odorat et goût: de la neurobiologie des sens chimiques aux applications*. Quae, Versailles, Paris, pp. 323–338.

Logan, J.G., Birkett, M.A., Clark, S.J., Powers, S., Seal, N.J., Wadhams, L.J., Mordue, A.J. and Pickett, J.A. (2008) Identification of human-derived volatile chemicals that interfere with attraction of *Aedes aegypti* mosquitoes. *Journal of Chemical Ecology* 34, 308–322. DOI: 10.1007/s10886-008-9436-0

Matsuo, T., Suguya, S., Yasukawa, J., Aigaki, T. and Fuyama, Y. (2007) Odorant-binding proteins OBP57d and OBP57e affect taste perception and host-plant preference in *Drosophila sechellia*. *PLoS Biology* 5, 985–995. DOI: 10.1371/journal.pbio.0050118

McBride, C.S. (2007) Rapid evolution of smell and taste receptor genes during host specialization in *Drosophila sechellia*. *Proceedings of the National Academy of Sciences USA* 104, 4996–5001. DOI: 10.1073/pnas.0608424104

Menashe, I., Abaffy, T., Hasin, Y., Goshen, S., Yahalom, V., Luetje, C.W. and Lancet, D. (2007) Genetic elucidation of human hyperosmia to isovaleric acid. *PLoS Biology* 5, 2462–2468. DOI: 10.1371/journal.pbio.0050284

Moncrieff, R.W. (1966) *Odour Preferences*. John Wiley, New York, pp. 11–29.

Natsch, A., Derrer, S., Flachsmann, F. and Schmidt, J. (2006) A broad diversity of volatile carboxylic acids, released by a bacterial amynoacylase from axilla secretions, as candidate molecules for the determination of human-body odour type. *Chemistry and Biodiversity* 3, 1–20. DOI: 10.1002/cbdv.200690015

Pajot-Augy, E. (2012) Diagnostic olfactif des pathologies. In: Salesse, R. and Gervais, R. (eds) *Odorat et goût: de la neurobiologie des sens chimiques aux applications*. Quae, Versailles, Paris, pp. 401–411.

Palmieri, B., Nardo, B., Lippi, G., Palmieri, L., Vadalà, M. and Laurino, C. (2016) Dogs' olfactory diagnostics applied on human species: state of the art and clinical perspectives. *La Clinica Terapeutica* 167, e78–e84. DOI: 10.7417/CT.2016.1943

Penn, D.J. and Potts, W.K. (1998) Chemical signals and parasite-mediated sexual selection. *Trends in Ecology and Evolution* 13, 391–396. DOI: 10.1016/S0169-5347(98)01473-6

Pourtier, L. (2012) Evaluation des nuisances olfactives et désodorisation. In: Salesse, R. and Gervais, R. (eds) *Odorat et goût: de la neurobiologie des sens chimiques aux applications*. Quae, Versailles, Paris,, pp. 451–457.

Prugnolle, F., Lefèvre, T., Renaud, F., Møller, A.P., Misse, D. and Thomas, F. (2009) Infection and body odours: evolutionary and medical perspectives. *Infection, Genetics and Evolution* 9, 1006–1009. DOI: 10.1016/j.meegid.2009.04.018

Rankin, C.H. (2015) A review of transgenerational epigenetics for RNAi, longevity, germline maintenance and olfactory imprinting in *Caenorhabditis elegans*. *Journal of Experimental Biology* 218, 41–49. DOI: 10.1242/jeb.108340

Remy, J.J. (2010) Stable inheritance of an acquired behaviour in *Caenorhabditis elegans*. *Current Biology* 20, R877–R878. DOI: 10.1016/j.cub.2010.08.013

R'Kha, S., Capy, P. and David, J.R. (1991) Host-plant specialization in the *Drosophila melanogaster* species complex: a physiological, behavioural, and genetical analysis. *Proceedings of the National Academy of Sciences USA* 88, 1835–1839.

Schmidt, K. and Podmore, I. (2015) Current challenges in volatile organic compounds analysis as potential biomarkers of cancer. *Journal of Biomarkers* 2015, 981458. DOI: 10.1155/2015/981458

Shepherd, G.M. (2012) *Neurogastronomy*. Columbia University Press, Columbia, New York.

Stensmyr, M.C., Dekker, T. and Hansson, B.S. (2003) Evolution of the olfactory code in the *Drosophila melanogaster* subgroup. *Proceedings of the Royal Society B: Biological Sciences* 270, 2333–2340. DOI: 10.1098/rspb.2003.2512

Suarez, F.L., Springfield, J. and Levitt, M.D. (1998) Identification of gases responsible for the odour of human flatus and evaluation of a device purported to reduce this odour. *Gut* 43, 100–104.

Syed, Z. and Leal, W.S. (2009) Acute olfactory response of *Culex* mosquitoes to a human- and bird-derived attractant. *Proceedings of the National Academy of Sciences USA* 106, 18803–18808. DOI: 10.1073/pnas.0906932106

Wicker, B., Keysers, C., Plailly, J., Royet, J.P., Gallese, V. and Rizzolatti, G. (2003) Both of us disgusted in *My* insula: the common neural basis of seeing and feeling disgust. *Neuron* 40, 655–664. DOI: 10.1016/S0896-6273(03)00679-2

Xu, P.X., Atkinson, R.A., Jones, D.N.M. and Smith, D.P. (2005) *Drosophila* OBP LUSH is required for activity of pheromone-sensitive neurons. *Neuron* 45, 193–200. DOI: 10.1016/j.neuron.2004.12.031

Yang, Z. and Huffman, S.L. (2013) Nutrition in pregnancy and early childhood and associations with obesity in developing countries. *Maternal and Child Nutrition* 9, 105–119. DOI: 10.1111/mcn.12010

Zarzo, M. (2008) Psychologic dimensions in the perception of everyday odours: pleasantness and edibility. *Journal of Sensory Studies* 23, 354–376. DOI: 10.1111/j.1745-459X.2008.00160.x

Zhang, Z.M., Cai, J.J., Ruan, G.H. and Li, G.K. (2005) The study of fingerprint characteristics of the emanations from human arm skin using original sampling system by SPME-GC/MS. *Journal of Chromatography B* 822, 244–252. DOI: 10.1016/j.jchromb.2005.06.026

6 The Role of Olfaction in Feeding and Foraging

MARIJE OOSTINDJER

Norwegian University of Life Sciences, Ås, Norway

That the sense of smell is important in the context of food should come as no surprise. Everyone who has had a bad cold, or suffers from olfactory impairment for another reason – for example, due to old age, or due to exposure to chemotherapy (Rolls, 1999; Brisbois *et al.*, 2006), will recognize that food becomes tasteless when the sense of smell is absent. Although it is hard to pinpoint exactly what part of the ability to taste food comes from the sense of smell, there is consensus that olfaction (particularly in the passageways up from the back of the mouth, also known as retronasal olfaction) plays a dominant role in taste (Spence, 2015). Taste receptors on the tongue also play a role, by distinguishing five different tastes (sweet, salty, bitter, sour and umami). Other sensory roles in food behaviour are played by taste receptors in the small intestine, by chemosensory receptors in the airways that react strongly to bitter compounds and other irritants, and even by sight and sound (Moyer, 2015). The different sensory inputs are then integrated in the brain, where taste is translated to liking or disliking (Rolls, 2005; but see also Chapter 5).

It is actually important to like and dislike food in order to stay healthy and to maintain energy homeostasis (Woods, 2004). It acts as a mechanism of motivation and protection that helps to decide which foods one should consume more of, and which foods should be avoided. Bitter taste, for example, is often related to poisonous or spoiled foods, while sweet taste is often associated with sugar and therefore with calories. Even when the tongue fails to pick up the taste, the taste receptors in the gut can provide information about the nutritional value of the food (Rozengurt and Sternini, 2007). However, food needs to be ingested first before it can be tasted in the gut, and this presents a potential danger if the food is poisonous. The way a food item looks and smells are the first signs used to judge whether or not the food is going to be healthy and nutritious, before eating it. This is why humans and other animals make associations between the smell of food, and how the food made them feel after ingestion (satisfied, or a stomach ache), and memorize this association (Sclafani, 1995). That way, next time the food is encountered, there is no need to consider the food for very long before deciding to eat it or avoid it. These olfactory food memories are helpful at various stages of feeding and foraging, for example, during the search for food (if the preferred food has a strong odour), and during the decision to ingest and digest the food. Post-ingestive consequences, including sensory input from the stomach and the intestine, as well as hormonal signals and blood sugar response, can help (re-)shape the olfactory food memories to result in aversion or preference for the food.

In Fig. 6.1, an overview is provided of how olfaction is involved in feeding and foraging. This chapter shows examples of studies that focus on different stages of feeding and foraging: when locating food, when experiencing post-ingestive consequences and when memorizing locations of previously found food sources. It also discusses studies that investigated how flexible olfactory food memories are, and how olfaction plays a role in learning from other animals with regard to what foods are good to eat.

Role of Olfaction in Locating Food

For many animals, finding food is a challenge. Depending on the density of the food source, and whether the food is hidden or easily available, animals may spend a large share of their time searching for food. Olfaction, in addition to other senses such as vision or even sound, can play an important role in locating food, particularly when the food

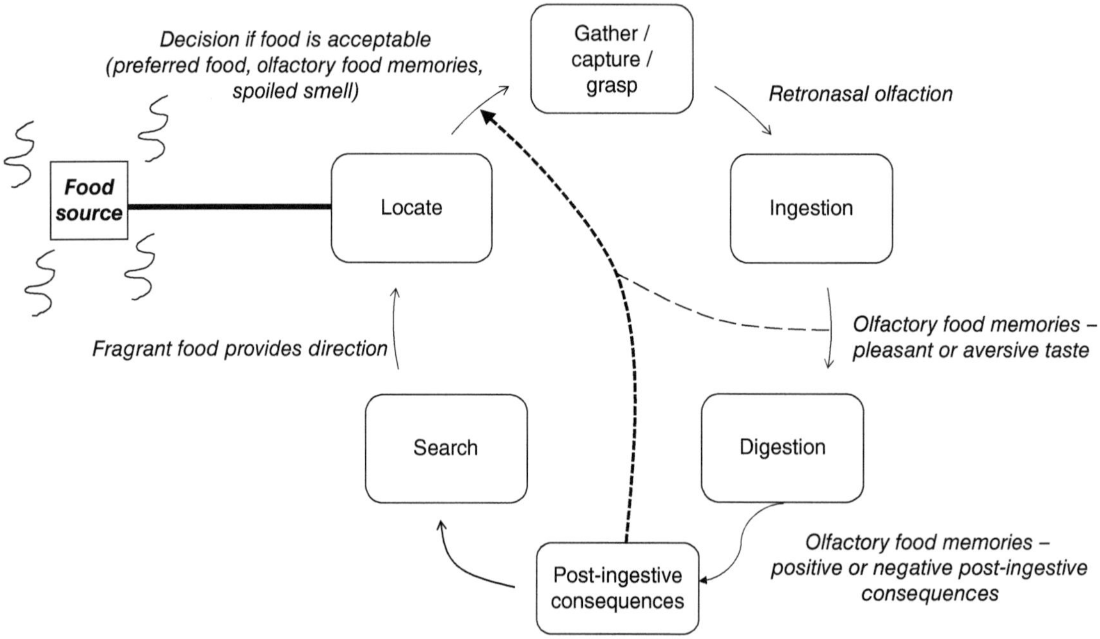

Fig. 6.1. Overview of the potential role of olfaction in feeding and foraging.

source gives off a particular odour that is recognizable for the animal. An example of this is the case of truffle searching pigs: the truffle gives off a particular odour (dimethyl sulfide) that pigs (and dogs) can use to locate the truffle in the soil (Talou *et al.*, 1990). In addition, odours may be used to avoid certain food sources. For example, cows will avoid grazing in areas where the smell of dead cattle is present, which helps prevent the ingestion of harmful pathogens or microbes. Some plants exploit this avoidance mechanism by emitting carrion odour that attracts pollinators, but at the same time repels herbivores (Lev-Yadun and Gutman, 2013).

Hara (2006) conducted a study with four fish species that demonstrates how fish can use olfaction to locate food, based on key amino acids. In this study, rainbow trout, lake charr, lake whitefish and goldfish were exposed to selected amino acids that are part of food sources for these species, and the effects of this exposure on feeding behaviour were studied. Fish were individually housed in tanks that had a flow-through feed trough made out of transparent plastic, in which the test stimuli could be provided in a fixed concentration. The amino acids that were used in this study were alanine, arginine, cysteine, glutamic acid, lysine, proline and serine. In addition, an extract from standard trout pellets and a mixture of all amino acids were used in the test.

Fish were first acclimatized to the tanks, which took 1 week for goldfish and 3 weeks for lake charr. The fish were then tested with different amino acids on successive days, so several stimuli could be tested per fish, while preventing confusion for the fish and overlapping stimuli. Each stimulus was tested on at least five fish per species. During a test session, the fish was observed in the tank for 15 min as a baseline, then observed for 15 min while the stimulus was presented in the trough, and then observed for 15 min after the stimulus had disappeared. Feeding behaviour is different between the four species in this study, but all four fish species display an increase in locomotor activity, with more swimming with 180° turns when being exposed to food. Other species-specific feeding behaviours included: bottom searching in rainbow trout, exploratory and escape behaviour against the window in lake whitefish, surfacing and jumping in lake charr, and gravel pecking in goldfish.

All fish increased their locomotor activity in response to the cysteine stimulus, often even more so than in response to the trout pellet extract. Most fish also showed increased locomotion in response to glutamic acids and proline. Species-specific feeding behaviour was mostly absent when no stimulus was present (with the exception of pecking in goldfish), but was elicited by several of the stimuli, though the food extract was the most effective in eliciting feeding behaviour. All behaviours that were increased by the stimuli decreased again when the stimuli disappeared. These results, together with results from an earlier electrophysiology study on amino acid response through olfaction and gustation (Hara, 2005), indicate that these four fish species can use single amino acids to help them locate food through olfaction. Once the food is located, additional stimuli and senses are used to start feeding, including visual and taste cues.

Olfaction is modulated by post-ingestive consequences

Olfaction can help to locate food sources, but once food is consumed and the animal is satiated, food searching is no longer necessary. The behaviour of food searching is supported by a feedback mechanism, where post-ingestive consequences (feeling full, increase in blood sugar or even feeling sick) are registered by the brain (Woods and D'Alessio, 2008), and influence the decision whether to keep searching for food, or to stop searching and instead spend time digesting the food (Berthoud, 2008). There has been a notion for a long time that this feedback mechanism of satiety is also linked to how sensitive an individual is to odours. For example, humans have been shown to be less sensitive to smells right after lunch (e.g. Goetzl and Stone, 1947). However, studies on the topic of olfactory sensitivity are tricky, and more recent studies in humans have not been able to show a clear effect of satiation on olfactory sensitivity (e.g. Koelega, 1994). In animal studies, it is much easier to control food intake, and therefore it is easier to pick up subtle effects. Earlier studies have shown that the olfactory neural system is more reactive to smells when the animal is food deprived, and less reactive when the animal is satiated (Pager *et al.*, 1972; Apelbaum *et al.*, 2005). To see whether this change in reactivity of the neural system translates to behaviour, Aimé *et al.* (2007) set out to study whether satiation and fasting affected the actual ability to detect odours in rats.

In this study, the rats were trained on day 0 to associate a neutral odour with a negative experience (see Fig. 6.2 for an overview of the study). In a test room, rats were provided with drinking water, odorized with the neutral odour (isoamyl acetate, ISO), at a concentration of 1 µl/100 ml water (10^{-5} dilution). Fifteen minutes after they drank the water, rats were given an injection with lithium chloride, which makes them sick for a short time. Rats were exposed to the odorized water three times, and had therefore three sessions to learn to avoid water that emitted this odour. During the test phase, rats were exposed to two tubes of water: one with odorized water and one with normal tap water. Rats had been water-deprived prior to the test and were therefore thirsty. At the start of the test they were always placed under the odorized water tube, and most rats would sample the odorized water first, because they were very motivated to drink. The rats were tested on five consecutive days with two sessions per day starting with a 10^{-10} odorized water dilution on day 1, progressing to a 10^{-7} dilution on day 4. Finally, on day 5 the aversion was retested in the same way as used during training on day 0, by using a 10^{-5} dilution in two sessions. A pre-test showed that the ability of rats to detect the odour was the same when tested with the same dilution in two consecutive sessions.

To test how satiation and fasting affects the detection of odours, rats were put on a food deprivation schedule in addition to the water deprivation, so that they now had access to food for 4 h per day only. Rats were then tested two times per day: once after 16 h of fasting and once 2 h after the meal (when the animals were still satiated). The main measurements in the test were the water intake from the odorized water and intake from the tap water.

Intake of the odorized water at a 10^{-5} dilution was very low for all rats, both when fasted and when satiated (on average less than 10% of total water intake), indicating that the odour was easily detected at this concentration. Satiated rats had difficulties detecting the odour in the water when the dilution was 10^{-10}, 10^{-9} or 10^{-8}, with average intakes around 70% of total intake (Fig. 6.2). This was significantly higher than the odorized water intake of fasted rats at the same dilutions, which was around 45% of total water intake. Since rats were very motivated to drink and the odorized water was the first water they encountered, many fasted rats did drink from the odorized water, but they also switched between the two waters more often than satiated rats, particularly at the 10^{-8} dilution. In this study, fasted rats were indeed better at detecting odour in water than satiated rats, indicating that animals are most sensitive to smells

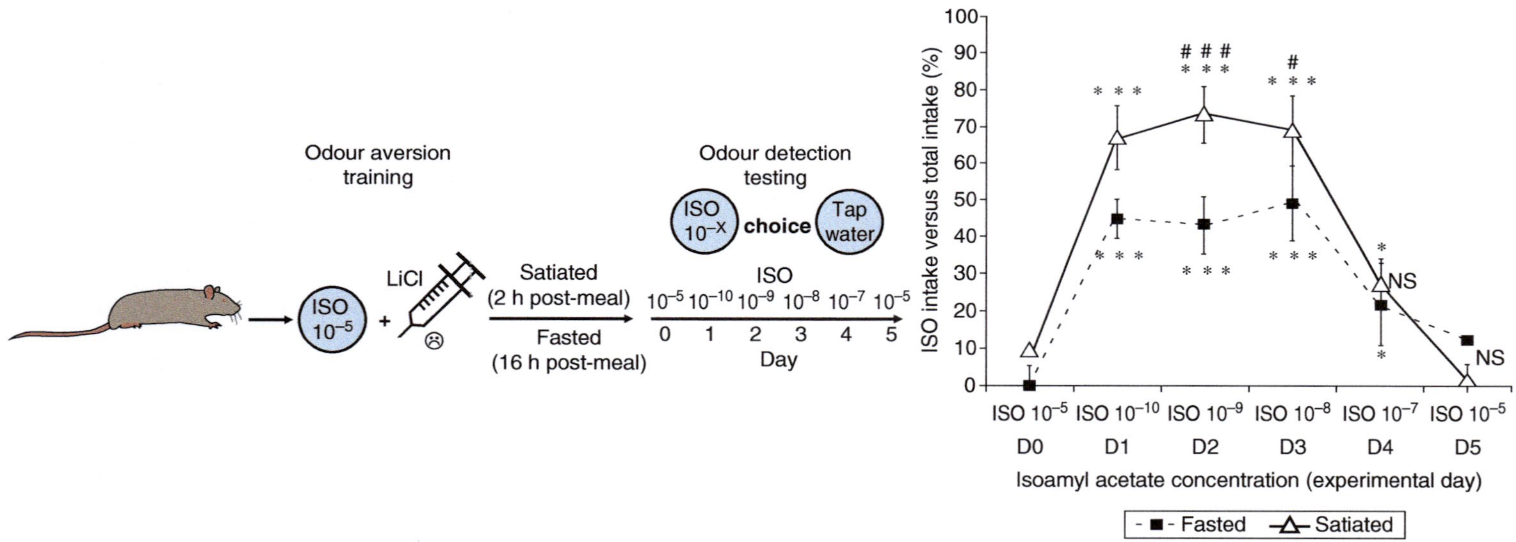

Fig. 6.2. Overview of study conducted by Aimé *et al*. (2007). Rats were trained to show an aversive response to a neutral odour (ISO) by use of lithium chloride (LiCl). After successful training, a series of odour detection tests was carried out, with concentrations from 10^{-10} to 10^{-5}, both when rats were satiated and when fasted. Fasted rats were better at detecting the ISO at low concentrations (10^{-10} to 10^{-8}) than satiated rats. (Graph from Aimé *et al*., 2007.)

when they need it the most in the context of feeding, which is when they are hungry.

Role of Olfaction in Memorizing Food Source Locations

For some animals, it is useful to remember where a certain food source is, so that the food source can be revisited in the future. This is particularly true when food sources are immobile and cannot be depleted by one individual in a single visit, or when the food source replenishes itself. Honey bees visit food sources to collect pollen or nectar that they then bring back to the hive. Often it takes several visits by one or more individuals to deplete a patch of flowers: hence, the famous honeybee round and waggle dances that indicate to other individuals of the hive in which direction and at what distance from the hive food may be found (Couvillon *et al.*, 2014). In addition, bees encounter many flowers of similar appearance, some of which provide a rich source of food, and some of which provide no food at all (Seeley *et al.*, 1991). Being able to memorize which types of flowers provide food will obviously help the bees to forage more efficiently, which is important because foragers are short-lived and the hive needs to be provided with sufficient nutrition to ensure survival of the hive and rearing of the brood (Bell, 1990; Rueppell *et al.*, 2009).

Olfaction helps bees to remember which locations provide a food reward, and which ones do not. Nectar, which is a rewarding food for bees that is rapidly metabolized (it contains mostly sugars), has been shown previously to be associated rapidly with cues from the environment, including location and scent of the flower from which the nectar comes (Scheiner *et al.*, 1999). When the scent is then encountered, bees will head in the direction of the scent to obtain the food reward (Chaffiol *et al.*, 2005; Arenas *et al.*, 2007). Pollen, however, is less rewarding to bees because it is mostly made of slow-digestible protein, and is often brought back in its entirety to the hive (or accidentally deposited on other plants). This means that there is a delay between experiencing the scent of the flower, and obtaining nutrients from pollen, which may prevent bees from forming olfactory food memories associated with the scent of flowers that provide good quality pollen. Understanding how bees forage, together with their preferences for flowers and for pollen is useful because bees play such a large role in the world's food supply: an estimated one-third of all food crops relies on insect pollination, of which bees account for 80% (Kleijn *et al.*, 2015).

To investigate how bees develop pollen preferences, and how this is related to flower scent and to flower location, Arenas and Farina (2012) conducted four different experiments. The first experiment studied whether bees are able to associate specific odours with the presence of pollen and whether the memory of these scents is helpful when looking for pollen in the future. First, bees were trained by being exposed to a feeder with one of two neutral odours used in this study, and then tested by giving them access to two feeders: one with the odour used during training, and one with the other neutral odour. All landings at both feeders were recorded, and any landing was taken to indicate preference for that feeder. Indeed, most landings were at the feeder where the conditioned scent was present, regardless of whether pollen was available at the feeders or not. The scent could already be learned in one single visit, but pollen had to be available at the feeder for the bees to start proper feeding behaviour: bees were unwilling to collect a non-food item (chalk powder). So, the learned odour-association can guide the bees to the food source but in order for feeding to occur, multiple senses are likely to be used, and this includes perceiving the smell of the pollen itself.

The second experiment included bees that were trained with a neutral scent linked to pollen in a feeder, which were in turn brought back into the lab and tested in one of two tests: a Y-maze test or a proboscis extension response (PER) test (Fig. 6.3). The aim was to see whether learned responses were transferable to other contexts. The Y-maze consists of a starting arm, a central part and two arms that can be used for two different cues (Fig. 6.3A). In this study, the trained scent was in one arm and the neutral scent was in the other arm. The bees' first choice and the time spent in each arm were measured. The PER test (Takeda, 1961) involves bees being immobilized and exposed to different odours. If the bee has learned to associate a specific odorant with a pollen source, the bee should extend the proboscis (its tongue) when the trained scent is present, as it should signal food. In the Y-maze, bees spent more time in the arm where the trained scent was present. However, in the PER test, bees failed to extend their proboscis in response to scents associated with pollen; they only showed an extension response when a scent previously linked to sugar was presented.

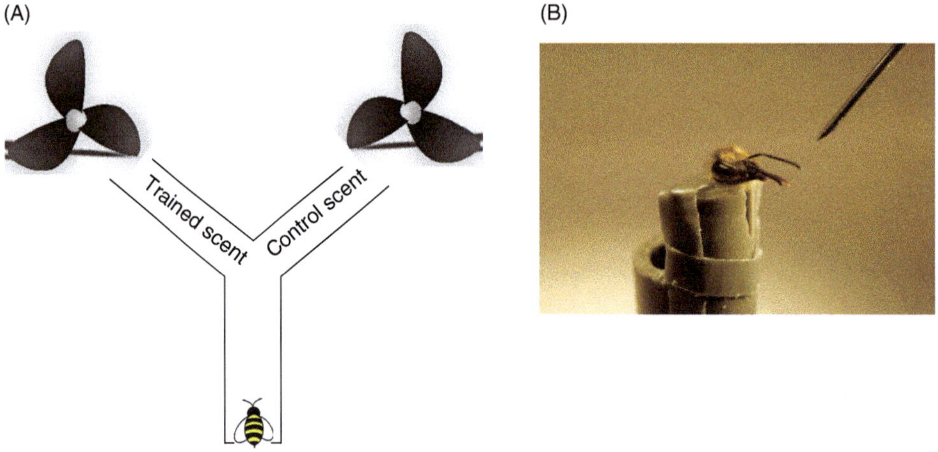

Fig. 6.3. Two behavioural tests performed by Arenas and Farina (2012). Panel A shows a diagram of the Y-maze test, in which the bee can fly around freely in three arms: the starting arm (bottom) and two scented arms (top), of which one contains the trained scent and one the control scent. Panel B shows the proboscis extension response (PER) test, in which the immobilized bee is exposed to a scent via a continuous airflow from a syringe. (Credits PER test photo: Daniel Münch.)

The third experiment tested whether bees would return to a feeder after re-exposure to a scent associated with the presence of pollen. Bees were first trained on a feeding platform that contained both a scent and pollen. Then, the feeder was cleared of both scent and pollen for 3 h, and bees would stop visiting the feeder. To test whether visiting the feeder could be reinstated, the scent was blown into the hive, and revisits of bees were counted. Indeed, about three times as many bees re-visited the feeder after the trained scent was blown into the hive, compared with when air or a neutral scent were blown into the hive.

Finally, Arenas and Farina (2012) conducted a fourth experiment as a follow-up from experiment two, in which bees failed to transfer learned associations between pollen and scent from outside the lab to a PER test in the lab. As bees did respond to sugar, many different substrates including sugar, pollen, fatty acids and amino acids (since pollen contains mostly protein) were used to test which substrates do and do not result in successful conditioning in the PER test. A distinction was made between pollen foragers and non-pollen foragers. Conditioning of both types of bees was successful when using sugar and to some extent fatty acids, but not when using an amino acid mixture. In the lab, pollen foragers could be conditioned to associate an odour with (hand-picked) pollen, while non-pollen foragers would show an initial response, but could not be conditioned.

This study showed that bees can use odours to locate a food source (pollen), and form memories of the location of this pollen for later visits. When the pollen source disappears, bees stop visiting the location, but when the odour associated with the previous pollen location returns, bees will start visiting the pollen location again. The olfactory pollen memories are strong and not context-dependent, as they could be retrieved in a lab setting as well. However, it is not the protein inside the pollen, but rather the sugar and fatty acids that are on the outside of the pollen that help form the memory.

Salience of olfactory food memories

Some olfactory memories are hard to forget, and this is true also for food. Most people have experienced how the smell of food can bring back memories from the past, as far back as childhood, often associated with the feeling of safety and comfort (like grandma's home baked cookies; Herz, 2002; Willander and Larsson, 2006). This strong recall of a mood or memory in response to a food odour is also thought to be involved in the current public health problems of excessive body weight and obesity. While the causes of surplus weight and obesity are many and complex, it basically comes down to an imbalance between energy intake (overconsumption, consumption of energy-dense foods with low nutritional value) and energy expenditure (physical activity; WHO, 2015). Energy dense foods are often very palatable due to their high content of fat and/or sugar, and even when people manage to stop eating such foods for some time, for example during a diet, they often relapse. To investigate whether olfactory memories may be involved in the likelihood of relapsing into unhealthy food behaviours, La Mela *et al.* (2010) performed an experiment in mice, in which chocolate was used as the energy-dense food. The experiment was based on the idea of cue reactivity, which is a trained response to a cue (such as a neutral odour) that elicits a certain behaviour (e.g. intake of alcohol or drug use; Carter and Tiffany, 1999) in order to obtain a positive experience (sedation or a high in the case of alcohol or drug use, and a pleasant taste or satiety in the case of palatable food intake). In the study, male mice were trained in a place-conditioning set-up, in which the location of two objects would signal either the presence or absence of a chocolate treat, in one of the two rooms of the set-up. A control group was tested in the same set-up, but where the objects would signal only the presence or absence of a chocolate smell. Indeed, animals in the test group successfully learned that when the objects were in a certain location it would signal the presence of chocolate. After training was successful, animals went through an extinction phase, in which they successfully learned that the chocolate was no longer linked to the location of the two objects. Then, to reinstate the previously learned object-location and chocolate presence combination, the smell of chocolate was presented to the mice, which were then tested in a similar setup. The smell of chocolate was indeed enough for the mice to show food seeking behaviour again. The authors conclude that olfactory cues are powerful with regard to reinstating the behaviour of palatable food searching. As we are tempted by many food odours throughout the day in many places and from different sources, it may be no surprise that most people have trouble eating healthily and in moderation.

Role of Olfaction in Learning from Other Individuals What to Eat

Olfaction can help animals learn what foods are healthy for them. In fact, olfactory food memories can already form before birth based on the information coming from the mother. This process is described as perinatal flavour learning, i.e. learning about flavours in the period around birth. The young animal is exposed to flavours that come from what the mother is eating. This process is called flavour learning rather than odour learning, as there are many different molecules in the mother's diet that may be experienced as smells, as tastes or both, and it is the combined experience that results in learning. When flavours from the mother's diet are experienced frequently, the young animal may develop a preference for these flavours. This is a mechanism of learning from an experienced animal through the following paradigm: given that the mother animal has reached maturity and procreated, she must have a lot of experience with eating healthy and nutritious food, and therefore flavours to which the young animal are exposed frequently via the mother must be linked to healthy and nutritious foods: if it's good for Mum, it's good for the offspring. This flavour exposure can happen before birth, while the animal is still in the uterus (mammals) or in the egg (fish, birds), or after birth, through maternal milk (mammals) or from the birth environment. Flavour learning as a mechanism has been shown in a variety of animals, including humans (Marlier *et al.*, 1998; Mennella *et al.*, 2001), rabbits (Bilkó *et al.*, 1994), rats (Galef and Henderson, 1972), dogs (Wells and Hepper, 2006), pigs (Langendijk *et al.*, 2007; Oostindjer *et al.*, 2010), sheep (Schaal *et al.*, 1995), goats (Hai *et al.*, 2012), cats (Becques *et al.*, 2009) and even chickens (Sneddon *et al.*, 1998) and frogs (Hepper and Waldman, 1992). Learning what to eat is a challenge for a young animal (Galef and Giraldeau, 2001), and this early flavour learning helps the animal choose nutritious foods from early in life.

A number of the studies on flavour learning are fundamental in nature, but some aim to improve the welfare of the species under investigation. Research in humans, for example, may ultimately aim to understand how early exposure to flavours can result in a healthy diet later in life (Forestell and Mennella, 2007). Flavour learning may also be useful to improve farm animal welfare. One study that illustrates this is a study on flavour learning in pigs, which I carried out with colleagues in the Netherlands (Oostindjer *et al.*, 2011). Young pigs face a challenge under husbandry conditions: they are separated (weaned) much earlier from the mother than would occur under (semi-)natural conditions (Jensen and Stangel, 1992), and many piglets have never tried solid food at the time of weaning (Berkeveld *et al.*, 2007; Kuller *et al.*, 2007). This leads to a lot of problems and stress in the first weeks after weaning, with low food intakes in the first days after weaning, low growth or even weight loss, and intestinal problems including diarrhoea (Brooks and Tsourgiannis, 2003). Establishing (solid) food intake right after weaning is key, and perinatal flavour learning may be a useful catalyst to establish food intake early.

The study by Oostindjer *et al.* (2011) aimed to investigate whether perinatal flavour learning in pigs can help reduce problems after weaning and – if so – through which mechanism(s) the flavour learning helps the piglets. Is it through an increased preference for food with a certain flavour, or is the smell of the odour part of the familiar flavour, independent of the food, enough to help make the weaning process smoother for piglets? For this purpose, piglets from 12 sows were exposed to a moderately liked flavour, anise (also known as aniseed, provided here as the molecule anethol), through the maternal diet before and after birth. Exposure took place in the late stage of pregnancy (days 98 to 115 of gestation, sows usually give birth on day 115), and between birth and weaning (days 2 to 25 after birth, to ensure a good food intake of sows in the first 2 days post-partum). Piglets from 12 other sows were never exposed to the study flavour. At weaning, half of the piglets from each sow were placed in pens where normal piglet food as well as food with anise flavour were available. The other half of the piglets were placed in pens where only normal piglet food was available to eat, and where the food with anise flavour was provided in a bucket with small holes, to allow the smell of anise to escape, while inaccessible for the piglets to eat. This resulted in four groups of piglets in 12 pens per group: with or without perinatal flavour exposure, and with the same flavour either in the food or in the air post-weaning (see Fig. 6.4 for an overview). Over a period of 2 weeks, food intake and weight was monitored, behaviours that indicate that piglets are stressed (aggression, biting or manipulating other piglets' body parts, including belly nosing) or relaxed (play) were observed and diarrhoea prevalence was noted. In addition, on the day of weaning, stress levels were assessed through cortisol measured in the saliva.

Piglets that were exposed to the flavour before weaning ate sooner after weaning than control piglets: around 9 h earlier on average, where average latency to eat for all piglets was 30 h after weaning. In the 2 weeks after weaning, food intake was the same for all groups, as was growth, but there was less variation in growth within the pens of flavour-exposed piglets. Flavour-exposed piglets were less stressed than control piglets: they vocalized less on the day of weaning, had lower cortisol levels and showed less manipulation behaviour in the 2 weeks after weaning. It did not matter whether the flavour after weaning was provided in the food or whether only the smell was present, except that piglets exposed to the flavour perinatally and subsequently in the air post-weaning played more than the other groups. Otherwise, the presence of the familiar odour was enough to generate the positive effects of flavour learning on piglet welfare. These results match studies on familiar odours and pheromones in terms of their impact on stress, even when these are not associated with food, such as maternal pheromones (McGlone and Anderson, 2002). In pigs, the perinatal flavour exposure did not result in a preference for the food with the familiar flavour, as it can do in other species. However, flavour learning from the mother helped the young pig to establish an early food intake.

Perinatal flavour learning is just one way of social learning in which olfaction plays a role. Animals may also benefit from a very direct type of social learning that involves olfaction, where one animal can learn about foods that another animal has eaten by smelling the mouth or another body part of the other animal. This type of olfactory learning about food, called flavour enhancement, can result in a preference for that food type, as has been shown in a study in dogs by Lupfer-Johnson and Ross, 2007. In this study, 12 pairs of dogs were studied: in each pair, there was a demonstrator dog and an observer dog of approximately equal size. The demonstrator dog was fed a commercial dog food with added basil or thyme. After the demonstrator dog had eaten, the observer dog was allowed to interact with the observer dog for 10 min. The interactions during this period were recorded to ensure that the observer dog had sniffed the mouth of the demonstrator dog. After the interaction, the observer dog was provided with two flavoured feeds in equal amounts: one flavoured with thyme, and one with basil. The amount of each feed eaten by the observer dog was measured to see if observer dogs had a preference for the same type of feed as that eaten by the demonstrator dog. This was indeed the case: on average 68% of the food that the observer dogs ate was the same type as that the demonstrator dog had eaten. The authors suggest that this type of direct learning about food through olfaction was very useful for dogs during the process of domestication. During that time, dogs had to learn which foods provided

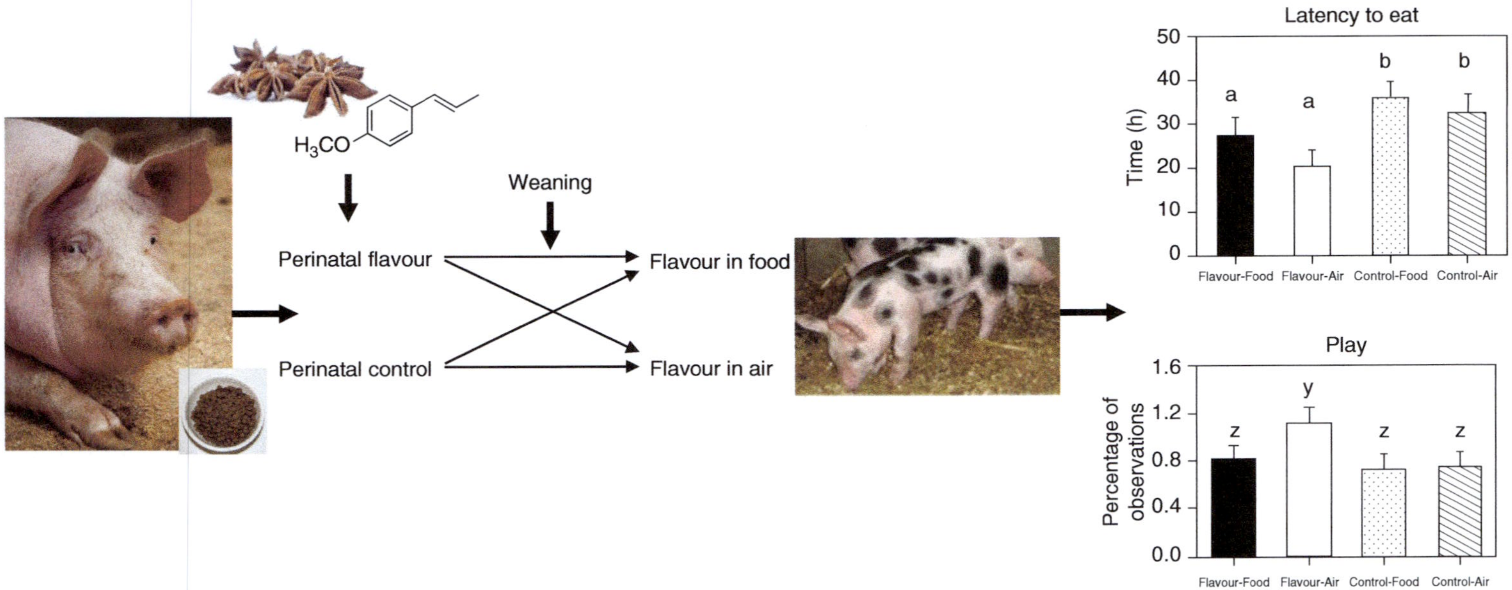

Fig. 6.4. Effects of perinatal flavour learning on food intake and behaviour of pigs in the first 2 weeks after weaning. Sows either received an anethol-flavoured diet or a control diet around birth. After weaning, piglets received the anethol flavour either in the food, or only in the air. Piglets exposed to the flavour before weaning showed a shorter latency to start eating after weaning (top graph), and piglets that were exposed to flavour before weaning and received the flavour in the air after weaning showed more play behaviour (bottom graph). Different letters indicate significant differences at $P < 0.05$. (Graphs modified from Oostindjer *et al.*, 2011; piglet photo courtesy of Monique Suer.)

by humans or from the human environment were nutritious, and which ones were harmful. Similar learning occurs in wild rats, another species that often eats food from the human environment (Galef *et al.*, 1988), but one that tends to be particularly neophobic as they cannot vomit (and thus not expel poisoned food easily; Yamamoto *et al.*, 2016). This type of learning may be less useful for wild dogs or wolves, as their diet is less diverse (Coppinger and Coppinger, 2001), and they encounter fewer unfamiliar foods in the wild than the dogs did in the human environment.

Role of Olfaction in Feeding and Foraging: Can It Be Used to Improve Animal Welfare?

Olfaction plays an important role in feeding and foraging. Understanding this role better may ultimately be useful to improve animal welfare. One example of this was described earlier in this chapter, where perinatal flavour learning helped piglets to adapt better to the situation after weaning. There may be other applications as well, particularly with regard to animals that are faced with challenges of over- or undereating. Both overeating (a problem with many pets) and undereating (a problem caused by management practices in breeding sows and broiler breeders) pose threats to animal welfare. Breeding sows are restrictedly fed through most of the breeding cycle (with the exception of peak lactation), in order to maintain optimal body composition for fertility. Broiler breeders are fed restrictively to slow down growth, so as to lower mortality and increase egg production. On the other hand, many dogs and cats are overweight or obese, partly due to an overabundance of food present in the home, including treats and human foods (Sandøe *et al.*, 2014). Restrictive feeding means that the animals are always motivated to find more food, while animals that are overweight may also seek and eat more food than necessary. Olfaction, as discussed in this chapter, plays a role in food searching behaviour, as well as in satiety. It may be possible to reduce the motivation of animals to search for food by using certain smells that reduce appetite. Rats, for example, show reduced appetite when exposed to the smell of grapefruit (Shen *et al.*, 2005a), while lavender increases appetite (Shen *et al.*, 2005b). Research in humans show that the smell of dark chocolate alone is enough to suppress appetite in women (Massolt *et al.*, 2010). Indeed, some patents can be found that exploit the sense of smell to reduce appetite, either by providing an odour that suppresses appetite (e.g. Nietupski, 1985) or by making the olfactory system less sensitive, which in turn may help to suppress appetite (e.g. More, 2006). Yet these ideas have not yet resulted in ready-to-use products. There may not be any odour-based methods ready for appetite suppression in farm animals and pets, but it provides an interesting opportunity for future research and applications.

References

Aimé, P., Duchamp-Viret, P., Chaput, M.A., Savigner, A., Mahfouz, M. and Julliard, A.K. (2007) Fasting increases and satiation decreases olfactory detection for a neutral odor in rats. *Behavioural Brain Research* 179, 258–264. DOI: 10.1016/j.bbr.2007.02.012

Apelbaum, A.F., Perrut, A. and Chaput, M. (2005) Orexin A effects on the olfactory bulb spontaneous activity and odor responsiveness in freely breathing rats. *Regulatory Peptides* 129, 49–61. DOI: 10.1016/j.regpep.2005.01.003

Arenas, A. and Farina, W.M. (2012) Learned olfactory cues affect pollen-foraging preferences in honeybees, *Apis mellifera*. *Animal Behaviour* 83, 1023–1033. DOI: 10.1016/j.anbehav.2012.01.026

Arenas, A., Fernández, V.M. and Farina, W.M. (2007) Floral odor learning within the hive affects honeybees' foraging decisions. *Naturwissenschaften* 94, 218–222. DOI: 10.1007/s00114-006-0176-0

Becques, A., Larose, C., Gouat, P. and Serra, J. (2009) Effects of pre-and postnatal olfactogustatory experience on early preferences at birth and dietary selection at weaning in kittens. *Chemical Senses* 35, 41–45. DOI: 10.1093/chemse/bjp080

Bell, W.J. (1990) Searching behavior patterns in insects. *Annual Review of Entomology* 35, 447–467. DOI: 10.1146/annurev.en.35.010190.002311

Berkeveld, M., Langendijk, P., van Beers-Schreurs, H.M., Koets, A.P., Taverne, M.A. and Verheijden, J.H. (2007) Postweaning growth check in pigs is markedly reduced by intermittent suckling and extended lactation. *Journal of Animal Science* 85, 258–266. DOI: 10.2527/jas.2006-143

Berthoud, H.R. (2008) Vagal and hormonal gut–brain communication: from satiation to satisfaction. *Neurogastroenterology and Motility* 20, 64–72. DOI: 10.1111/j.1365-2982.2008.01104.x

Bilkó, Á., Altbäcker, V. and Hudson, R. (1994) Transmission of food preference in the rabbit: the means of information transfer. *Physiology and Behavior* 56, 907–912. DOI: 10.1016/0031-9384(94)90322-0

Brisbois, T.D., Hutton, J.L., Baracos, V.E. and Wismer, W.V. (2006) Taste and smell abnormalities as an

independent cause of failure of food intake in patients with advanced cancer – an argument for the application of sensory science. *Journal of Palliative Care* 22, 111–114.
Brooks, P.H. and Tsourgiannis, C.A. (2003) Factors affecting the voluntary feed intake of the weaned pig. In: Pluske, J.R., Le Dividich, J. and Verstegen, M.W.A. (eds) *Weaning the Pig, Concepts and Consequences*. Wageningen Academic Publishers, Wageningen, The Netherlands, pp. 81–116. DOI: 10.3920/978-90-8686-513-0
Carter, B.L. and Tiffany, S.T. (1999) Meta-analysis of cue-reactivity in addiction research. *Addiction* 94, 327–340. DOI: 10.1046/j.1360-0443.1999.9433273.x
Chaffiol, A., Laloi, D. and Pham-Delègue, M.H. (2005) Prior classical olfactory conditioning improves odour-cued flight orientation of honey bees in a wind tunnel. *Journal of Experimental Biology* 208, 3731–3737. DOI: 10.1242/jeb.01796
Coppinger, R. and Coppinger, L. (2001) *Dogs: A Startling New Understanding of Canine Origin, Behavior and Evolution*. Scribner, New York.
Couvillon, M.J., Schürch, R. and Ratnieks, F.L. (2014) Waggle dance distances as integrative indicators of seasonal foraging challenges. *PLoS ONE* 9, e93495. DOI: 10.1371/journal.pone.0093495
Forestell, C.A. and Mennella, J.A. (2007) Early determinants of fruit and vegetable acceptance. *Pediatrics* 120, 1247–1254. DOI: 10.1542/peds.2007-0858
Galef, B.G. and Giraldeau, L.A. (2001) Social influences on foraging in vertebrates: causal mechanisms and adaptive functions. *Animal Behaviour* 61, 3–15. DOI: 10.1006/anbe.2000.1557
Galef, B.G. and Henderson, P.W. (1972) Mother's milk: a determinant of the feeding preferences of weaning rat pups. *Journal of Comparative and Physiological Psychology* 78, 213. DOI: 10.1037/h0032186
Galef, B.G., Mason, J.R., Preti, G. and Bean, N.J. (1988) Carbon disulfide: a semiochemical mediating socially-induced diet choice in rats. *Physiology and Behavior* 42, 119–124. DOI: 10.1016/0031-9384(88)90285-5
Goetzl, F.R. and Stone, F. (1947) Diurnal variations in acuity of olfaction and food intake. *Gastroenterology* 9, 444–453.
Hai, P.V., Everts, H., Van Tien, D., Schonewille, J.T. and Hendriks, W.H. (2012) Feeding *Chromonaela odorata* during pregnancy to goat dams affects acceptance of this feedstuff by their offspring. *Applied Animal Behaviour Science* 137, 30–35. DOI: 10.1016/j.applanim.2012.01.010
Hara, T.J. (2005) Olfactory responses to amino acids in rainbow trout: revisited. In: Reutter, K. and Kapoor, B.G (eds) *Fish Chemosenses*. Science Publishers, Enfield, UK, pp. 32–64.
Hara, T.J. (2006) Feeding behaviour in some teleosts is triggered by single amino acids primarily through olfaction. *Journal of Fish Biology* 68, 810–825. DOI: 10.1111/j.0022-1112.2006.00967.x
Hepper, P.G. and Waldman, B. (1992) Embryonic olfactory learning in frogs. *Quarterly Journal of Experimental Psychology: Section B* 44, 179–197. DOI: 10.1080/02724999208250611
Herz, R.S. (2002) Influences of odors on mood and affective cognition. In: Rouby, C., Schaal, B., Dubois, D., Gervais, R. and Holley, A. (eds) *Olfaction, Taste, and Cognition*. Cambridge University Press, Cambridge, UK, pp. 160–177.
Jensen, P. and Stangel, G. (1992) Behaviour of piglets during weaning in a seminatural enclosure. *Applied Animal Behaviour Science* 33, 227–238. DOI: 10.1016/S0168-1591(05)80010-3
Kleijn, D., Winfree, R., Bartomeus, I., Carvalheiro, L.G., Henry, M., Isaacs, R., Klein, A.M., Kremen, C., M'gonigle, L.K., Rader, R. and Ricketts, T.H. (2015) Delivery of crop pollination services is an insufficient argument for wild pollinator conservation. *Nature Communications* 6, 7414. DOI: 10.1038/ncomms8414
Koelega, H.S. (1994) Diurnal variations in olfactory sensitivity and the relationship to food intake. *Perceptual and Motor Skills* 78, 215–226. DOI: 10.2466/pms.1994.78.1.215
Kuller, W.I., Soede, N.M., van Beers-Schreurs, H.M.G., Langendijk, P., Taverne, M.A.M., Kemp, B. and Verheijden, J.H.M. (2007) Effects of intermittent suckling and creep feed intake on pig performance from birth to slaughter. *Journal of Animal Science* 85, 1295–1301. DOI: 10.2527/jas.2006-177
La Mela, I., Latagliata, E.C., Patrono, E., Puglisi-Allegra, S. and Ventura, R. (2010) Olfactory priming reinstates extinguished chocolate-induced conditioned place preference. *Appetite* 54, 237–240. DOI: 10.1016/j.appet.2009.12.008
Langendijk, P., Bolhuis, J.E. and Laurenssen, B.F.A. (2007) Effects of pre-and postnatal exposure to garlic and aniseed flavour on pre-and postweaning feed intake in pigs. *Livestock Science* 108, 284–287. DOI: 10.1016/j.livsci.2007.01.083
Lev-Yadun, S. and Gutman, M. (2013) Carrion odor and cattle grazing: evidence for plant defense by carrion odor. *Communicative and Integrative Biology* 6, p.e26111. DOI:10.4161/cib.26111
Lupfer-Johnson, G. and Ross, J. (2007) Dogs acquire food preferences from interacting with recently fed conspecifics. *Behavioural Processes* 74, 104–106. DOI: 10.1016/j.beproc.2006.09.006
Marlier, L., Schaal, B. and Soussignan, R. (1998) Bottle-fed neonates prefer an odor experienced in utero to an odor experienced postnatally in the feeding context. *Developmental Psychobiology* 33, 133–145. DOI: 10.1002/(SICI)1098-2302(199809)33:2<133::AID-DEV4>3.0.CO;2-K
Massolt, E.T., van Haard, P.M., Rehfeld, J.F., Posthuma, E.F., van der Veer, E. and Schweitzer, D.H. (2010) Appetite suppression through smelling of dark chocolate correlates with changes in ghrelin in young women.

Regulatory Peptides 161, 81–86. DOI: 10.1016/j.regpep.2010.01.005

McGlone, J.J. and Anderson, D.L. (2002) Synthetic maternal pheromone stimulates feeding behavior and weight gain in weaned pigs. *Journal of Animal Science* 80, 3179–3183. DOI: 10.2527/2002.80123179x

Mennella, J.A., Jagnow, C.P. and Beauchamp, G.K. (2001) Prenatal and postnatal flavor learning by human infants. *Pediatrics* 107, e88–e88. DOI: 10.1542/peds.107.6.e88

More, R.J. (2006) Orally and nasally administered appetite suppressant. US Patent Application 11/449,118.

Moyer, M. (2015) What makes food taste so good? *Scientific American* 24, 4–8. DOI: 10.1038/scientificamericanfood0615-4

Nietupski, R.S. (1985) Appetite suppressant and method of application thereof. US Patent 4,521,427.

Oostindjer, M., Bolhuis, J.E., van den Brand, H., Roura, E. and Kemp, B. (2010) Prenatal flavor exposure affects growth, health and behavior of newly weaned piglets. *Physiology and Behavior* 99, 579–586. DOI: 10.1016/j.physbeh.2010.01.031

Oostindjer, M., Bolhuis, J.E., Simon, K., van den Brand, H. and Kemp, B. (2011) Perinatal flavour learning and adaptation to being weaned: all the pig needs is smell. *PLoS ONE* 6, e25318. DOI: 10.1371/journal.pone.0025318

Pager, J., Giachetti, I., Holley, A. and Le Magnen, J. (1972) A selective control of olfactory bulb electrical activity in relation to food deprivation and satiety in rats. *Physiology and Behavior* 9, 573–579. DOI: 10.1016/0031-9384(72)90014-5

Rolls, B.J. (1999) Do chemosensory changes influence food intake in the elderly? *Physiology and Behavior* 66, 193–197. DOI: 10.1016/S0031-9384(98)00264-9

Rolls, E.T. (2005) Taste, olfactory, and food texture processing in the brain, and the control of food intake. *Physiology and Behavior* 85, 45–56. DOI: 10.1016/j.physbeh.2005.04.012

Rozengurt, E. and Sternini, C. (2007) Taste receptor signaling in the mammalian gut. *Current Opinion in Pharmacology* 7, 557–562. DOI: 10.1016/j.coph.2007.10.002

Rueppell, O., Kaftanouglu, O. and Page, R.E. (2009) Honey bee (*Apis mellifera*) workers live longer in small than in large colonies. *Experimental Gerontology* 44, 447–452. DOI: 10.1016/j.exger.2009.04.003

Sandøe, P., Corr, S. and Palmer, C. (2014) Fat companions: understanding the welfare effects of obesity in cats and dogs. In: Appleby, M.C. and Weary, D.M (eds) *Dilemmas in Animal Welfare*. CAB International, Wallingford, UK, pp. 28–45. DOI: 10.1079/9781780642161.0028

Schaal, B., Orgeur, P. and Arnould, C. (1995) Olfactory preferences in newborn lambs: possible influence of prenatal experience. *Behaviour* 132, 351–365. DOI: 10.1163/156853995X00603

Scheiner, R., Erber, J. and Page, R.E., Jr (1999) Tactile learning and the individual evaluation of the reward in honey bees (*Apis mellifera* L.). *Journal of Comparative Physiology A* 185, 1–10. DOI: 10.1007/s003590050360

Sclafani, A. (1995) How food preferences are learned: laboratory animal models. *Proceedings of the Nutrition Society* 54, 419–427. DOI: 10.1079/PNS19950011

Seeley, T.D., Camazine, S. and Sneyd, J. (1991) Collective decision-making in honey bees: how colonies choose among nectar sources. *Behavioral Ecology and Sociobiology* 28, 277–290. DOI: 10.1007/BF00175101

Shen, J., Niijima, A., Tanida, M., Horii, Y., Maeda, K. and Nagai, K. (2005a) Olfactory stimulation with scent of grapefruit oil affects autonomic nerves, lipolysis and appetite in rats. *Neuroscience Letters* 380, 289–294. DOI: 10.1016/j.neulet.2005.01.058

Shen, J., Niijima, A., Tanida, M., Horii, Y., Maeda, K. and Nagai, K. (2005b) Olfactory stimulation with scent of lavender oil affects autonomic nerves, lipolysis and appetite in rats. *Neuroscience Letters* 383, 188–193. DOI: 10.1016/j.neulet.2005.04.010

Sneddon, H., Hadden, R. and Hepper, P.G. (1998) Chemosensory learning in the chicken embryo. *Physiology and Behavior* 64, 133–139. DOI: 10.1016/S0031-9384(98)00037-7

Spence, C. (2015) Just how much of what we taste derives from the sense of smell? *Flavour* 4, 1. DOI: 10.1186/s13411-015-0040-2

Takeda, K. (1961) Classical conditioned response in the honey bee. *Journal of Insect Physiology* 6, 168–179. DOI: 10.1016/0022-1910(61)90060-9

Talou, T., Gaset, A., Delmas, M., Kulifaj, M. and Montant, C. (1990) Dimethyl sulphide: the secret for black truffle hunting by animals? *Mycological Research* 94, 277–278. DOI: 10.1016/S0953-7562(09)80630-8

Wells, D.L. and Hepper, P.G. (2006) Prenatal olfactory learning in the domestic dog. *Animal Behaviour* 72, 681–686. DOI: 10.1016/j.anbehav.2005.12.008

WHO (2015) Obesity and overweight. Fact sheet N° 311. Available at: http://www.who.int/mediacentre/factsheets/fs311/en/ (accessed 17 January 2017).

Willander, J. and Larsson, M. (2006) Smell your way back to childhood: autobiographical odor memory. *Psychonomic Bulletin and Review* 13, 240–244. DOI: 10.3758/BF03193837

Woods, S.C. (2004) Gastrointestinal satiety signals I. An overview of gastrointestinal signals that influence food intake. *American Journal of Physiology – Gastrointestinal and Liver Physiology* 286, G7–G13. DOI: 10.1152/ajpgi.00448.2003

Woods, S.C. and D'Alessio, D.A. (2008) Central control of body weight and appetite. *The Journal of Clinical Endocrinology and Metabolism* 93, s37–s50. DOI: 10.1210/jc.2008-1630

Yamamoto, K., Yamamoto, E., Sugimoto, T., Sagakami, T. and Yamatodani, A. (2016) Sevoflurane-induced pica in female rats. *Journal of Pharmacological Sciences* 131, 68–71. DOI: 10.1016/j.jphs.2016.03.015

7 The Role of Olfaction in Mate Selection and Reproductive Behaviour

LUISA AMO

Department of Evolutionary Ecology, Museo Nacional de Ciencias Naturales, CSIC, Madrid, Spain

Chemical communication is probably the most ancient and widespread form of communication (Wyatt, 2003). The importance of chemical cues in intraspecific communication has been described in many animal taxa, from invertebrates to vertebrates (Johansson and Jones, 2007). The chemical compounds that animals use for social communication can range from a single compound that induces a fixed response, such as the pheromone released by many insect females to attract males, to complex mixtures of chemical compounds that provide information about the sender to which conspecifics may react or not depending on their motivation or physiological state. For example, two components of mouse urine (2-*sec*-butyl-4,5-dihydrothiazole and 2,3-dehydro-*exo*-brevicomin) can elicit aggressive displays if the receiver is a male (Novotny *et al.*, 1985) or induce oestrus if the receiver is a female (Jemiolo *et al.*, 1986). This variation in chemical signals used by animals in social contexts is the result of both natural and sexual selective forces that have shaped the evolution of chemical cues (Wyatt, 2003).

Mate Selection

When selecting partners for mating, females may choose high quality males that will provide the best resources, the best genes or both (Bateman, 1948; Trivers, 1972). For this purpose, females may rely on the characteristics that provide honest information about male quality (Grafen, 1990). Therefore, sexual selection has favoured the evolution of elaborated sexual signals in males of many animal species. Among these sexual signals, plumage coloration has provided good examples of sexual selection in birds (Hill, 2002). In other, less colourful and visually conspicuous animals, researchers have focused on other sensory systems to study mate choice preferences, such as acoustic cues in amphibians (Ron, 2008) or chemical cues in mammals (Hurst, 2009).

Darwin (1871) suggested that the exaggerated displays of males, including 'glands for emitting odours' have evolved through sexual selection. Scent-producing glands, scent marking behaviour and the chemical signals themselves are often different between males and females, even in species with non-conspicuous visual dimorphism, as in the case of many small mammals. So the scent of a male mouse will be the equivalent of the plumage of a peacock (Fig. 7.1), as they both function to attract females and they both are costly to produce (Penn and Potts, 1998a). Sexual chemical signals may thus allow potential partners or rivals to evaluate an individual's quality. Theoretical models have predicted that signals can only be evolutionarily stable if they are condition-dependent, or costly to the signaller, and if the cost is correlated with the signaller's quality (Zahavi, 1975; Grafen, 1990; Zahavi and Zahavi, 1997). Therefore, individuals cannot afford to cheat, i.e. to signal at too high a level (Krebs and Dawkins, 1983), and conspecifics can rely on the information provided by these honest chemical signals.

Scent marking behaviour and mate attraction

Pheromones or chemical signals are frequently deposited as scent marks left on various substrates with the purpose of attracting mates or signalling

Fig. 7.1. Both the chemical cues that a male mouse emits and the plumage of a peacock are costly to produce and signal quality to females and rival males (Penn and Potts, 1998a). (Figure from Penn and Potts, 1998a, with permission.)

territory ownership to rival conspecifics. When chemical cues emitted by an animal are condition-dependent, they provide reliable information about the donor of the scent, and therefore can be used both by potential mates (Johansson and Jones, 2007; Thomas, 2011) or by rivals (e.g. Martín and López, 2007; Arakawa *et al.*, 2008) to assess the quality of the transmitter of such chemical cues.

Similar to mammals, many reptile species produce chemical signals (Mason and Parker, 2010) that they use to mark their territory. For example, male Carpetan rock lizards (*Iberolacerta cyreni*) scent mark substrates with secretions from the femoral glands (found on their inner thighs), which contain proteins and lipids such as fatty acids and steroids (López *et al.*, 2006), including ergosterol. These chemical cues provide information on sex, body size, age, familiarity and health condition (Martín *et al.*, 2007). Because the production of ergosterol in the gland secretions is costly, only high quality males can afford it. Therefore this scent mark can be considered as an honest sexual display. These chemical signals could be useful for male lizards to attract females to their territories, thereby increasing the probability of mating.

In a field experiment, Martín and López (2012) tested this hypothesis. They experimentally simulated scent marking by Carpetan rock lizard (*I. cyreni*) males of good quality by adding ergosterol to rocks inside home ranges occupied by male lizards, and examined the effects of this manipulation on the density of females in the home range. They also added a control scent on rocks of the home range of other male lizards. They found increased relative densities of females in the territories of males where the ergosterol was added. They also observed a higher number of females associated with males in those territories (Fig. 7.2), which probably increased mating opportunities for males in these areas. The authors conclude that female rock lizards choose to settle in home ranges based on the presence of characteristics of scent marks from conspecific males. Male rock lizards may therefore attract more females and obtain more matings by increasing the proportion of ergosterol when scent marking their home ranges.

Discrimination between individual characteristics when looking for partners

When an animal is searching for a partner, first it needs to discriminate a conspecific from an individual of another species. When closely related species are in sympatry (live in the same area), this first step becomes especially important in maintaining reproductive isolation between species. Mihailova and collaborators (2014) performed a study to investigate whether the crimson rosella, *Platycercus elegans*, a species of parrot, could discriminate between the odours of subspecies. Once the females had completed laying and started incubating, the authors placed one of the following inside the nest-box: (i) feathers from an individual of the same subspecies (*Platycercus elegans elegans*); (ii) feathers from an individual of a different subspecies (*Platycercus elegans flaveolus*); (iii) feathers from an individual of a different species; and (iv) no feathers, as odourless control. The feathers were placed inside a black cotton cloth pouch, so the birds could smell but not see them. The authors filmed the behaviour of females and investigated whether there were differences between the treatments in the latency to enter the nest-box and the time spent inside the nest-box. The authors found that females were able to differentiate between the scent of the feather of individuals from the same subspecies and that of individuals from a different subspecies or species. When female *P. e. elegans* individuals were exposed to the scent of an individual from the same subspecies, they returned to and entered the nest-box quicker (Fig. 7.3A) and spent more time on the nest-box (Fig. 7.3B) than when they were exposed to other scents.

Fig. 7.2. Number of males of the Carpetan rock lizard with females in their territory after adding ergosterol or a control scent to their territory. (Based on data from Martín and López, 2012, and photo of a female Carpetan rock lizard by Luisa Amo.)

Despite the study being undertaken during the incubation period, when females are not actively searching for partners, the results suggest that *P. elegans* females could use subspecies odour recognition during mate choice. It is thus due to olfaction that *P. elegans* subspecies can prevent hybridization and maintain divergence and isolation between the phenotypically distinct subspecies, as has been observed in other taxa such as insects and mammals (Nevo *et al.*, 1976; Nosil *et al.*, 2007; Smadja and Butlin, 2008).

The chemical compounds that seem to play a role in intraspecific communication in different taxa, from insects to mammals, are similar between species. However, the mixture of chemical compounds that animals emit for intraspecific communication has evolved to be specific to a species, i.e. although many of the odorant compounds that animals produce can be found in more than one species, the proportion of such compounds in a signal is unique for a particular species. Species recognition may constitute the first step in the use of scents in mate recognition.

A second step in choosing a mate is to correctly assess the sex of the conspecific. Especially in species with scarce visual sexual dimorphism, the use of

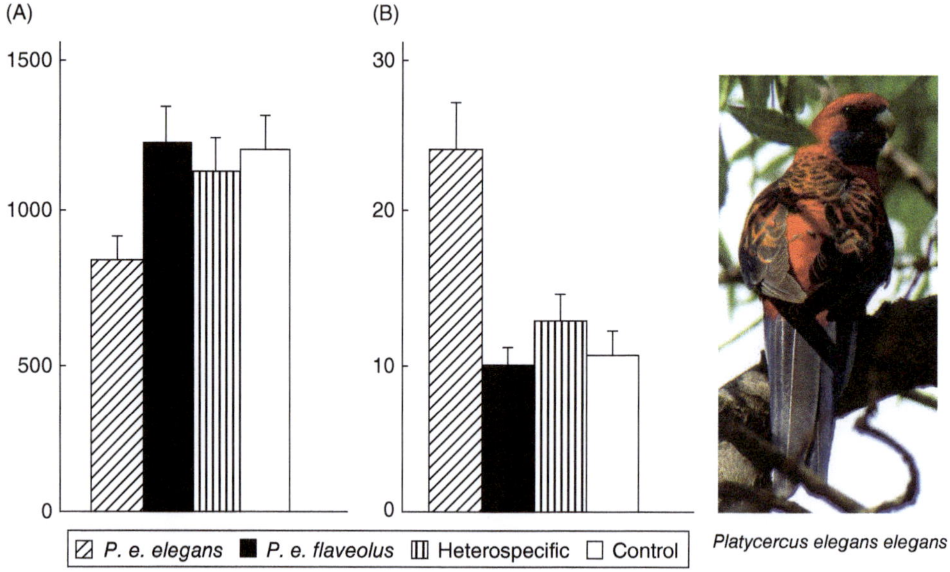

Fig. 7.3. Mean (± SE) (A) time (s) to enter the nest-box for the first time and (B) time (s) on the nest-box when female *Platycercus elegans elegans* were exposed to the scent of the feathers from an individual of the same subspecies (diagonal lines), from an individual of a different subspecies (black) and from an individual of a different species (vertical lines) as well as an odourless control (white). (Data from Mihailova *et al.*, 2014; photograph by Guillermo Amo de Paz.)

olfaction to discriminate between males and females is highly relevant. For example, Amo and collaborators (2012) investigated whether a passerine bird, the spotless starling, *Sturnus unicolor*, was able to discriminate the sex of conspecifics by using olfactory cues and whether the size and secretion composition of the uropygial gland (also called the preen gland) conveyed information on sex, age and reproductive status in this species. Secretions from the uropygial gland, placed just above the tail, are considered as the main odour source in birds. This secretion is a mixture of monoester and diester waxes, triglycerides, fatty acids and hydrocarbons, although its composition varies widely among avian groups (Jacob and Ziswiler, 1982). It contains both volatile and non-volatile compounds in the form of waxy fluids that birds collect and spread on their feathers during preening (Jacob and Ziswiler, 1982). Therefore, the chemical components of the uropygial secretion are also present on the feathers of birds (Soini *et al.*, 2007; Mardon *et al.*, 2011).

The finding that the gland secretory activity as well as the chemical components of uropygial secretions vary between seasons (Jacob *et al.*, 1979; Reneerkens *et al.*, 2002), sexes (Jacob *et al.*, 1979; Piersma *et al.*, 1999; Whittaker *et al.*, 2010), age classes, diets (Sandilands *et al.*, 2004a,b) and hormone levels (Whelan *et al.*, 2010) suggests that these secretions may provide important information during intraspecific interactions, particularly in sex recognition (Whittaker *et al.*, 2011), individual recognition (Bonadonna and Nevitt, 2004) and mate choice (Slade *et al.*, 2016).

Amo and collaborators (2012) captured spotless starlings (*S. unicolor*) during the mating period and performed a choice experiment in an olfactometry apparatus (Fig. 7.4), offering males and females the scent of a male and a female. The focal bird was introduced in the central chamber, where it received two air flows coming from the scent donor birds during 5 min. A small fan created the scent flows by extracting the air from the chamber. After 5 min of exposure to the scents, the doors of the central chamber were opened so the bird could select one of the choice chambers. The experimenter was blind to the treatments, i.e. she did not know the location of the scent donor male or female inside the chamber.

They found that spotless starlings were able to discriminate the sex of conspecifics by using chemical cues alone. Both male and female starlings preferred male scents (Fig. 7.5). The results found for female starlings are in accordance with the hypothesis that

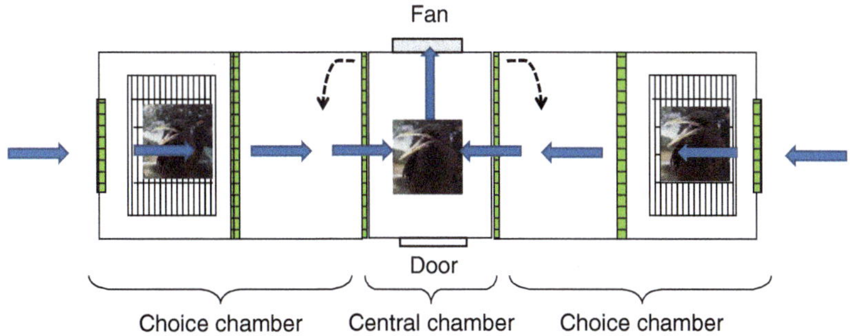

Fig. 7.4. Olfactometry chamber where the experiment on spotless starlings was performed. Scent donor birds were in small cages to limit movement. The chamber was in darkness to prevent birds from emitting calls. The solid arrows indicate the direction of air flow within the chamber, whereas the dashed arrows indicate the direction of opening of the two doors connecting the two choice chambers. (Data from Amo *et al.*, 2012.)

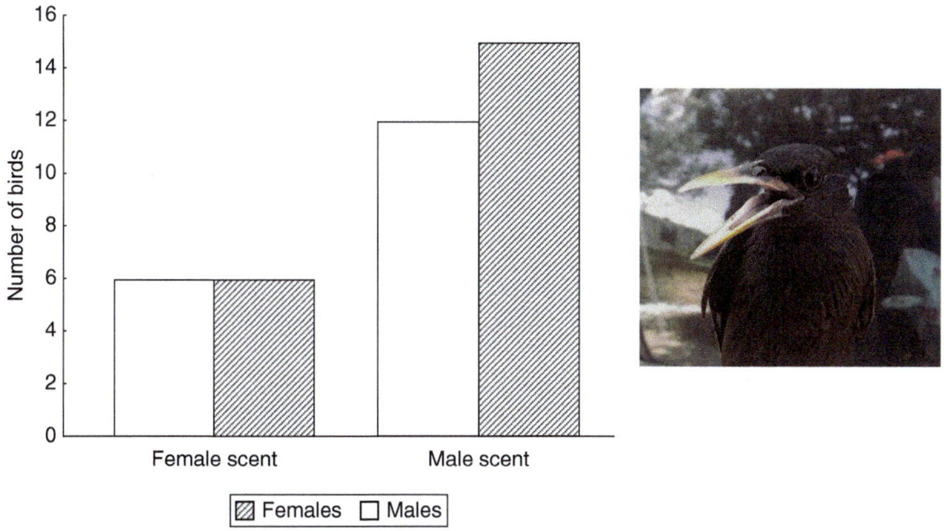

Fig. 7.5. Number of male (white) and female (lined) adult spotless starlings that chose the side of the chamber containing the scent of a male or a female starling, respectively ($Z = 2.87$, $P = 0.004$). (Data from Amo *et al.*, 2012; photograph of a female spotless starling by Gustavo Tomás Gutiérrez.)

females may recognize and respond to the scent of a potential partner, as has previously been observed in budgerigars, *Melopsittacus undulatus* (Zhang *et al.*, 2010). However, males also oriented towards male scents. This was contrary to the hypothesis that males may prefer the scent of a female over the scent of a male. In this study, starlings were captured at the beginning of reproduction, when males often engage in aggressive intrasexual encounters to obtain a cavity for breeding. The authors therefore explained the preference of males for the scent of another male in terms of intrasexual competition. Similar results were obtained by Jones *et al.* (2004) in a study with crested auklets (*Aethia cristatella*), where it was found that although both sexes approached scented male models more closely than unscented male models, males responded more to scented male models than females did, which the authors assume was a sign of intrasexual aggression. Male mice (*Mus musculus*) are also attracted to scent marks of other males because they provide useful information about the social dominance of rival males (Arakawa *et al.*, 2008).

In the study by Amo *et al.* (2012), analysis of the chemical composition of the uropygial gland secretion by gas chromatography–mass spectrometry (GC–MS) revealed differences between sexes, ages and reproductive status of adult spotless starlings. The authors concluded that the uropygial gland secretion may potentially function as a chemical signal used in mate choice and/or intrasexual competition in this as well as in other bird species (Whittaker *et al.*, 2011).

Assessment of body condition, health condition and parasite load of potential partners

Along with obtaining information about species, gender or age, it has been demonstrated that animals can use chemical cues to detect complex information about potential mates, such as body size and body condition, health state and parasite load (Johansson and Jones, 2007; Arakawa *et al.*, 2008). Several studies have shown that females not only discriminate between the scent of uninfected and infected males, but are also more attracted to uninfected males (for a review, see Beltran-Bech and Richard, 2014).

Penn and collaborators (1998) tested whether female house mice, *Mus musculus domesticus*, were able to assess a male's infection status through olfactory cues. The authors presented females with the urine of uninfected males and males experimentally infected with influenza, a respiratory virus. Females spent more time investigating the scent of males before (Fig. 7.6A) and after (Fig. 7.6B) infection than when they were infected. Additionally, most females preferred to nest in boxes containing the

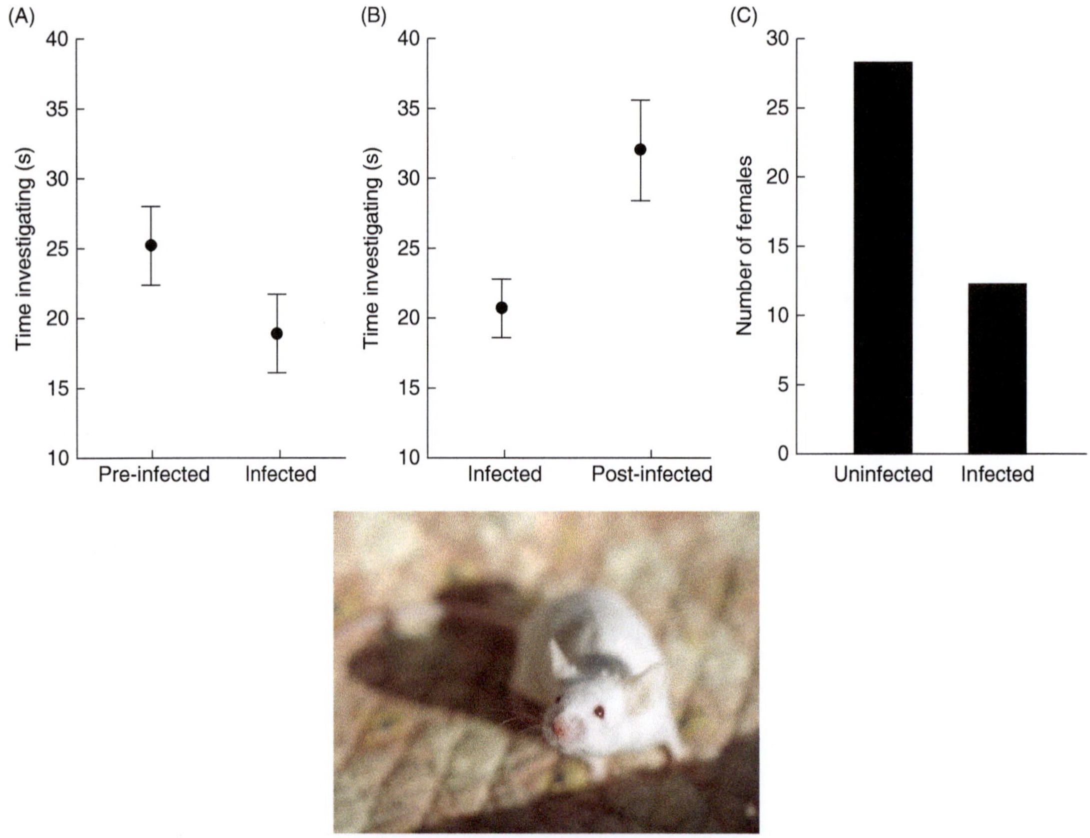

Fig. 7.6. (A) Mean (± SE) time female house mice (*Mus musculus domesticus*) spent investigating the urine of males before infection with influenza (pre-infection) versus during infection. (B) Mean (± SE) time females spent investigating the urine of males during infection versus after infection (post-infection). (C) Number of females settled in nest-boxes containing urine collected while males were uninfected (pre- and post-infection combined) versus infected. (Data from Penn *et al.*, 1998; photograph of mouse © Tomo.yun' (www.yunphoto.net/es/).)

scent of uninfected rather than infected males (Fig. 7.6C). This study was the first evidence that mice can discriminate virally infected individuals through chemical signals and the first evidence that infection causes odour changes in the urine, as was previously observed when male mice were infected with coccidian gut parasites or nematodes (Kavaliers and Colwell, 1995a,b).

Detecting the reproductive status of partners

Chemical signals also provide valuable information about the reproductive status of conspecifics for animals searching for a partner. This may be especially useful for males in order to avoid pregnant females, where mating is futile, or to determine if a mating is worth sperm investment in species where females display multiple matings. Males of many species are able to distinguish between mated and unmated females using chemical cues, and they usually exhibit a preference for unmated females (Thomas, 2011). For example, in invertebrates, *Agelenopsis aperta* spider males and *Tenebrio molitor* beetle males preferentially respond to chemical signals produced by unmated over mated females (Riechert and Singer, 1995; Carazo *et al.*, 2004). Males of the lizard *Podarcis hispanica* also prefer the scent of non-gravid over that of gravid females (Cooper and Pérez-Mellado, 2002). Males can discriminate between mated and unmated females through three mechanisms. First, females may stop the emission of pheromones for attracting males. Second, after mating, females may emit pheromones for repelling males. And third, during mating, males may apply chemical cues to females in order to modify their attractiveness or to signal that she has already copulated. Males can transfer these chemical signals during courtship and during or after copulation. Chemicals can also be transferred with the ejaculate or the mating plug (secretion that males deposit into a female genital tract during mating, that later hardens into a plug, to prevent sperm being expelled by females or sperm competition with other male sperm; Thomas, 2011). These mechanisms are not mutually exclusive, and are often used in combination. For example, in many moth species the female stops her pheromone production due to neural or mechanical stimulation during mating as well as to the substances that the male transfers in the ejaculate (Wedell, 2005).

Kin recognition and inbreeding avoidance

In philopatric species (those that establish in the same territory where they were born instead of dispersing; Pusey and Wolf, 1996), the possibility of meeting a related individual is high. Therefore, females may be able to discriminate between related and unrelated individuals when searching for a partner, in order to avoid inbreeding. Olfaction is the mechanism responsible for kin discrimination in many taxa, such as invertebrates (Gamboa, 1996), fish (Neff and Sherman, 2003), birds (Coffin *et al.*, 2011) and mammals (Mateo, 2003).

Coffin and collaborators (2011) performed an experiment to examine whether the Humboldt penguin (*Spheniscus humboldti*) in captivity was able to discriminate between the scent of familiar individuals and related individuals (kin). They offered captive penguins two kennels containing (i) the scent of a familiar non-kin penguin versus an unfamiliar non-kin penguin, or (ii) the scent of an unfamiliar kin versus an unfamiliar non-kin penguin. The time spent by the penguins inspecting the kennels was then recorded over a period of 10 min.

Humboldt penguins use olfactory cues to recognize and discriminate between familiar and unfamiliar individuals (Coffin *et al.*, 2011). When simultaneously presented with the scents from a familiar and an unfamiliar penguin, they investigated the kennels with unfamiliar odours first, but then spent more time inside kennels with familiar odours (Fig. 7.7A). Penguins have the ability to discriminate odours based on genetic differences, allowing the discrimination between kin and non-kin individuals (Fig. 7.7B). Although the study was not performed during the mating period, the discrimination between kin and non-kin individuals as well as the preference for non-kin scents suggest that penguins are likely to use olfaction to avoid inbreeding during mate searching.

Major histocompatibility complex-associated mate choice

The major histocompatibility complex (MHC) has been proposed to be a signaller of genetic relatedness that may facilitate inbreeding avoidance across vertebrates (Hurst, 2009). The MHC is a large chromosomal region containing several highly polymorphic genes (MHC class I and II loci) that play a central role in controlling immunological self/non-self recognition (Klein, 1986; Janeway, 1993) and encode

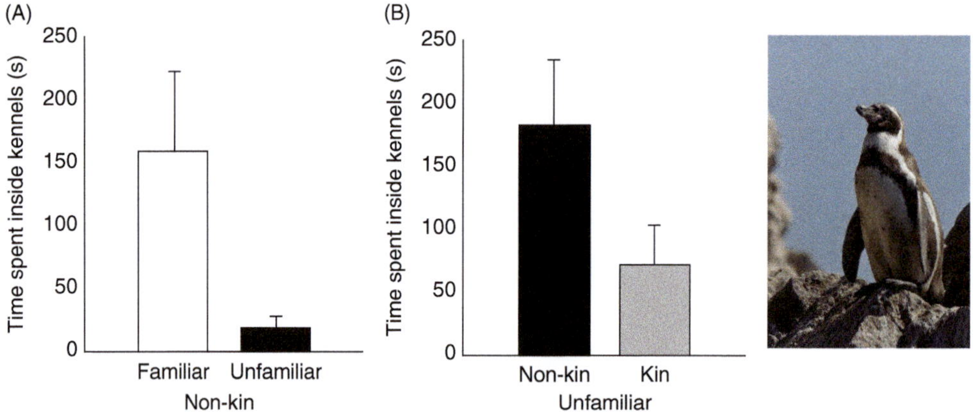

Fig. 7.7. Mean time (± SE) spent by Humboldt penguins (*Spheniscus humboldti*) inside kennels containing the scent of (A) a familiar vs unfamiliar non-kin penguin, and (B) an unfamiliar kin and non-kin penguin. (Figure adapted from Coffin *et al.*, 2011; photograph by Adam Kumiszcza.)

cell-surface glycoproteins that control antigen presentation to T-lymphocytes. Therefore, MHC genes play a fundamental role in immune recognition of pathogens and parasites. As the MHC is highly polymorphic, individuals that share the same genotype are likely to be closely related, and therefore, females that avoid mating with partners with similar MHC-based scents are decreasing the probability of inbreeding. By choosing males with a high number of MHC alleles, females are maximizing the number of MHC alleles in offspring so they have an ability to recognize a wider range of antigens (Penn and Potts, 1999; Drury, 2010). Therefore, both parasite-mediated selection (Eizaguirre *et al.*, 2012) and sexual selection (Wedekind and Penn, 2000; for a review, see Piertney and Oliver, 2006) may explain the maintenance of MHC polymorphism.

Although the evidence is not as abundant as in mammals (Yamazaki *et al.*, 1976; Radwan *et al.*, 2008; Cutrera *et al.*, 2012), including humans (Wedekind *et al.*, 1995; Havlicek and Roberts, 2009), evidence for MHC-associated mate choice has also been found in fish (Reusch *et al.*, 2001; Milinski *et al.*, 2005; Forsberg *et al.*, 2007), reptiles (Olsson *et al.*, 2003) and birds (Freeman-Gallant *et al.*, 2003; Strandh *et al.*, 2012).

Wedekind and collaborators (1995) tested whether MHC influenced body odour and body odour preferences in humans. They characterized the MHC of 49 female students and 44 male students, and they asked men to wear a T-shirt for two consecutive nights. The next day, they offered six of the T-shirts to women, and they were asked to score the scents of the T-shirts for intensity and for pleasantness and sexiness. They were tested near the time of ovulation, as women appear to be most odour-sensitive at this time (Doty *et al.*, 1981). The pleasantness and sexiness of men's body odours scored by different women depended on their own MHC. Female students ranked the scent of the T-shirts of males with MHC dissimilar to their own as more pleasant than those of male students whose MHC was more similar to the female student (Fig. 7.8A). Also, the scent of the T-shirts worn by MHC-dissimilar men more often reminded the women of their own current or former partners than did the scent of MHC-similar men. In contrast, when females were taking oral contraceptives, they did not prefer the scent of men with similar MHC (Fig. 7.8B). Oral contraceptives increase the levels of circulating steroids that are naturally released during pregnancy. The finding that women taking oral contraceptives do not exhibit a preference for the scent of men with dissimilar MHC but a preference for similar MHC males indicates that they prefer odours similar to those of relatives instead of potential partners.

The detection of individual MHC characteristics is performed via olfaction in most taxa. In birds, however, the mechanisms involved in the recognition of MHC genetic compatibility of potential mates remain unknown, probably because olfaction has been neglected in most studies of avian sexual selection. Also, it is still unknown how MHC genes influence scent, although several hypotheses to explain the mechanisms have been suggested (Penn and Potts, 1998b). As MHC proteins appear in

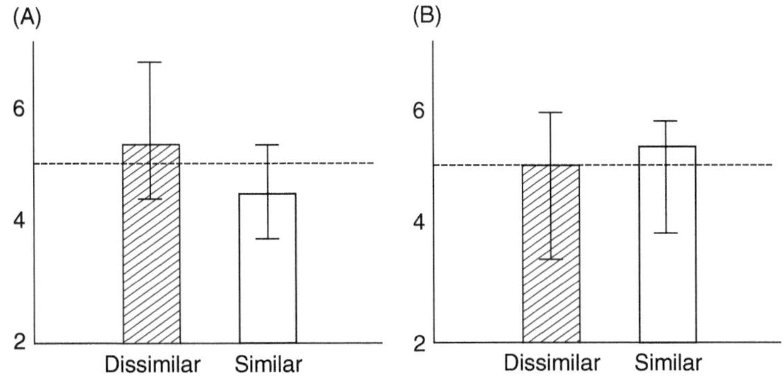

Fig. 7.8. Median and quartiles of average score per female regarding the pleasantness of the body odour of males that have a similar or dissimilar MHC to the scoring females in (A) women who do not take the contraceptive pill and (B) women who do. (Based on data from Wedekind *et al.*, 1995.)

urine and sweat, it has been proposed that these molecules could be the odorants (Singh *et al.*, 1987). However, MHC proteins are non-volatiles and scent recognition occurs when these proteins have been denaturalized (Singer *et al.*, 1993). Other hypotheses propose that the peptides bound by MHC molecules are the precursors of the volatile odorants (Singer *et al.*, 1997), or that those peptides can be detected by the vomeronasal organ (Leinders-Zufall *et al.*, 2004). Another hypothesis is that MHC genes may influence the bacteria profile of an individual, which in turn may influence the scent (Howard, 1977).

By smelling potential partners, females may choose the ones that increase the heterozygosity (genetic diversity) of offspring and also avoid the deleterious consequences of genetic incompatibility. For example, in humans, couples that share MHC alleles have lower possibilities of successful pregnancy than couples that differ in MHC alleles (Ober *et al.*, 1988; Weckstein *et al.*, 1991; Laitinen, 1993). The behaviour of women is also affected by the similarity in MHC alleles with their partner. Women with a partner with similar MHC alleles are less responsive to their partners, have a greater number of extra-pair sexual partners and increase their attraction to men other than their partners, particularly during the fertile phase of their cycles (Garver-Apgar *et al.*, 2006). With this strategy, women may increase the heterozygosity of their offspring. In birds, similar strategies have been found when females share MHC alleles with their partners. For example, blue tit (*Cyanistes caeruleus*) females are known to use song (Kempenaers *et al.*, 1992) and plumage coloration (Andersson *et al.*, 1998) when selecting mates. However, females select extra-pair partners in relation to their genetic dissimilarity, as nestlings born from extra-pair copulations were more heterozygous than the nestlings of the social father (Foerster *et al.*, 2003). Since these more heterozygous nestlings were more likely to survive (Foerster *et al.*, 2003), females may increase their fitness by selecting genetically dissimilar mates when looking for extra-pair copulations. Females may ensure nestling survival by choosing the best parental male thanks to characteristics such as plumage coloration (Andersson *et al.*, 1998). However, since chemical cues of birds are related to genetic heterozygosity (Leclaire *et al.*, 2011), olfaction is likely to be the mechanism for assessing the genetic characteristics of potential partners.

Therefore, both types of female selection are not exclusive and females can evaluate mates based simultaneously on resource-acquisition abilities and genetic quality. For example, it is already known that female mice prefer dominant males that scent mark more frequently, and they also prefer males with dissimilar MHC alleles. When confronted with males that differ in dominance status and MHC alleles, females preferred dominant males but they chose MHC dissimilar males when male dominance status was similar (Roberts and Gosling, 2003). In this context, multiple signals that are emitted by animals may provide different information to potential mates and the scent of an individual has been largely proven to provide information about genetic compatibility (Penn and Potts, 1998b).

Female MHC diversity has also been shown to influence mating strategies of males. For example, male mice may choose females based on their

MHC, as female settlements in male territories, controlled by males, were not random with respect to MHC (Potts *et al.*, 1992). Red junglefowl (*Gallus gallus*) males decrease their investment in reproduction, i.e. by allocating less sperm, when copulating with females with similar MHC alleles (Gillingham *et al.*, 2009).

Reproductive Behaviour

Chemical signals not only provide valuable information about the sender, but may also induce changes in the physiology of the receptor. Mammals have provided many examples of the physiological changes induced by chemical signals of conspecifics (Hurst, 2009). Among these changes, odours can influence the reproductive development of young animals by accelerating or delaying their development. Scents can cause a synchrony or an inhibition of oestrus among females, or even cause the interruption of pregnancy establishment. In males, scents are known to modulate luteinizing hormone levels, affect spermatogenesis and influence sperm density and motility (Brown, 1985; Koyama, 2004).

Induction of reproduction via olfaction

Among the first evidence of the induction of reproduction due to chemical signals of conspecifics is what became known as the Whitten effect (Whitten, 1966), in which the presence of a male induces and synchronizes the oestrus in unisexually grouped mouse (*Mus musculus*) females. Females exposed to males or to their urine will exhibit shorter and more regular cycles than females kept in the absence of male stimuli (Whitten, 1966; Whitten *et al.*, 1968). Indeed, housing female rodents in the absence of male odour can induce a state of permanent pseudo-oestrus, where the females stop cycling altogether (McClintock and Adler, 1978). The volatiles, contained in the male urine, responsible for such physiological change have been described as 2-*sec*-butyl-4,5-dihydrothiazole and 2,3-dehydro-*exo*-brevicomin (Schwende *et al.*, 1986). The secretion of the latter compound appears to be regulated by testosterone as its concentration drastically decreases following male castration, and returns to normal levels after testosterone supplementation (Novotny *et al.*, 1985).

Jemiolo and collaborators (1986) designed an experiment aiming to evaluate the possible role of these two compounds in the regulation of the fertility of female mice. They exposed group-caged females to Petri dishes containing: (i) water (control), (ii) castrate urine, (iii) castrate urine plus 2-(*sec*-butyl) thiazoline and dehydro-*exo*-brevicomin, or (iv) urine from intact males. The samples were presented in Petri dishes covered with a mesh to ensure that females could smell the volatiles released by the samples but not be in physical contact with the samples. Females were offered the Petri dish when they were housed in different conditions: individually, in groups of four or eight females per cage. The authors exposed females to the samples daily for 15 min per day for 21 days. They took vaginal smears of females and examined them under the microscope to determine the state of the oestrous cycle during the experiment.

In accordance with previous evidence, the authors found that females housed in large groups experienced suppression of the ovarian cycle activity, decreasing the number of oestrous cycles (Fig. 7.9). However, after exposure to male scent, they increased the number of oestrous cycles and they synchronized their cycles 3 or 4 days later. Females exposed to male scent had shorter and more regular cycles than females unexposed to male stimuli, and the latter could even experience a suppression of cycles (McClintock and Adler, 1978). The results from this study indicate that females responded to the exposure to the odorants 2-(*sec*-butyl) thiazoline and dehydro-*exo*-brevicomin diluted in castrated males' urine in a manner similar to those exposed to urine of intact males. The authors also exposed females living in groups of ten to the volatiles diluted in water and found a similar effect to that found when the volatiles were diluted in the urine of a castrated male. This result suggests that exposure to these odorants on their own is enough to induce the female physiological response, modulating receptivity and inducing oestrus in female mice (Jemiolo *et al.*, 1986).

Suppression of reproduction mediated by conspecific chemical cues

Chemical cues from other conspecifics may also affect reproduction in a negative way. For example, female scent induced suppression of the ovarian activity in adult female mice (Whitten, 1959; Champlin, 1971; Clee *et al.*, 1975). When female mice are grouped with other females, they synchronize their oestrous cycle (Jemiolo *et al.*, 1986); they also exhibit an extension of the cycle length (Lamond,

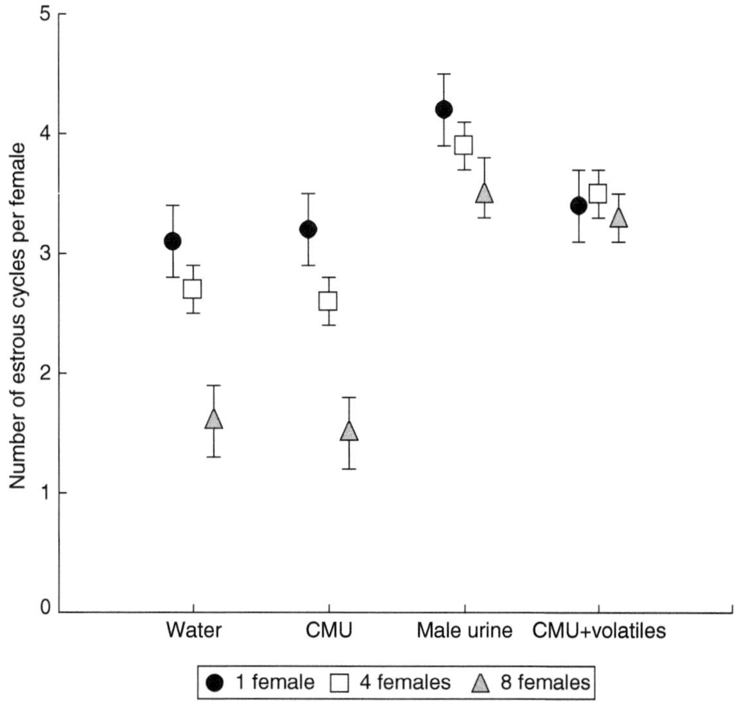

Fig. 7.9. Mean number (± SE) of oestrous cycles in 21 days during the exposure to different stimuli: water, urine from castrated male mice (CMU), urine from intact males (Male urine), or CMU + volatiles (2-(*sec*-butyl) thiazoline and dehydro-*exo*-brevicomin) in females housed in groups of one (squares), four (circles) or eight (triangles) individual(s) per cage. (Data from Jemiolo *et al*., 1986.)

1958, 1959; Jemiolo *et al*., 1986), have a prolongation of the di-oestrous phase of the oestrous cycle (Whitten, 1959) or even suffer spontaneous pseudopregnancies (van der Lee and Boot, 1955, 1956).

Chemical cues from a male that did not impregnate female mice can cause the interruption of early pregnancy. During mating, the female learns the scent of her mate, and the exposure to his urine does not cause this effect. However, if the female is exposed to the urine of an unfamiliar male at the beginning of the pregnancy, she can experience abortion and is brought back into oestrus. The duration of sensitivity to the male urine varies between species, ranging from 4 to 5 days post-mating in species of the *Mus* genus (Parkes and Bruce, 1961), and up to 17 days post-mating in microtine species (Stehn and Jannett, 1981). This phenomenon is known as the Bruce effect (Bruce, 1959). The signals that mediate this pregnancy failure are the MHC class-I peptides found in male urine (Leinders-Zufall *et al*., 2004), which are also used by the female to identify the male that produced the urine. When the recently pregnant female is exposed to unfamiliar MHC peptides added to the mating male's urine, she will undergo an abortion. Early pregnancy failure can therefore be mediated by olfaction. Males obtain benefits from interrupting a female's pregnancy by other males. For example, dominant males that normally scent mark their territory can block the pregnancies of competitors. As females may return to oestrus after abortion, they may increase their fitness (Schwagmeyer, 1979), as from a female's perspective, the Bruce effect may help them to exert a post-copulatory mate choice. Females can control this effect by seeking or avoiding the male scent during their most susceptible period; this way they can opt to reserve their reproductive resources for the highest quality male if he was not their initial partner (Becker and Hurst, 2009). Also, females may avoid investing in pregnancy if the possibility of survival of their offspring is very low, e.g. when the risk of infanticide by a new dominant male is high (Pillay and Kinahan, 2009).

Selective response to chemical cues of conspecifics in relation to physiological state

Chemical cues of conspecifics can have strong effects on the physiology of receivers of such signals. However, the response to such chemical signals is also mediated by the physiological state of the receiver, as a fixed response to the signals may not be adaptive in many cases. For example, male chemical cues may elicit aggression in male mice (Novotny *et al.*, 1985), and induce oestrus in females (Jemiolo *et al.*, 1986). Also, females are attracted towards major urinary proteins (MUPs) presented in the urine of male mice during oestrus but they do not respond to such signals during di-oestrus (Dey *et al.*, 2015). MUPs provide an individual genetic identity signature and are used by females for individual recognition as well as for kin recognition, assessment of male competitive ability and genetic heterozygosity (Hurst, 2009). The MUPs are ligands that stimulate the vomeronasal sensory neurons, i.e. they are detected thanks to the vomeronasal organ (Dey *et al.*, 2015). When female mice are not searching for a partner, i.e. they are in the di-oestrous phase of the ovarian cycle, they have increased levels of progesterone, which blocks the receptors of the MUPs presented in male urine. Therefore, they do not detect and exhibit an attraction response to these male chemical cues (Dey *et al.*, 2015). The effect of progesterone is very specific, as this hormone only affects the receptors of MUPs but not of other ligands such as FELD4, a predator chemical cue emitted by cats (Papes *et al.*, 2010). So female mice remain continuously responding to predator chemical cues irrespective of their reproductive state, but their response to chemical cues of potential partners is finely modulated by their internal physiological state (Dey *et al.*, 2015).

As we have seen in this chapter, olfaction plays an important role in sexual selection and reproductive behaviour of many animals. Until recently, the role of olfaction in bird biology has been largely neglected. It is therefore important to emphasize that olfaction should not be disregarded even if we do not have sufficient information about the olfactory capabilities of the animal in question. It is via chemical signals that animals assess and choose their partners, and the scent of conspecifics can have positive as well as negative effects on the reproduction of the odour recipient. All these aspects should be taken into account, not only to ensure the welfare of animals kept on farms (see Chapter 12), in zoos (see Chapter 13), laboratories (see Chapter 14) or at home (see Chapter 15), but also when trying to develop programmes of reproduction for endangered species or to improve the productivity of farm animals. Potential uses of animal chemical cues include the identification of oestrus as well as the stimulation of early puberty and the stimulation and synchronization of oestrus in the non-breeding season. For example, chemical cues from the males have been found to hasten the onset of puberty and oestrus in pigs (Signoret and Lindsay, 1982; Izard, 1983), sheep (Underwood *et al.*, 1944), goats (Shelton, 1960) and cows (Roberson *et al.*, 1991). The scent of the boar at the time of insemination of the sow improves sperm transport and ovulation (Soede, 1993). The scent of a bull has been reported to accelerate the resumption of ovarian activity in cattle following parturition (Zalesky *et al.*, 1984; Scott and Montgomery, 1987). The use of chemical cues therefore offers a potentially useful and practical way to improve reproductive efficiency in livestock species (Rekwot *et al.*, 2001).

References

Amo, L., Avilés, J.M., Parejo, D., Peña, A., Rodríguez, J. and Tomás, G. (2012) Sex recognition by odour and variation in the uropygial gland secretion in starlings. *Journal of Animal Ecology* 81, 605–613. DOI: 10.1111/j.1365-2656.2011.01940.x

Andersson, S., Örnborg, J. and Andersson, M. (1998) Ultraviolet sexual dimorphism and assortative mating in blue tits. *Proceedings of the Royal Society B: Biological Sciences* 265, 445–450. DOI: 10.1098/rspb.1998.0315

Arakawa, H., Blanchard, D.C., Arakawa, K., Dunlap, C. and Blanchard, R.J. (2008) Scent marking behavior as an odorant communication in mice. *Neuroscience and Biobehavioral Reviews* 32, 1236–1248. DOI: 10.1016/j.neubiorev.2008.05.012

Bateman, A.J. (1948) Intra-sexual selection in *Drosophila*. *Heredity* 2, 349–368. DOI: 10.1038/hdy.1948.21

Becker, S.D. and Hurst, J.L. (2009) Female behaviour plays a critical role in controlling murine pregnancy block. *Proceedings of the Royal Society B: Biological Sciences* 276, 1723–1729. DOI: 10.1098/rspb.2008.1780

Beltran-Bech, S. and Richard, F.J. (2014) Impact of infection on mate choice. *Animal Behaviour* 90, 159–170. DOI: 10.1016/j.anbehav.2014.01.026

Bonadonna, F. and Nevitt, G.A. (2004) Partner-specific odor recognition in an Antarctic seabird. *Science* 306, 835. DOI: 10.1126/science.1103001

Brown, R.E. (1985) The rodents I: effects of odours on reproductive physiology (primer effects). In: Brown R.E. and Macdonald, D.W. (eds) *Social Odours in Mammals*. Clarendon Press, Oxford, UK, pp. 245–344.
Bruce, H.M. (1959) An exteroceptive block to pregnancy in the mouse. *Nature* 184, 105. DOI: 10.1038/184105a0
Carazo, P., Sanchez, E., Font, E. and Desfilis, E. (2004) Chemosensory cues allow male *Tenebrio molitor* beetles to assess the reproductive status of potential mates. *Animal Behaviour* 68, 123–129. DOI: 10.1016/j.anbehav.2003.10.014
Champlin, A.K. (1971) Suppression of oestrus in grouped mice: the effects of various densities and the possible nature cf the stimulus. *Journal of Reproduction and Fertility* 27, 233–241. DOI: 10.1530/jrf.0.0270233
Clee, M.D., Humphreys, E.M. and Russell, J.A. (1975) The suppression of ovarian cyclical activity in groups of mice, and its dependence on ovarian hormones. *Journal of Reproduction and Fertility* 45, 395–398. DOI: 10.1530/jrf.0.0450395
Coffin, H.R., Watters, J.V. and Mateo, J.M. (2011) Odor-based recognition of familiar and related conspecifics: a first test conducted on captive Humboldt penguins (*Spheniscus humboldti*). *PLoS ONE* 6, e25002. DOI: 10.1371/journal.pone.0025002
Cooper, W.E. and Pérez-Mellado, V. (2002) Pheromonal discriminations of sex, reproductive condition, and species by the lacertid lizard *Podarcis hispanica*. *Journal of Experimental Zoology* 292, 523–527. DOI: 10.1002/jez.10089
Cutrera, A.P., Fanjul, M.S. and Zenuto, R.R. (2012) Females prefer good genes: MHC-associated mate choice in wild and captive tuco-tucos. *Animal Behaviour* 83, 847–856. DOI: 10.1016/j.anbehav.2012.01.006
Darwin, C. (1871) *The Descent of Man, and Sexual Selection in Relation to Sex*. John Murray, London.
Dey, S., Chamero, P., Pru, J.K., Chien, M.S., Ibarra-Soria, X., Spencer, K.R., Logan, D.W., Matsunami, H., Peluso, J.J. and Stowers, L. (2015) Cyclic regulation of sensory perception by a female hormone alters behavior. *Cell* 161, 1334–1344. DOI: 10.1016/j.cell.2015.04.052
Doty, R.L., Snyder, P.J., Huggins, G.R. and Lowry, L.D. (1981) Endocrine, cardiovascular, and psychological correlated of olfactory sensivity changes during the human menstrual cycle. *Journal of Comparative Physiology and Psychology* 95, 45–60. DOI: 10.1037/h0077755
Drury, J.P. (2010) Immunity and mate choice: a new outlook. *Animal Behaviour* 79, 539–545. DOI: 10.1016/j.anbehav.2009.12.023
Eizaguirre, C., Lenz, T.L., Kalbe, M. and Milinski, M. (2012) Rapid and adaptive evolution of MHC genes under parasite selection in experimental vertebrate populations. *Nature* 3, 621. DOI: 10.1038/ncomms1632
Foerster, K., Delhey, K., Johnsen, A., Lifjeld, J.T. and Kempenaers, B. (2003) Females increase offspring heterozygosity and fitness through extra-pair matings. *Nature* 425, 714–717. DOI: 10.1038/nature01969
Forsberg, L.A., Dannewitz, J., Petersson, E. and Grahn, M. (2007) Influence of genetic dissimilarity in the reproductive success and mate choice of brown trout – females fishing for optimal MHC dissimilarity. *Journal of Evolutionary Biology* 20, 1859–69. DOI: 10.1111/j.1420-9101.2007.01380.x
Freeman-Gallant, C.R., Meguerdichian, M., Wheelwright, N.T. and Sollecito, S.V. (2003) Social pairing and female mating fidelity predicted by restriction fragment length polymorphism similarity at the major histocompatibility complex in a songbird. *Molecular Ecology* 12, 3077–3083. DOI: 10.1046/j.1365-294X.2003.01968.x
Gamboa, G.J. (1996) Kin recognition in social wasps. In: Turillazzi, S. and West-Eberhard, M.J. (eds) *Natural History and Evolution of Paper Wasps*. Oxford University Press, Oxford, UK, pp. 161–177.
Garver-Apgar, C.E., Gangestad, S.W., Thornhill, R., Miller, R.D. and Olp, J.J. (2006) Major histocompatibility complex alleles, sexual responsivity, and unfaithfulness in romantic couples. *Psychological Science* 17, 830–835. DOI: 10.1111/j.1467-9280.2006.01789.x
Gillingham, M.A.F., Richardson, D.S., Løvlie, H., Moynihan, A., Worley, K. and Pizzari, T. (2009) Cryptic preference for MHC-dissimilar females in male red junglefowl, *Gallus gallus*. *Proceedings of the Royal Society B: Biological Sciences* 276, 1083–1092. DOI: 10.1098/rspb.2008.1549
Grafen, A. (1990) Biological signals as handicaps. *Journal of Theoretical Biology* 144, 517–546. DOI: 10.1016/S0022-5193(05)80088-8
Havlicek, J. and Roberts, S.C. (2009) MHC-correlated mate choice in humans: a review. *Psychoneuroendocrinology* 34, 497–512. DOI: 10.1016/j.psyneuen.2008.10.007
Hill, G.E. (2002) *A Red Bird in a Brown Bag: The Function and Evolution of Colorful Plumage in the House Finch.* Oxford University Press, Oxford, UK.
Howard, J.C. (1977) H-2 and mating preferences. *Nature* 266, 406–408. DOI: 10.1126/science.3375818
Hurst, J.L. (2009) Female recognition and assessment of males through scent. *Behavioural Brain Research* 200, 295–303. DOI: 10.1016/j.bbr.2008.12.020
Izard, M.K. (1983) Pheromones and reproduction in domestic animals. In Vandenbergh, J.G. (eds) *Pheromones and Reproduction in Mammals*. Academic Press, New York, pp. 253–285.
Jacob, J.P. and Ziswiler, V. (1982) The uropygial gland. In: Farner, D.S., King, J.R. and Parkes, K.C. (eds) *Avian Biology*. Academic Press, New York, pp. 199–324.
Jacob, J., Balthazart, J. and Schoffeniels, E. (1979) Sex differences in the chemical composition of uropygial gland waxes in domestic ducks. *Biochemical Systematics and Ecology* 7, 149–153. DOI: 10.1016/0305-1978(79)90024-3
Janeway, C.A., Jr (1993) How the immune system recognizes invaders. *Scientific American* 269, 73–79.

Jemiolo, B., Harvey, S. and Novotny, M. (1986) Promotion of the Whitten effect in female mice by synthetic analogs of male urinary constituents. *Proceedings of the National Academy of Sciences USA* 83, 4576–4579. DOI: 10.1073/pnas.83.12.4576

Johansson, B.G. and Jones, T.M. (2007) The role of chemical communication in mate choice. *Biological Reviews* 82, 265–289. DOI: 10.1111/j.1469-185X.2007.00009.x

Jones, I.L., Hagelin, J.C., Major, H.L. and Rasmussen, L.E.L. (2004) An experimental field study of the function of crested auklet feather odor. *The Condor* 106, 71–78. DOI: 10.1650/7363

Kavaliers, M. and Colwell, D.D. (1995a) Discrimination by female mice between the odours of parasitized and non-parasitized males. *Proceedings of the Royal Society B: Biological Sciences* 261, 31–35. DOI: 10.1098/rspb.1995.0113

Kavaliers, M. and Colwell, D.D. (1995b) Odours of parasitized males induce aversive response in female mice. *Animal Behaviour* 50, 1164–1169. DOI: 10.1016/0003-3472(95)80032-8

Kempenaers, B.,Verheyen, G.R. and Dhondt, A.A. (1992) Extrapair paternity results from female preference for high-quality males in the blue tit. *Nature* 357, 494–496. DOI: 10.1038/357494a0

Klein, J. (1986) *Natural History of the Histocompatibility Complex*. Wiley, New York.

Koyama, S. (2004) Primer effects by conspecific odors in house mice: a new perspective in the study of primer effects on reproductive activities. *Hormones and Behaviour* 46, 303–310. DOI: 10.1016/j.yhbeh.2004.03.002

Krebs, J.R. and Dawkins, R. (1983) Animal signals: mind-reading and manipulation. In: Krebs, J.R. and Davies, N.B. (eds) *Behavioural Ecology: An Evolutionary Approach*, 2nd edn. Blackwell, Oxford, UK, pp. 380–402.

Laitinen, T. (1993) A set of MHC haplotypes found among Finnish couples suffering from recurrent spontaneous abortions. *American Journal of Reproductive Immunology* 29, 148–154. DOI: 10.1111/j.1600-0897.1993.tb00580.x

Lamond, D.R. (1958) Spontaneous anoestrus in mice. *Proceedings of the Australian Society of Animal Production* 2, 97–101.

Lamond, D.R. (1959) Effect of stimulation derived from other animals of the same species on oestrous cycles in mice. *Journal of Endocrinology* 18, 343–349. DOI: 10.1677/joe.0.0180343

Leclaire, S., Merkling, T., Raynaud, C., Mulard, H., Bessière, J.M., Lhuillier, E., Hatch, S.A. and Danchin, E. (2011) Semiochemical compounds of preen secretion reflect genetic make-up in a seabird species. *Proceedings of the Royal Society B: Biological Sciences* 279, 1185–1193. DOI: 10.1098/rspb.2011.1611

Leinders-Zufall, T., Brennan, P., Widmayer, P., Prashanth Chandramani, S., Maul-Pavicic, A., Jager, M. and Li, X.H. (2004) MHC class I peptides as chemosensory signals in the vomeronasal organ. *Science* 306, 1033–1037. DOI: 10.1126/science.1102818

López, P., Amo, L. and Martín, J. (2006) Reliable signaling by chemical cues of male traits and health state in male lizards, *Lacerta monticola*. *Journal of Chemical Ecology* 32, 473–488. DOI: 10.1007/s10886-005-9012-9

Mardon, J., Saunders, S.M. and Bonadonna, F. (2011) From preen secretions to plumage: the chemical trajectory of blue petrels' *Halobaena caerulea* social scent. *Journal of Avian Biology* 42, 29–38. DOI: 10.1111/j.1600-048X.2010.05113.x

Martín, J. and López, P. (2007) Scent may signal fighting ability in male Iberian rock lizards. *Biology Letters* 3, 125–127. DOI: 10.1098/rsbl.2006.0589

Martín, J. and López, P. (2012) Supplementation of male pheromone on rock substrates attracts female rock lizards to the territories of males: a field experiment. *PLoS ONE* 7: e30108. DOI: 10.1371/journal.pone.0030108

Martín, J., Civantos, E., Amo, L. and López, P. (2007) Chemical ornaments of male lizards *Psammodromus algirus* may reveal their parasite load and health state to females. *Behavioral Ecology and Sociobiology* 62, 173–179. DOI: 10.1007/s00265-007-0451-x

Mason, R.T. and Parker, M.R. (2010) Social behavior and pheromonal communication in reptiles. *Journal of Comparative Physiology A – Neuroethology Sensory Neural and Behavioral Physiology* 196, 729–749. DOI: 10.1007/s00359-010-0551-3

Mateo, J.M. (2003) Kin recognition in ground squirrels and other rodents. *Journal of Mammalogy* 84, 1163–1181. DOI: 10.1644/BLe-011

McClintock, M.K. and Adler, N.T. (1978) Induction of persistent estrus by airborne chemical communication among female rats. *Hormones and Behaviour* 11, 414–418. DOI: 10.1016/0018-506X(78)90041-7

Mihailova, M., Berg, M.L., Buchanan, K.L. and Bennett, A.T.D. (2014) Odour-based discrimination of subspecies, species and sexes in an avian species complex, the crimson rosella. *Animal Behaviour* 95, 155–164. DOI: 10.1016/j.anbehav.2014.07.012

Milinski, M., Griffiths, S., Wegner, K.M., Reusch, T.B.H., Haas-Assenbaum, A. and Boehm, T. (2005) Mate choice decisions of stickleback females predictably modified by MHC peptide ligands. *Proceedings of the National Academy of Sciences USA* 102, 4414–4418. DOI: 10.1073/pnas.0408264102

Neff, B.D. and Sherman, P.W. (2003) Nestling recognition via direct cues by parental male bluegill sunfish (*Lepomis macrochirus*). *Animal Cognition* 6, 87–92. DOI: 10.1007/s10071-003-0166-y

Nevo, E., Bodmer, M. and Heth, G. (1976) Olfactory discrimination as an isolating mechanism in speciating mole rats. *Experientia* 32, 1511–1512. DOI: 10.1007/BF01924423.

Nosil, P., Crespi, B., Gries, R. and Gries, G. (2007) Natural selection and divergence in mate preference during speciation. *Genetica* 129, 309–327. DOI: 10.1007/s10709-006-0013-6.

Novotny, M., Harvey, S., Jemiolo, B. and Alberts, J. (1985) Synthetic pheromones that promote inter-male aggression in mice. *Proceedings of the National Academy of Sciences USA* 82, 2059–2061. DOI: 10.1073/pnas.82.7.2059

Ober, C., Elias, S., O´Brien, E., Kostyu, D.D., Hauck, W.W. and Bombard, A. (1988) HLA sharing and fertility in Hutterite couples: evidence for prenatal selection against compatible fetuses. *American Journal of Reproductive Immunology* 18, 111–115. DOI: 10.1111/j.1600-0897.1988.tb00245.x

Olsson, M., Madsen, T., Nordby, J., Wapstra, E., Ujvari, B. and Wittsellet, H. (2003) Major histocompatibility complex and mate choice in sand lizards. *Proceedings of the Royal Society B: Biological Sciences* 270, S254–S256. DOI: 10.1098/rsbl.2003.0079

Papes, F., Logan, D.W. and Stowers, L. (2010) The vomeronasal organ mediates interspecies defensive behaviors through detection of protein pheromonehomologs. *Cell* 141, 692–703. DOI: 10.1016/j.cell.2010.03.037

Parkes, A.S. and Bruce, H.M. (1961) Olfactory stimuli in mammalian reproduction. *Science* 134, 1049–1054. DOI: 10.1126/science.134.3485.1049

Penn, D. and Potts, W.K. (1998a) Chemical signals and parasite-mediated sexual selection. *Trends in Ecology and Evolution* 13, 391–396. DOI: 10.1016/S0169-5347(98)01473-6

Penn, D. and Potts, W. (1998b) How do major histocompatibility complex genes influence odor and mating preferences? *Advances in Immunology* 69, 411–436. DOI: 10.1016/S0065-2776(08)60612-4

Penn, D.J. and Potts, W.K. (1999) The evolution of mating preferences and major histocompatibility complex genes. *The American Naturalist* 153, 145–164. DOI: 10.1086/303166

Penn, D., Schneider, G., White, K., Slev, P. and Potts, W. (1998) Influenza infection neutralizes the attractiveness of male odour to female mice (*Mus musculus*). *Ethology* 104, 685–694. DOI: 10.1111/j.1439-0310.1998.tb00102.x

Piersma, T., Dekker, M. and Sinninghe Damsté, J.S. (1999) An avian equivalent of make-up? *Ecology Letters* 2, 201–203. DOI: 10.1046/j.1461-0248.1999.00078.x

Piertney, S.B. and Oliver, M.K. (2006) The evolutionary ecology of the major histocompatibility complex. *Heredity* 96, 7–21. DOI: 10.1038/sj.hdy.6800724

Pillay, N.A. and Kinahan, A.A. (2009) Mating strategy predicts the occurrence of the Bruce effect in the vlei rat *Otomys irroratus*. *Behaviour* 146, 139–151. DOI: 10.1163/156853908X390968

Potts, W.K., Manning, C.J. and Wakeland, E.K. (1992) MHC-based mating preferences in *Mus* operate through both settlement patterns and female controlled extra-territorial matings. In Doty, R.L. and Müller-Schwarze D. (eds) *Chemical Signals in Vertebrates VI.* Plenum Press, New York, pp. 183–187.

Pusey, A. and Wolf, M. (1996) Inbreeding avoidance in animals. *Trends in Ecology and Evolution* 11, 201–206. DOI: 10.1016/0169-5347(96)10028-8

Radwan, J., Tkacz, A. and Kloch, A. (2008) MHC and preferences for male odour in the bank vole. *Ethology* 114, 827–833. DOI: 10.1111/j.1439-0310.2008.01528.x

Rekwot, P.I., Ogwu, D., Oyedipe E.O. and Sekoni, V.O. (2001) The role of pheromones and biostimulation in animal reproduction. *Animal Reproduction Science* 65, 157–170. DOI: 10.1016/S0378-4320(00)00223-2

Reneerkens, J., Piersma, T. and Sinninghe Damsté, J.S. (2002) Sandpipers (Scolopacidae) switch from monoester to diester preen waxes during courtship and incubation, but why? *Proceedings of the Royal Society B: Biological Sciences* 269, 2135–2139. DOI: 10.1098/rspb.2002.2132

Reusch, T.B.H., Häberli, M.A., Aeschlimann, P.B. and Milinski, M. (2001) Female sticklebacks count alleles in a strategy of sexual selection explaining MHC polymorphism. *Nature* 414, 300–302. DOI: 10.1038/35104547

Riechert, S.E. and Singer, F.D. (1995) Investigation of potential male mate choice in a monogamous spider. *Animal Behaviour* 49, 715–723. DOI: 10.1016/0003-3472(95)80204-5

Roberson, M.S., Wolfe, M.W., Stumpf, T.T., Werth, L.A., Cupp, A.S., Kojima, N.D., Wolfe, P.L., Kittok, R.J. and Kinder, J.E. (1991) Influence of growth rate and exposure to bulls on age at puberty in beef heifers. *Journal of Animal Science* 69, 292–298. DOI: 10.2527/1991.6952092x

Roberts, S.C. and Gosling, L.M. (2003) Genetic similarity and quality interact in mate choice decisions by female mice. *Nature Genetics* 35, 103–106. DOI: 10.1038/ng1231

Ron, S.R. (2008) The evolution of female mate choice for complex calls in túngara frogs. *Animal Behaviour* 76, 1783–1794. DOI: 10.1016/j.anbehav.2008.07.024

Sandilands, V., Powell, K., Keeling, L. and Savory, C.J. (2004a) Preen gland function in layer fowls: factors affecting preen oil fatty acid composition. *British Poultry Science* 45, 109–115. DOI: 10.1080/00071660410001668932

Sandilands, V., Savory, J. and Powell, K. (2004b) Preen gland function in layer fowls: factors affecting morphology and feather lipid levels. *Comparative Biochemistry and Physiology A* 137, 217–225. DOI: 10.1016/j.cbpb.2003.10.004

Schwagmeyer, P.L. (1979) The Bruce effect: an evaluation of male/female advantages. *The American Naturalist* 114, 932–938. DOI: 10.1086/283541

Schwende, F.J., Wiesler, D., Jorgenson, J.W., Carmack, M. and Novotny, M. (1986) Urinary volatile constituents of the house mouse, *Mus musculus*, and their endocrine dependency. *Journal of Chemical Ecology* 12, 277–296. DOI: 10.1007/BF01045611

Scott, L.C. and Montgomery, G.W. (1987) Introduction of bulls induces return of cyclic ovarian function in postpartum beef cows. *New Zealand Journal of Agricultural Research* 30, 189–194. DOI: 10.1080/00288233.1987.10430495

Shelton, M. (1960) Influence of the presence of a male goat on the initiation of oestrous cycling and ovulation of Angora does. *Journal of Animal Science* 19, 368–375. DOI: 10.2527/jas1960.192368x

Signoret, J.P. and Lindsay, D.R. (1982) The male effect in domestic mammals. Effect on LH secretion and ovulation. Importance of olfactory ewes. In: Breiphol, W. (ed.) *Olfaction and Endocrine Regulation*. IRL Press, London, pp. 63–72.

Singer, A.G., Tsuchiya, H., Wellington, J.L., Beauchamp, G.K. and Yamazaki, K. (1993) Chemistry of odor types in mice – fractionation and bioassay. *Journal of Chemical Ecology* 19, 569–579. DOI: 10.1007/BF00994326

Singer, A.G., Beauchamp, G.K. and Yamazaki, K. (1997) Volatile signals of the major histocompatibility complex in male mouse urine. *Proceedings of the National Academy of Sciences USA* 94, 2210–2214. DOI: 10.1007/s10886-006-9230-9

Singh, P.B., Brown, R.E. and Roser, B. (1987) MHC antigens in urine as olfactory recognition cues. *Nature* 327, 161–164. DOI: 10.1038/327161a0

Slade, J.W.G., Watson, M.J., Kelly, T.R., Gloor, G.B., Bernards, M.A. and MacDougall-Shackleton, E.A. (2016) Chemical composition of preen wax reflects major histocompatibility complex similarity in songbirds. *Proceedings of the Royal Society B: Biological Sciences* 283, 20161966. DOI: 10.1098/rspb.2016.1966

Smadja, C. and Butlin, R.K. (2008) On the scent of speciation: the chemosensory system and its role in premating isolation. *Heredity* 102, 77–97. DOI: 10.1038/hdy.2008.55

Soede, N.M. (1993) Boar stimuli around insemination affect reproductive processes in pigs: a review. *Animal Reproduction Science* 32, 107–125. DOI: 10.1016/0378-4320(93)90062-V

Soini, H.A., Schrock, S.E., Bruce, K.E., Wiesler, D., Ketterson, E.D. and Novotny, M.V. (2007) Seasonal variation in volatile compound profiles of preen gland secretions of the dark-eyed junco (*Junco hyemalis*). *Journal of Chemical Ecology* 33, 183–198. DOI: 10.1007/s10886-006-9210-0

Stehn, R.A. and Jannett, F.J. (1981) Male-induced abortion in various microtine rodents. *Journal of Mammalogy* 62, 369–372. DOI: 10.2307/1380713

Strandh, M., Westerdahl, H., Pontarp, M., Canbäck, B., Dubois, M.P., Miquel, C., Taberlet, P. and Bonadonna, F. (2012) Major histocompatibility complex class II compatibility, but not class I, predicts mate choice in a bird with highly developed olfaction. *Proceedings of the Royal Society B: Biological Sciences* 279, 4457–4463. DOI: 10.1098/rspb.2012.1562

Thomas, M.L. (2011) Detection of female mating status using chemical signals and cues. *Biological Reviews* 86, 1–13. DOI: 10.1111/j.1469-185X.2010.00130.x.

Trivers, R. (1972) Parental investment and sexual selection. In: Campbell, B. (ed.) *Sexual Selection and the Descent of Man*. Aldine, Chicago, Illinois, pp. 136–179.

Underwood, E.J., Shier, F.L. and Davenport, N. (1944) Studies in sheep husbandry. The breeding season of Merino crossbreeds and British breeds ewes in the agricultural districts. *Journal of Agriculture* 11, 135–143.

van der Lee, S. and Boot, L.M. (1955) Spontaneous pseudopregnancy in mice. *Acta Physiology and Pharmacology Neerlandica* 4, 442–444.

van der Lee, S. and Boot, L.M. (1956) Spontaneous pseudopregnancy in mice II. *Acta Physiology and Pharmacology Neerlandica* 5, 213–214.

Weckstein, L.N., Patrizio, P., Balmaceda, J.P., Asch, R.H. and Branch, D.W. (1991) Human leukocite antigen compatibility and failure to achieve a viable pregnancy with assisted reproductive technology. *Acta Europaea Fertilitatis* 22, 103–107.

Wedekind, C. and Penn, D. (2000) MHC genes, body odours, and odour preferences. *Nephrology Dialysis Transplantation* 15, 1269–1271. DOI: 10.1093/ndt/15.9.1269

Wedekind, C., Seebeck, T., Bettens, F. and Paepke, A.J. (1995) MHC-dependent mate preference in humans. *Proceedings of the Royal Society of London B: Biological Sciences* 260, 245–249. DOI: 10.1098/rspb.1995.0087

Wedell, N. (2005) Female receptivity in butterflies and moths. *Journal of Experimental Biology* 208, 3433–3440. DOI: 10.1242/jeb.01774

Whelan, R.J., Levin, T.C., Owen, J.C. and Garvin, M.C. (2010) Short-chain carboxylic acids from gray catbird (*Dumetella carolinensis*) uropygial secretions vary with testosterone levels and photoperiod. *Comparative Biochemistry and Physiology B* 156, 183–188. DOI: 10.1016/j.cbpb.2010.03.005

Whittaker, D.J., Soini, H.A., Atwell, J.W., Hollars, C., Novotny, M.V. and Ketterson, E.D. (2010) Songbird chemosignals: volatile compounds in preen gland secretions vary among individuals, sexes, and populations. *Behavioral Ecology* 21, 608–614. DOI: 10.1093/beheco/arq033

Whittaker, D.J., Richmond, K.M., Miller, A.K., Kiley, R., Burns, C.B., Atwell, J.W. and Ketterson, E.D. (2011) Intraspecific preen oil odor preferences in dark-eyed juncos (*Junco hyemalis*). *Behavioral Ecology* 22, 1256–1263. DOI: 10.1093/beheco/arr122

Whitten, W.K. (1959) Occurrence of anoestrus in mice caged in groups. *Journal of Endocrinology* 8, 102–107. DOI: 10.1677/joe.0.0180102

Whitten, W.K. (1966) Pheromones and mammalian reproduction. In: McLaren, A. (ed.) *Advances in Reproductive Physiology*, Vol. 1. Academic, London, pp. 155–177.

Whitten, W.K., Bronson, F.H. and Greenstein, J.A. (1968) Estrus-inducing pheromone of male mice: transport by movement of air. *Science* 161, 584–585. DOI: 10.1126/science.161.3841.584

Wyatt, T.D. (2003) *Pheromones and Animal Behaviour: Communication by Smell and Taste*. Cambridge University Press, Cambridge, UK.

Yamazaki, K., Boyse, E.A., Mike, V., Thaler, H.T. and Mathieson, B.J. (1976) Control of mating preferences in mice by genes in the major histocompatibility complex. *Journal of Experimental Medicine* 144, 1324–1335. DOI: 10.1084/jem.144.5.1324

Zahavi, A. (1975) Mate selection – a selection for a handicap. *Journal of Theoretical Biology* 53, 205–214. DOI: 10.1016/0022-5193(75)90111-3

Zahavi, A. and Zahavi, A. (1997) *The Handicap Principle*. Oxford Press, New York.

Zalesky, D.D., Day, M.L., Garcia-Winder, M., Imakawa, K., Kittok, R.J., D'Occhio, M.J. and Kinder, J.E. (1984) Influence of exposure to bulls on resumption of oestrous cycles following parturition in beef cows. *Journal of Animal Science* 59, 1135–1139. DOI: 10.2527/jas1984.5951135x

Zhang, J.X., Wei, W., Zhang, J.H. and Yang, W.H. (2010) Uropygial gland-secreted alkanols contribute to olfactory sex signals in budgerigars. *Chemical Senses* 35, 375–382. DOI: 10.1093/chemse/bjq025

8 The Role of Olfaction in Maternal Care and Offspring Survival

Frédéric Lévy and Raymond Nowak

Physiologie de la Reproduction et des Comportements, INRA, Université F. Rabelais, Nouzilly, France

Introduction

Behaviours associated with the birth and care of young are essential for the survival of mammals as the mother provides food, warmth, shelter and protection from predators and conspecifics. Right after parturition, the mother generally provides licking or grooming, which stimulates the offspring and in some species promotes urination and defecation. She also adopts a specific posture that facilitates nursing. The young are not passive and have to adapt rapidly to the mother's behaviour to be able to reach the udder. Indeed, suckling must occur shortly after birth since it is a question of offspring survival. As the body energy reserves are very limited, any delay in the ingestion of milk puts the neonate in danger of hypothermia. In addition, failure to suckle within a few hours of birth results in decreased absorption of antibodies and reduces protection against neonatal infection (Korhonen *et al.*, 2000a,b).

The success of mother–young interactions depends on the behavioural synchronization between the dam and her offspring, which is accomplished by finely tuned physiological and sensory changes. Among the various types of sensory information exchanged by the parent and offspring, the importance of olfaction differs among mammalian species according to the maternal style and the maturity of the young at birth (Lévy *et al.*, 2004; Lévy and Keller, 2009). In so-called altricial/nidicolous species (e.g. mice, rats, rabbits, dogs and cats), mothers give birth to a large litter of still largely immature young with limited sensory and locomotor abilities. The mother builds a nest in which mother–young interactions take place. It plays a key role in the thermoregulation of the offspring, and the mother retrieves the young if they wander away. In so-called precocial/nidifugous species (which includes most ungulates), the females give birth to offspring with fully developed sensory and motor capacities, and nest building is non-existent. Mothers in these species develop discriminative care, allowing only their own young to suckle, and a clear bond based on mutual recognition develops between the mother and her offspring soon after birth. Between the two extremes, a continuum of mothering styles and neonatal development can be observed. In these 'intermediate' species, like pigs, neonates have functional sensory systems but limited thermoregulation. A nest is provided by the mother but, although mutual recognition takes place to some extent, it does not comprise a bonding process. Olfaction is far more significant in altricial species, while in precocial and intermediate species mother–young interactions rely on multisensory cues. The former is even more obvious in the young because their developmental stage at birth is so immature.

This chapter will compare the relative importance of the sense of smell for the development of both maternal behaviour and neonatal behaviour across the different styles of mother–young interactions defined above. Although the laboratory rat is the best-studied mammal for analysing the importance of olfaction in mother–young interactions and therefore will receive the most attention in this review, we will also refer to other rodents (mice, hamsters, gerbils), lagomorphs (rabbits), cavies (guinea pigs) and ungulates (sheep and goats). First, we will describe how infantile odours become meaningful stimuli at parturition, allowing the normal development of maternal care but also providing a basis for individual recognition of the offspring in some species. Next, we will review the role of olfaction for three behavioural aspects that play a key role in survival of the offspring: nipple search behaviour, recognition of the postnatal environment (mother and nest), and the development of food preference through chemosensory signals perceived via maternal influence.

Maternal Behaviour

Olfaction and the onset of maternal behaviour

Olfaction regulates avoidance of the neonate in non-pregnant females

Outside parturition, most mammalian females are not responsive to young and instead avoid them. This is due to the majority of virgin or non-pregnant, non-lactating females being repelled by the placenta and/or amniotic fluids (AFs) that cover the neonate (Kristal and Graber, 1976; Kristal, 1980; Lévy *et al.*, 1983). This avoidance is reported in rats and other species of rodents like golden hamsters or Mongolian gerbils (Rowell, 1960; Richards, 1966; Marques and Valenstein, 1976; Elwood, 1977, 1981; Swanson and Campbell, 1979).

In other species, the emergence of maternal behaviour appears to be subject to a stronger inhibition than that observed in rodents. In rabbits, virgin females are absolutely unresponsive to foster young, even after 2 weeks of exposure (González-Mariscal, 2001, 2004) and in sheep, maternal responsiveness has never been reported in virgin females (Poindron *et al.*, 2007). Lesions of both the accessory and the main olfactory systems (see Chapter 1) eliminate perception of aversive cues coming from the neonate and consequently females exhibit a rapid onset of maternal behaviour (Fleming and Rosenblatt, 1974a,b; Carretero *et al.*, 2003). However, disruption of both chemosensory systems induces the shortest onset latency (Fleming *et al.*, 1979). Similarly in the hamster, half of virgin females cannibalize young (Marques and Valenstein, 1976), whereas after lesion of both the main and the accessory olfactory systems they retrieve them (Marques, 1979). In the rabbit, removal of the accessory olfactory bulb in virgins activates maternal responsiveness in half of the animals within a few days (Chirino *et al.*, 2007). In sheep, AFs that cover the neonate are found repulsive outside parturition inducing avoidance of the lamb. Also, input from the main olfactory system plays an inhibitory role. In non-pregnant ewes, anosmic females that are hormonally primed and receive a genital stimulation that mimics parturition are more likely to accept a new-born lamb than intact primed females (Poindron *et al.*, 1988). In contrast to the species listed above, the mouse is a special case since pup odours are not repulsive to virgins and instead are necessary to trigger maternal behaviour. Olfactory bulbectomy induces disruption of maternal behaviour and cannibalism in virgin mice (Gandelman *et al.*, 1971). More recently, a study in virgin mice lacking type 3 adenylyl cyclase (AC3), an enzyme involved in olfactory signalling in the main olfactory epithelium, reported that the mice had deficits in licking, nursing, nest building and pup retrieval. Although AC3 is also expressed in several brain regions including those involved in maternal behaviour, $AC3^{-/-}$ virgins could not detect pup odours, indicating that the main olfactory system is involved (Wang and Storm, 2011).

Olfaction is key for the onset of maternal behaviour at parturition

At parturition, the mother is highly responsive to cues from the young that change her behaviour towards the neonate from avoidance to approach. Some of these cues are of an olfactory nature. In rats and sheep, virtually all parturient females consume the placenta (placentophagia) and lick AF at parturition (Kristal and Graber, 1976; Kristal, 1980). In sheep, anosmic females are neither repelled nor clearly attracted to AF at parturition (Lévy *et al.*, 1983). Lesions of the vomeronasal organ are without effect on the attraction to AF, indicating a mediation by the primary olfactory system (Lévy *et al.*, 1995). Also, virgins and early-to-late-pregnant females do not show a preference between nesting material taken from the nest of an unfamiliar mother and her pups and material taken from a virgin's nest. By contrast, a marked preference for a nest built by maternal females occurs during the peripartum period (Bauer, 1983; Fleming *et al.*, 1989; Kinsley and Bridges, 1990). Similar shifts have been reported in rabbits, where an absence of placentophagia is observed during oestrus and pregnancy, whereas most of the females become placentophagic at parturition (Melo and González-Mariscal, 2003).

Depending on the species, olfaction plays a more or less important role in the organization of maternal care. For instance, in mice the onset of maternal behaviour relies heavily on offspring odours.

Anosmia induced by irrigation of the nostrils with zinc sulfate, or by depletion of noradrenaline within the main olfactory bulb, results in the majority of the females eating their offspring (Seegal and Denenberg, 1974; Dickinson and Keverne, 1988). However cannibalism could be the consequence of side effects of nasal irrigation with zinc sulfate: deleting genes in genetic/transgenic mouse models involved in olfactory signal transduction in the main olfactory epithelium generates a deficit in pup retrieval but no cannibalism is reported (Weiss *et al.*, 2011; Fraser and Shah, 2014). On the other hand, removal of the vomeronasal organ does not preclude the expression of maternal behaviour (Lepri *et al.*, 1985). In addition, mice without Trpc2 (a gene coding for ion channels in the vomeronasal organ), which essentially lack odour-evoked activity in the vomeronasal organ, express normal pup retrieval behaviour (Fraser and Shah, 2014). However, experienced mother mice are able to use other sensory information to compensate for the loss of olfactory cues. Anosmic multiparous mice do not kill their pups (Seegal and Denenberg, 1974), and depletion of noradrenaline within the olfactory bulb has no consequence on cannibalism in multiparous mice (Dickinson and Keverne, 1988).

The onset of maternal behaviour in precocial ungulates like sheep also depends primarily on olfaction. Studies were undertaken to observe, at parturition, the effects of deprivation of olfactory information in the mother, by either lesions of the main and the accessory olfactory systems, or by preventing access to the olfactory cues from the neonate. In primiparous ewes, anosmia induces deficits in licking of the neonate, in maternal bleats and increases protest bleats expressed at parturition (Lévy *et al.*, 1995). Such behavioural disturbances are not observed in multiparous anosmic mothers. On the other hand, females with only lesions of the vomeronasal organ show few disturbances in maternal care. These results underline the importance of the main olfactory system in the establishment of maternal behaviour. The olfactory cues that the ewe responds to at parturition and that allow her to be attracted to any new-born lamb are mainly contained in AF. This is well illustrated by the behaviour of primiparous ewes whose lambs have had their AF washed off at birth (Lévy and Poindron, 1987). Washing the neonate prevents the manifestation of any acceptance behaviour while aggressive behaviour is increased. AF is also sufficient by itself to induce maternal acceptance in a context wherein females reject young. Parturient experienced ewes do not accept 1-day-old lambs as readily as they accept new-borns (Poindron and Le Neindre, 1980). However, presentation of lambs coated with AF increases their rate of acceptance (Lévy and Poindron, 1984). Similar results were found in a study in which fostering of alien lambs onto parturient ewes was facilitated by covering the lambs with jackets soaked in AF (Basiouni and Gonyou, 1988; see also Fig. 12.2 in Chapter 12). Thus, olfactory cues provided by AF are necessary to ensure appropriate maternal behaviour at parturition, especially in naïve ewes. AF appears to be a potent organizer of maternal behaviour in focusing the attention of the dam from licking her body to licking her neonate (Schneirla *et al.*, 1963).

The importance of olfaction for the establishment of maternal behaviour is not a common mechanism across mammalian species. In the rat, neither the main nor the accessory olfactory systems appear necessary. Normal onset of maternal behaviour is displayed in primiparous parturient rats rendered anosmic by destruction of the olfactory epithelium before parturition (Benuck and Rowe, 1975; Jirik-Babb *et al.*, 1984; Kolunie and Stern, 1995). Similarly, removal of the vomeronasal organ or section of the vomeronasal nerves do not result in deficits of maternal behaviour (Jirik-Babb *et al.*, 1984; Fleming *et al.*, 1992; Kolunie and Stern, 1995). On the other hand, bilateral bulbectomy performed before parturition induces deficits in pup retrieval and incomplete placentophagia (Benuck and Rowe, 1975; Schwartz and Rowe, 1976), together with profound disturbances of licking behaviour and time spent with pups (Kolunie and Stern, 1995). Nevertheless, in addition to deprivation of olfaction, bulbectomy in female rats produces numerous non-sensory effects such as irritability, hyperactivity and infanticide (Leonard and Tuite, 1981), together with presumably physiological disturbances. Finally, other studies have reported a normal onset of maternal behaviour after bulbectomy (Beach and Jaynes, 1956b; Fleming and Rosenblatt, 1974b). Therefore, although olfaction certainly facilitates mother–young contact in rats it is not critical for the successful care of the neonate.

In summary, in sheep and mice, olfactory information is critical for inexperienced mothers. Once they have experienced motherhood, they can compensate for the loss of the sense of smell at parturition. Also, preventing mothers from having access to olfactory cues has effects much more dramatic than anosmia, suggesting that the loss or impairment of the sensory function may lead to an underestimation of the actual role of olfaction in the intact animal.

Olfaction regulates specific component of maternal behaviour

Although olfaction is not necessary for the initiation of maternal behaviour in parturient rats, olfaction is critical for the organization of some dam-specific behaviours. For instance, a specific olfactory cue coming from the pups was characterized that regulates anogenital licking, a behaviour crucial for rat pup survival because it stimulates defecation and urination (Rosenblatt *et al.*, 1979; Moore, 1981) and contributes to the mother's ability to distinguish the sex of her pups (Moore, 1981, 1985). This olfactory substance, attractive to the mother, was found in the rat pups' anogenital area (Fig. 8.1; Brouette-Lahlou *et al.*, 1991b). When the pups' heads were rubbed with anogenital smears while their anogenital areas were cleaned, the dams licked only the heads. Anogenital licking was completely disorganized after ablation of exocrine glands of pups that are located close to the genitals, the preputial glands. Using combined gas chromatography and mass spectrometric (GC–MS) analysis, an active compound responsible for attraction to anogenital smears was isolated from the pups' preputial glands: dodecyl propionate (Brouette-Lahlou *et al.*, 1991a). When given a two-choice test between several components isolated from rat pups' preputial glands and this molecule, dodecyl propionate was the only compound for which the attractiveness was equal to preputial secretions themselves. Interestingly, dodecyl propionate concentration in preputial secretions is higher in male than in female pups (Brouette-Lahlou, 1989) and this difference may help mothers to discriminate the gender of their pups. Mothers without vomeronasal organs displayed a disorganized anogenital licking behaviour similar to the one observed after preputialectomy, and reduced the time spent licking the pups' heads odourized with dodecyl propionate, whereas such a difference was not observed in anosmic mothers (Brouette-Lahlou *et al.*, 1999). Thus, the accessory, but not the main olfactory system, is necessary for the perception of this compound. Overall, dodecyl propionate is an olfactory cue that helps the mother to identify pups that need

Fig. 8.1. Olfactory substances from the anogenital area of rat pups have been found to guide the anogenital licking by the mother. (With the permission of A.S. Fleming; Brouette-Lahlou *et al.*, 1991a,b.)

to be licked. In this sense, it regulates, but does not trigger, anogenital licking behaviour.

Olfaction and recognition of the young

Offspring recognition in altricial species

Laboratory strains of rats and mice do not form a selective bond with their own young and they also retrieve alien young from outside the nest. Nevertheless, when given the choice, mother rats retrieve their own young first before taking care of alien pups, a preference abolished by olfactory bulbectomy (Beach and Jaynes, 1956a; Rosenblatt and Lehrman, 1963). Similarly, wild guinea pigs nurse specifically their own offspring whereas laboratory strains care for alien young as well (Leon, 1983). Nonetheless in a choice test, domestic guinea pigs prefer their own litter to an unfamiliar one within 2 days after birth (Porter *et al.*, 1973).

The existence of scent marking and its involvement in pup recognition has been reported in gerbil mothers. Gerbils mark the pups with sebum, an oily substance secreted by microscopic glands in the skin, and cleaned pups are marked more vigorously than pups that already have sebum on them. Moreover, after the first week postpartum, the presence of the mother's scent marking on a strange pup enhances the pup's probability of being retrieved to the nest in comparison with an unscented pup (Wallace *et al.*, 1973). These data suggest that scent marking helps the mother identify her pups.

While rabbit mothers can nurse alien young (González-Mariscal *et al.*, 1998), they also mark their own young and are aggressive towards alien offspring that are not colony members once they are out of the burrow (Mykytowycz and Dudzinski, 1972). Females identify kittens by smell, sniffing most commonly at the hindquarters and head regions in which the skin glands are located. Mothers attack their own young if they have been marked with glandular secretions or urine from females living in another colony (Mykytowycz, 1985). However, mothers are more tolerant towards young from their own social group, suggesting the existence of group specific odours (Mykytowycz and Dudzinski, 1972).

Offspring recognition in precocial species

Individual recognition of young associated with exclusive nursing has been demonstrated in sheep, goats, cattle and horses (Herscher *et al.*, 1963; Klopfer *et al.*, 1964; Smith *et al.*, 1966; Hudson and Mullord, 1977; Maletinska *et al.*, 2002). Nonetheless, it is in sheep and goats that the role of olfaction in young recognition has been most extensively studied (Lévy *et al.*, 1996; González-Mariscal and Poindron, 2002).

When olfactory deficits are induced experimentally prior to parturition via bulbectomy (Baldwin and Shillito, 1974), sectioning of the olfactory nerves (Morgan *et al.*, 1975) or by irrigation of the olfactory mucosa with zinc sulfate solution (Romeyer *et al.*, 1994; Lévy *et al.*, 1995), ewes or does subsequently show no evidence of recognizing their neonate, but accept alien young as well as their own. It should be pointed out that the accessory olfactory system is not involved in lamb recognition, since severing its nerves does not prevent ewes from being selective (Lévy *et al.*, 1995). Experiments in which the lambs' sensory cues are manipulated provide additional evidence that the main olfactory system of the ewe is sufficient for the development of recognition of the neonate at parturition. Ewes that are exposed for 4 or 12 h to their lambs whilst these are confined in a 5-cm-thick double-wire-mesh cage (receiving olfactory, visual and auditory stimuli, but not physical contact), develop a selective bond with them (Poindron and Le Neindre, 1980). If the vomeronasal organ plays a critical role in maternal selectivity, such lamb confinement should at least impair its establishment since activation of neurons from the accessory olfactory bulb requires the nose of the subject to be in direct contact with the stimulus animal (Luo *et al.*, 2003). However, in goats unlike in sheep, depriving the mother of tactile contact but allowing them to perceive olfactory, visual and auditory stimuli for the first 4 h postpartum leads to rejection of the kid in some dams (Romeyer *et al.*, 1993a). This suggests participation of the vomeronasal organ in the onset of maternal behaviour and/or establishment of selective nursing in this species. Similarly, in pinnipeds like Australian sea lions, although vocal cues are important for long-distance recognition, females discriminate between the scent of their own pups over that of an alien pup, suggesting that olfactory cues are used for confirming pup identity before allowing it to suckle (Pitcher *et al.*, 2010).

In intermediate species, like pigs, mothers build a nest and do not develop selectivity towards their piglets immediately after parturition. However, cross-fostering later than 2 days postpartum is difficult because most sows already identify their own piglets and reject alien ones (Price *et al.*, 1994). This discrimination is based on olfaction since bulbectomy results in suppression of aggressive behaviour towards alien piglets (Meese and Baldwin, 1975). A more recent study shows that sows prefer their own anaesthetized piglets to alien anaesthetized piglets after the first day postpartum (Maletinska *et al.*, 2002).

Genotypic and environmental influences of olfactory signatures

Although the chemical nature of individual olfactory signatures of offspring is largely unknown, some experimental evidence indicates both genetic and environmental influences. Evidence for genetic influences of the olfactory signature is clearly demonstrated by research with mice that differ genetically only in the major histocompatibility complex (MHC) region, a set of cell surface proteins essential for the acquired immune system to recognize foreign molecules, which in turn determines histocompatibility (see also Chapters 3 and 7). MHC odour types are expressed in early life and these odours can be a basis for mothers to recognize their young. Adult mice are able to distinguish the urine of MHC-dissimilar pups of the same age (Yamazaki *et al.*, 1992). This raises the possibility that urine, ingested by the mother during anogenital licking, may be sufficient to discriminate her own from other H-2-dissimilar infant mice. This hypothesis is further supported by the finding that mothers retrieve pups of their own (familiar) MHC type in preference to otherwise identical pups bearing a different MHC type (Yamazaki *et al.*, 2000). Therefore, MHC-determined odours play a role in maternal recognition and since these odour types have been demonstrated also in rats and humans, investigation of such a role across mammalian species seems warranted.

Given the polygenic contributions to olfactory signatures, one would expect a positive correlation between the degree of genetic relatedness of individuals and the similarity of their odour types. This was investigated in sheep by testing if mothers are able to detect olfactory resemblance between their twins (Romeyer *et al.*, 1993b). It appears that the odour signatures of monozygotic twins are more similar than those of dizygotic twins, enabling the mothers to discriminate less effectively between monozygotic twins.

Individually distinctive odour signatures may be the product of each individual's genotype, but they are also influenced by the environment. Substances that are inhaled or ingested are known to alter the scent of the skin surface, faeces, urine or breath. The importance of dietary constituents has been clearly demonstrated to influence offspring recognition in the spiny mouse (*Acomys* spp.). Lactating females retrieve more rapidly alien pups, whose mothers have been fed the same diet as the subject females, than pups from mothers fed on a distinctly different diet (Doane and Porter, 1978). Cannibalism rates by foster mothers also differ for pups from own-diet versus unfamiliar-diet females (Porter and Doane, 1977; Doane and Porter, 1978). These diet-specific odours could be deposited onto the pups as their mothers lick or interact with them, or could be transferred to the pups through the milk.

A possible role of acquired maternal labels (i.e. olfactory marks) in offspring recognition through licking and nursing has also been proposed in goats (Gubernick, 1981). However, such maternal labels may not be necessary for the acceptance of familiar kids nor for rejection of aliens (Romeyer *et al.*, 1993a). Similarly, sheep mothers develop a selective bond with their new-born(s), even when direct physical contact is prevented, providing that they have access to the lamb's odour (Poindron and Le Neindre, 1980; Romeyer *et al.*, 1993a). Also, selective ewes reject alien young even if they have been housed with, and presumably labelled by their own mother (Lévy *et al.*, 1991; Porter *et al.*, 1991).

Maternal labels may not be necessary for lamb recognition, but if such cues are present they might be incorporated into the lamb's odour phenotype. It should be recalled that ewes did not discriminate between their monozygotic twins when they had no physical contact with the one they were familiarized with, but they did so if physical contact was allowed during the exposure for the first hours after parturition (Romeyer *et al.*, 1993b). This recognition may reflect a mosaic of chemical by-products mixing genetically mediated and acquired odours. By contrast, the odour profile of the isolated lamb during the first hours after parturition would not include the postnatal maternal contribution. When the first born twin is confined to a mesh cage in the mother's pen to prevent physical contact, neither that lamb nor its isolated monozygotic twin has the opportunity to acquire maternal labels. Thus, their (genetically influenced) odour signatures remain indistinguishable. This illustrates that environmental factors may contribute to the individual olfactory signature of lambs.

Infant Behaviour

Olfaction and nipple search behaviour

Suckling is displayed in a stereotyped manner in each species and, while olfaction helps in the initial localization of the nipple, getting access to milk is usually based on multisensory information. The importance of olfaction is usually related to the state of development of the new-born. In altricial species, localization of the mammary zone is mainly, if not uniquely, dependent on olfaction (e.g. rabbits, rats, Fig. 8.2A and B). By contrast, in precocial mammals (sheep), the fully developed neonate uses a wider panel of sensory cues (Fig. 8.2C). Intermediate species (pigs), where the neonate is fully developed but lacks thermoregulation, also use multisensory information (Fig. 8.2D).

Altricial species

Kangaroos and wallabies give birth to a small larva-like offspring. At parturition, the neonate crawls from the vaginal opening to the pouch through the abdominal fur. While the young invariably crawls upwards, suggesting that it is initially guided by gravity (Cannon *et al.*, 1976), once at the opening of the pouch, it needs to turn 180° to enter it and reach the nipple. This change in orientation behaviour is guided by olfactory cues provided by the mother's pouch, and gravity becomes irrelevant (Schneider *et al.*, 2009).

Rabbit kittens also rely on olfactory cues to find the nipple (Hudson and Distel, 1983, 1986). It takes the

Fig. 8.2. Comparison between rabbit kittens, rat pups, lambs and piglets in the sensory cues used in their initial search for a nipple (photos by R. Nowak). Note that the relative importance of olfaction varies between species. In the rabbit (A), the mammary pheromone 2-methylbut-2-enal (2MB2) is crucial, but prenatal odours can also be involved. In the rat (B), amniotic fluids (AFs) and then dimethyl disulfide (DMDS), a molecule found in saliva, act as attractants. The lamb (C) and the piglet (D) use several senses, but olfactory cues found in the AF, inguinal wax and saliva play a role.

kittens only a few seconds to attach to the nipple: after making contact with the doe's fur, they push their muzzle into it and display probing and rapid lateral head movements until a nipple is reached. By investigating the cues governing this behaviour it has been shown that an olfactory component produced by the nipples, the nipple search pheromone, is essential for the onset and the maintenance of searching behaviour and for nipple attachment. The kittens are very sensitive to the volatile cues that are present not only on the doe's ventrum but also in the milk (Coureaud *et al.*, 2001). When testing the reaction of kittens to fresh milk presented on a glass rod, it was found that even milk diluted 10,000-fold elicits searching and grasping (Keil *et al.*, 1990). Kittens remain inactive when presented in the same way with AF or blood from either pregnant or lactating females. Therefore, the reactivity to olfactory cues present in the milk is not programmed by prenatal exposure. A unique volatile substance extracted from rabbit milk by gas chromatography, 2-methylbut-2-enal, was shown to elicit the typical head-searching and grasping response, and Schaal *et al.* (2003) named it the mammary pheromone (Fig. 8.2A). Two-methylbut-2-enal corresponds unambiguously to the qualification of a pheromone as it fulfils all the operational criteria (Wyatt, 2014), including eliciting the same response as the natural stimulus and with no other compound giving rise to the same effect (see Chapter 3 for a more detailed discussion of pheromone characteristics). As far as we know, it is the only clear demonstration of the involvement of a pheromone in mammalian mother–young interactions. Whether the mammary pheromone is of the same origin and has the same structure as the nipple search pheromone described by Hudson and Distel (1983) still remains to be elucidated. Nonetheless, kittens delivered by caesarean section one day before term respond to a lactating rabbit doe (Hudson, 1985) and to the mammary pheromone (Schaal *et al.*, 2003) with normal teat searching activity and grasping. The fact that this pheromone is such a powerful releaser of neonatal teat searching behaviour does not exclude the involvement of other olfactory cues, in particular those from the uterine environment. Aroma preference acquired *in utero* from the mother's diet may provide additional information (Coureaud *et al.*, 2002) and prenatal learning of odours present in the amniotic fluid could help in the neonatal search for a nipple by providing additional facilitating cues. Nonetheless, the initial reactivity of the rabbit kittens to the pheromone is predictive of their survival. By testing 293 new-borns from 30 litters on days 1 and 3 after birth, Coureaud *et al.* (2007) showed that kittens that are unresponsive to the mammary pheromone have subsequent difficulties in finding the nipple, ingest less milk and are less likely to survive (Fig. 8.3A). This effect is modulated by

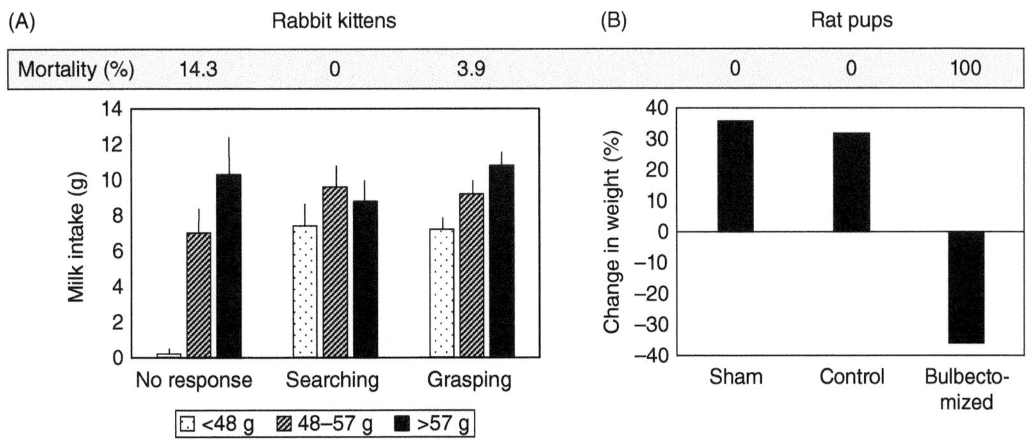

Fig. 8.3. Importance of olfactory cues in nipple search behaviour and survival in rabbit kittens and rat pups. Survival was measured up to day 7 postpartum in rabbits and up to day 4 in rats. (A) Milk intake in 1-day-old rabbit kittens according to their responsiveness to the mammary pheromone (no response, searching only, grasping); data are plotted according to birth weight. Olfactory unresponsiveness to milk affects more strongly the lightest neonates.
(B) Change in body weight of new-born rats 5 days after being subjected to bilateral bulbectomy, sham operation or nothing (control). Neonatal mortality is shown above the graphs as a percentage of sample size. (Adapted from Risser and Slotnick, 1987; Coureaud *et al.*, 2007.)

the birth weight since olfactory unresponsiveness affects the lightest kittens more strongly. As a consequence, mortality within the first week of life (but not beyond) is increased in lightweight unresponsive kittens.

Olfaction plays a very important role for suckling in rodents as well, and this has been shown by various methods. Contact with the fur elicits a rooting response through which rat and mouse pups explore the mother's body, moving their head laterally until a nipple is found. The pups then stop moving and probe the nipple area. In rats, elimination of the sense of smell via olfactory bulbectomy, destruction of the olfactory epithelium with intranasal infusion of a $ZnSO_4$ solution (Singh and Tobach, 1975; Singh *et al.*, 1976) or by washing the mother's ventrum impairs strongly the localization of the nipples (Hofer *et al.*, 1976; Teicher and Blass, 1976) and may lead to starvation (Risser and Slotnick, 1987; Fig. 8.3B). The olfactory cues that attract the pups in their initial search are nonetheless not produced by the nipple glandular system (Blass and Teicher, 1980). Rather, it is the AF and the saliva deposited by the mother around parturition as she licks herself, that orient the pups to the nipple (Teicher and Blass, 1976). Following their initial successful suckling, subsequent localization of the nipple is mediated by the odour of the pup's own saliva that was deposited in the previous feeding episodes. Washing the nipples disrupts suckling even in 30-day-old rats, when their eyes are opened, which shows that pup saliva is needed throughout the suckling period to elicit nipple attachment. Pedersen and Blass (1981) analysed saliva by GS–MS and identified a component that acted as an olfactory attractant: dimethyl disulfide (Fig. 8.2B). Dimethyl disulfide was shown to induce nipple attachment in 3- to 5-day-old rats and because it is not detectable in the AF, its behavioural salience is unlikely to depend on prenatal exposure. However, it has only half the potency of intact saliva, suggesting that other olfactory cues, still to be discovered, are involved. Surprisingly, rat milk has never been thoroughly tested in behavioural experiments.

Mice pups behave in a similar way. Washing the nipple and surrounding area prevents most of them from suckling within a 15-min trial (Logan *et al.*, 2012). AF, saliva and maternal milk are each sufficient to reinstate robust suckling when smeared on washed nipples, while saliva from virgin females and maternal urine do not promote suckling. Caesarean section-delivered pups, which have not experienced any suckling episode before being tested, only respond to nipples coated with AF, suggesting that mouse saliva, colostrum and milk do not contain a nipple search pheromone like in the rabbit (Logan *et al.*, 2012; see also Fig. 2.3A in Chapter 2). By using a very similar procedure, Al Ain *et al.* (2013) found that 5- to 7-h-old mouse pups without any suckling experience attach more easily to nipples of lactating females than to nipples of non-lactating females. They also found that the reactivity to the range of olfactory cues increases with suckling experience. AF, but also milk, induce nipple attachment in naïve pups, while pups that have a short suckling experience are attracted to nipples coated with AF, milk or saliva from lactating females, and pups that have a more extensive suckling experience respond to pup saliva as well. As can be expected, new-born pups with no functional olfaction fail to thrive. Mutant mice lacking the primary signal transduction (i.e. the transmission of a molecular signal) channel of the main olfactory epithelium (Cnga2) are unable to attach to the nipple. The importance of this defect is such that it results in severe mortality even though mother–young interactions appear normal (Brunet *et al.*, 1996).

Intermediate and precocial species

In pigs, the mother remains recumbent during parturition. This allows the young to follow the surface of her body until they reach the ventrum, sometimes after making contact with the sow's snout, suggesting an attractive role of her grunts. Olfactory, tactile and thermal cues provided by the sow are used by piglets to locate the udder (Parfet and Gonyou, 1991; Fig. 8.2D). However, new-born piglets take longer to attach to sows' teats that have been washed with organic solvent than to unwashed sows (Morrow-Tesch and McGlone, 1990). Olfactory deprivation of the piglets by anesthetizing their olfactory epithelium with lidocaine disrupts teat attachment almost completely. At least two maternal odour sources, AF and sow's milk, have been shown to attract piglets. It is believed that olfactory contamination of the bedding while the sow is giving birth, but also the trail left by first-born piglets, guide the next ones to the udder, and milk would serve as a final cue to locate the teats. However, piglets do not rely solely on olfaction, since anesthetizing the tongue and the snout affect teat attachment; obviously, tactile cues are used in the final act of grasping the teat into the mouth. First-born piglets spend less time searching for the udder than those born late, when there is considerably more competition for and fighting over the teats with increasing litter size. Once a teat is found, the piglet initially obtains milk continuously and can suckle from several teats as long as there is no competition. Twelve hours later, the release of milk becomes cyclic and the sow develops a regular nursing pattern. A teat order is consequently formed and each piglet fights to keep its own over the whole duration of lactation (de Passillé *et al.*, 1988). The stability of the teat order is maintained as long as milk supply is adequate, but will be broken if milk supply is scarce as the piglets will compete with each other. Very little work has investigated the cues that participate in the maintenance of the teat order. Jeppesen (1982) ran observational studies with piglets reared on an artificial sow fitted with rubber nipples. Alteration of visual features had little effect on the piglets' choice while fitting new nipples or moving familiar nipples to different positions on the udder disoriented them. Distinguishing nipples by individual olfactory signatures left by the piglets when they feed is the most plausible explanation. The fidelity for a specific teat is so strong that when 7-day-old piglets are cross-fostered, a practice commonly used on farms to move surplus piglets from large litters to sows that have non-occupied teats, they express distress behaviour and fight with resident piglets. Such individuals do eventually adapt to their novel situation but have reduced growth when weighed a week later (Horrell and Bennett, 1981; Horrell, 1982).

Lambs, and probably most ruminants, are attracted to their mother by sight and hearing (Vince, 1993) and the first contact is made with her chest or flank. Thereafter, the udder-seeking behaviour is under the guidance of thermotactile and olfactory cues (Vince, 1993). Touch on the face strongly activates oral exploration and orientation movements of the head, and lambs respond preferentially to warm, smooth, non-woolly, lightly yielding surfaces. There is also evidence that odours may act in combination with tactile stimuli to activate udder-seeking behaviour (Fig. 8.2C). One olfactory cue could be the inguinal wax produced by two lateral glands located at the outer surface of the udder. This wax covers the udder and spreads over the surrounding wool. Lambs taken from their mother at birth and tested with different olfactory cues presented on cotton swabs in front of their nose, respond with more head movement to the wax of their mother than to that of an alien ewe. Their respiration rate increases during stimulation with maternal wax, whereas there is little change when exposed to alien substances (Vince and Ward, 1984). Lambs do not seem to react to milk or to wool from the back of the ewe, suggesting that some odorous substances emanating from the inguinal wax may be perceived during prenatal experience. That prenatal factors influence postnatal suckling is also suggested by the finding that newly born lambs prior to any suckling experience are attracted to the smell of their own AF in comparison to alien AF (Schaal *et al.*, 1995). Anaesthesia of the olfactory epithelium at birth does not seem to have dramatic effects, even though fewer lambs found the teat in the time set by the behavioural test (Vince *et al.*, 1987).

In summary, it appears that olfactory cues are essential for nipple searching behaviour in intermediate species such as pigs. They seem to have a less crucial role in precocial mammals where other sensory inputs play a role.

Recognition of the postnatal environment

Altricial species

In altricial species, the postnatal environment includes the mother, nesting material and siblings. During the first few days or weeks after birth, the neonates are permanently in contact with the nest and the siblings, whereas the mother is encountered at nursing times and this sometimes very briefly (3 min/day in rabbits). In rabbits, it is the mammary pheromone that makes lactating does attractive to kittens. When submitted to two-choice tests where they can sense the presence of anesthetized stimulus animals through a wire mesh, 1- to 3-day-old kittens can detect the presence of females and males. They prefer the former to the latter, and lactating to non-lactating females (Coureaud and Schaal, 2000) but do not seem to discriminate between the mother and an unfamiliar female, if the latter is lactating (Patris *et al.*, 2008). On the other hand, when an odorant is paired with the mammary pheromone in the nursing context, 3- to 8-day-old kittens rapidly learn to orient to this new odorant. Later on when tested in the presence of two adult females, they show a preference for their scented mother compared to unscented females, but not systematically when having the choice between their own scented mothers and an unfamiliar female carrying the same odour. They are also equally attracted by the unscented mother and a scented unfamiliar lactating female. The new scent, although attractive, does not override other olfactory cues from the mother's ventrum, especially the mammary pheromone. Serra and Nowak (2008) also showed that kittens do not express any preference for their mother when exposed to two lactating does, nor when given a choice between two litters of three kittens, one of them being their own siblings. The lack of preference for a familiar social stimulus may be influenced by the warmth released by their bodies: kittens do prefer a warm (35°C) to a lukewarm surface (20°C). On the other hand, kittens display a clear preference for their own nesting materials compared to an unfamiliar nest, showing that they do recognize their postnatal environment based on olfactory characteristics. Overall, it appears that rabbit kittens are able to detect individual maternal odours but seem to react both to supra-individual cues (mammary pheromone, or body warmth) and individual odours that they have learned; their response depending on the degree of familiarity with such cues and on the context (Patris *et al.*, 2008; Val-Laillet and Nowak, 2008).

In Belding's ground squirrel, mothers rear their litter alone in underground burrows for about 25 days and defend their natal burrow against intrusion from conspecifics (Holmes, 1990). Under both field and laboratory conditions, the young distinguish between familiar and unfamiliar dams only at the age when they would emerge above ground in the field for the first time (at approximately 22 days) and litter-mate preferences are not fully developed at that stage either (Holmes, 1997; Mateo, 2006). Similar results have been reported in rats (Leon and Moltz, 1971, 1972; Carr *et al.*, 1979), house mice (Breen and Leshner, 1977) and Mongolian gerbils (Gerling and Yahr, 1982), suggesting that the processes of recognition develop at a stage when the young have to seek their mother more actively and initiate suckling outside of the nest. The inability to discriminate their mother at an early age seems quite logical in view of ecological constraints since as long as pups remain in the nest there is no need to respond preferentially to the mother; she will be the only female entering the nest. Nonetheless, with the exception of the rat, none of these studies dealt directly with olfactory recognition of the mother. It has been shown that rat pups discriminate and prefer the smell of lactating females compared to virgin females but do not seem to discriminate between their own and unfamiliar mothers. Authors have been tempted to use the expression 'maternal pheromone' to qualify the odorants that attract the pups but the phenomenon is better described by 'signature odours' (see also Chapter 3). For instance in Mongolian gerbils, development of attraction to maternal odours depends on prior exposure to them: pups raised by mothers lacking scent glands do not approach lactating females that have scent glands (Gerling and Yahr, 1982).

Rats are rather intriguing since the signature odour is not produced by a specific gland but found in specific maternal anal excreta: the caecotrophs. Caecotrophs are formed in the caecum and serve as a substrate for bacterial growth. They pass through the large intestine of the mother with relatively little water loss and when defecated are eaten by the adult directly from the anus. When rat pups are tested in a two chamber box, they prefer adult caecotrophs to the empty goal box, they find two lactating females equally attractive (one of them being their mother), while faeces are irrelevant to them. Moreover, the use of antibiotics such as neomycin, which wipes out the entire bacterial population in the adult, leads to complete loss of attractiveness by pups towards maternal females and their caecotrophs (Leon, 1974). While this shows that the signature odours could be of caecal origin, the chemical nature of the volatile compounds from bacterial origin has never been characterized, and the operational criteria to qualify them as a potential pheromone have not been tested.

Intermediate and precocial species

Both piglets and lambs can identify their mother very soon after birth on a multisensory basis. Tested in a T-maze, 1-day-old piglets are able to discriminate their home environment, including mother, bedding and siblings. When the testing procedure is performed in a novel room, they keep recognizing their own mother compared to an unfamiliar sow. To achieve this recognition, piglets are capable of using odours from the sow's faeces, urine and udder area in addition to auditory cues (Horrell and Hodgson, 1992). Quite interestingly, the mother's udder area is preferred from day 1 while other maternal odour sources are preferred at 7 days of age. Olfaction is therefore important for piglets to stay close to their mother, and the relevance of mammary odours at one day of age is very likely to reflect the importance for the neonate of achieving suckling. This helps them learn skin olfactory cues, perhaps including milk odours as well. The other odours will become familiar later as the piglets wander around the maternal body. Early recognition of the mother, in addition to the teat order, which relies on olfactory marking, explain why piglets show disturbed behaviour when fostered onto unfamiliar sows.

Lambs can discriminate their mother from unrelated ewes by the age of 12–24 h in a two-choice test where ewes are restrained behind metal hurdles (Nowak, 1991). However, while neonates do react differently when odours from their own mother and alien maternal females are presented in front of their nuzzle (Vince and Ward, 1984), olfactory cues are not important for choosing the mother in a choice test at the age of 24 h. Localization of the mother from a distance of several metres and selective close-proximity seeking is mainly based on visual and vocal information. Spraying lidocaine into the nostrils of 1-day-old lambs does not prevent them from correctly choosing their mother (Nowak, 1991). Rather, it is the visual and vocal display of maternal selectivity that guides the lambs to their mother (Terrazas *et al.*, 2002). Sheep are gregarious mammals, giving birth in open spaces. The fact that, soon after birth, most lambs have to follow their mother and find her amongst dozens (sometimes hundreds) of other nursing ewes makes visual and vocal cues more conspicuous than olfactory cues. Olfactory communication does exist as lambs frequently sniff their mother when close to her, but obviously the biological role of olfaction is to be found beyond mother recognition. One possibility is that the udder may have some appeasing properties. Stressed lambs, especially after being separated from their mother, are often seen going to the udder for a few seconds and this is not to satisfy their hunger.

Early olfactory learning and transmission of food preference

Altricial species

Much of the evidence concerning the role of maternal care in regulating infant olfactory behaviour comes from studies in altricial mammals. Newborn rabbit kittens can rapidly learn to associate new odours with nipple search behaviour. The learned response appears to be very similar to inborn responses. Hudson (1985) scented the rabbit doe's ventrum with one of four odorous substances (cologne Chypre, Chanel N°5, oil of camphor and citral) 15 min before the first nursing episode. When tested the next day on a mature female cat with well-developed nipples (chosen because it is of a similar size to the rabbit while being completely free of rabbit-specific odours), 2-day-old kittens showed strong conditional responses specific to the odour experienced during their previous suckling episode. In contrast to rat pups, the magnitude of the conditioned odour response declines rapidly with age, making it very difficult to induce by day 5 (Kindermann *et al.*, 1994). Under such conditions, oral non-nutritive stimulation appears to be rewarding since rabbit kittens that can suck nipples without obtaining milk show as clear signs of olfactory conditioning as pups that do obtain milk (Hudson *et al.*, 2002).

The apparent unimportance of milk is only relative since the mammary pheromone, present in milk, also acts as a learning promoter (Coureaud *et al.*, 2006), and when kittens suckle their odorized mother, the mammary pheromone and non-nutritive sucking act synergistically. Nonetheless, pairing the mammary pheromone on its own with a novel (neutral) odorant triggers, very efficiently on the following day, high reactivity and nipple search behaviour when the odorant is presented on its own to the kittens. The conditioned response is specific to the odorant paired with the pheromone and limited to the nursing period; there is no evidence that the pheromone facilitates any olfactory learning after weaning. Should it be mixed with novel solid food, it does not make this more palatable. The learning promoting properties of the mammary pheromone are therefore restricted to the nursing context. The biological function of such olfactory learning performed at the nipple is unknown. One may hypothesize that under natural conditions, environmental odours (plant, soil) contaminating the mother's ventrum while she wanders outside the nest, inform kittens of features they will encounter once they are weaned. Such odours perceived and learned by kittens at nursing times will provide valuable information, so that once they leave the nest without any parental guidance, the surroundings will not be unfamiliar.

The post-weaning transmission of food preference in the rabbit is achieved by transfer of food aromas from the mother to the kittens. The challenge in this species is even greater than for rat pups considering the rabbit's unusual pattern of maternal behaviour. The mother gives its young remarkably little care (one suckling bout per day lasting 3 min) and little or no direct assistance in making the transition to independent feeding. Transmission of food preference can be achieved in three ways. The first one is prenatal experience: kittens born to juniper-fed does but cross-fostered at birth to chow-fed mothers eat more juniper berries during their first post weaning meal than control animals (Bilko *et al.*, 1994). The second is the maternal faecal pellets left in the nest. Does are known to deposit three or four faecal pellets in the nest at each nursing visit. These pellets show clear signs of being nibbled by the kittens and some even disappear. Kittens raised by chow-fed mothers but receiving daily pellets from juniper-fed females, eat more juniper berries at weaning than controls (Bilko *et al.*, 1994). The last one is direct contact with the mother: kittens nursed by juniper-fed mothers find juniper berries more palatable at weaning than kittens nursed by chow-fed does. The three pathways, whether acting separately or in combination, are equally effective in terms of transmission of food preference. Such redundancy is highly relevant since the different aromas of the mother's diet have more chance to reach the young and be perceived by them should their concentrations and combinations differ for each pathway.

The possibility that rat pups can be fed with pulses of milk via an intra-oral cannula has also provided ample opportunities to study very early learning in the context of suckling. If a novel odour (initially aversive) is paired with an infusion of milk, pups come to prefer it and orient towards that odour in a two-choice chamber (Brake, 1981). These learning abilities are not induced by intra-oral infusion of water, indicating that milk has specific rewarding properties. Even when the smell/taste of the milk is masked by odorants mixed in the milk, it leads to the same behavioural outcome (Terry and Johanson, 1987). However, in the absence of milk, rat pups are also capable of associating a novel odour with the rewarding aspects of non-nutritive sucking (by attaching to the nipples of an anesthetized female) or of stroking (the anogenital region with a paint brush); hence an olfactory preference develops for the initially aversive odour (Brake, 1981). Milk, non-nutritive sucking and stroking are three elements of maternal care provided in the nest. They are perceived positively and as such facilitate olfactory learning for new odours provided by the mother's body.

Such early olfactory learning can have profound long-term consequences. Olfactory cues associated with the diet eaten by the mother are transmitted to her young and are sufficient to influence markedly the food preferences of her pups at weaning. This food habit transmission can be acquired directly by observing the behaviour of the mother or other conspecifics but also by acquiring information from her milk or body (Galef and Clark, 1971, 1972). When facing their first meals of solid food, weanlings actively seek and preferentially ingest the food eaten by their mother while they were nursed by them, even if it is relatively unpalatable (Galef and Henderson, 1972). In fact, early olfactory learning not only leads to selective odour preference at weaning and facilitates the transition to solid food consumption, it also impairs the development of conditioned olfactory aversion induced experimentally by an intra-peritoneal injection of lithium chloride (Sevelinges *et al.*, 2009). This effect is specific to the early-life context since after weaning a novel odour paired with an injection of lithium chloride induces a conditioned olfactory aversion. It seems that being exposed to an odour while suckling the mother gives it a strong safety value that is long lasting even though such an odour may be associated with sickness later on.

Intermediate and precocial species

There is some evidence of olfactory learning in intermediate and precocial mammals. Piglets can learn to eat a flavoured food while interacting with the sow at the feeding trough or just observing her. Surprisingly, interacting with the sow for 3 days is not more effective in reducing the latency to eat and increasing the amount of food consumed than observing her from a distance (Oostindjer *et al.*, 2011). Transfer of flavour in milk does not appear to occur (Oostindjer *et al.*, 2010), but aromas from flavoured food contained in the breath could potentially be perceived by all piglets. Yet, this information is not sufficient to induce a preference for the flavoured food in control piglets that did not have the opportunity to see or interact with their mother while she was feeding. Clearly, observing the sow eating, or interacting with her at the feeding place, provides the most important information about what is edible. This will, in a second step, be associated with the chemosensory characteristics of the food itself.

Lambs, too, learn which food to eat from their mother. When offered two types of food, lambs take more mouthfuls of the food ingested by the mother than of that avoided by her (Mirza and Provenza, 1990). The mother seems to be the most important demonstrator: lambs exposed at 6–7 weeks of age for 15 days to a novel food with their mother consume more of it at weaning than lambs exposed to this food with non-maternal ewes (Thorhallsdottir *et al.*, 1990). Further studies show that, while lambs eat as much thyme-flavoured food in the presence of their mother as in the presence of a non-maternal ewe, 1 month after weaning at 120 days of age, their food preference differs according to past experience and context (Saint-Dizier *et al.*, 2007). All the lambs preferred to ingest unflavoured rather than thyme-flavoured food. However, when they were given the choice between thyme- and cinnamon-flavoured foods, only those that experienced thyme-flavoured food in the presence of their mother expressed a preference for it. Lambs do not necessarily need to consume the diet with their mother. If they are next to her while she is feeding, being exposed to the odour without having access to what the mother is eating, the food preference still develops after weaning. On the other hand, lambs that were kept away from their mother while she was feeding did not develop any preference for thyme-flavoured food (Saint-Dizier *et al.*, 2007). This suggests that, like in pigs, neither the mother's breath nor her milk influence this type of food preference in lambs. The presence of the mother is the most effective way for the development of flavoured-food preferences in lambs, an effect that most likely finds its roots in the strong affectional bond that lambs have with their dam (Gaudin *et al.*, 2015).

Conclusion

Odours, or chemosensory signals, are one of the most ancient means of inter-individual communication. The examples presented in this chapter and summarized in Table 8.1 show that they have kept a high biological significance in mammalian mother–young interactions. Olfaction is involved in the onset of maternal responsiveness, recognition of the young, nipple search behaviour, recognition of maternal cues and development of food preference in the young. Nonetheless, the relative importance of olfaction varies not only across species but also between mothers and offspring.

What is striking is that olfaction is far more relevant in altricial species than in precocial or intermediate species, which use additional sensory cues. While detailed information is only available from a limited number of species, it seems nonetheless that the relative importance of olfaction relates to the biotope where the species has evolved. Mothers of nocturnal species living in burrows or hiding places, such as rodents and lagomorphs, rely more on odours to interact with their young than do ungulates. This is even more obvious in the young living in a nest. For rat and rabbit pups, finding the mother and subsequently the nipple is based on olfactory cues and, when this is somehow prevented, the future of the young is at risk. Even the development of food preference can develop through the sole presence of olfactory odorants in the perinatal environment. Lambs and, to a lesser extent, piglets use multisensory cues not only to locate the nipple but also to recognize their mother. It is the mother that is observed and imitated to learn and eat new foods, and olfaction appears less important compared with other senses. Yet it is functional at birth and newly born lambs react to maternal odours, although the biological significance remains to be elucidated. Whether such odours, through olfactory imprinting in early age, have reassuring properties when the young reunites with its mother after having been separated, or when participating in future social (e.g. choice of sexual partner) or maternal behaviour (e.g. attraction to amniotic fluids) would be worth investigating.

Interestingly, it can be seen from Table 8.1 that odours remain more relevant in mothers than in young. This is particularly clear in precocial species. Both in ewes and goats, mothers giving birth for the very first time need to perceive olfactory cues to develop maternal responsiveness. The lack of olfactory information has a limited influence on the establishment of maternal responsiveness in multiparous females that are able to rely on visual and auditory cues of their young. Nonetheless, and regardless of their maternal experience, both ewes and goats keep checking the lamb's olfactory signature before nursing it throughout lactation, and are deeply disturbed should their olfaction be impaired. To date, no study has identified the full chemical nature of the odours involved. Cues responsible for the attractiveness of any young in rats (dodecyl propionate; Brouette-Lahlou *et al.*, 1991a) or for eliciting nipple search behaviour in rabbits (Schaal *et al.*, 2003) seem to be based on

Table 8.1. Importance of olfaction in mother–young interactions across mammalian species.

	Mother		Young		
	Maternal responsiveness	Offspring recognition	Nipple search	Recognition of the mother	Food preference
Altricial species					
Rat	Inhibition in non-pregnant females/ unnecessary in parturient females/ regulation of ano-genital licking by dodecyl propionate	Preference for own young	Amniotic fluid/saliva/ dimethyl disulfide	Preference for own mother/attraction to caecotrophs	Olfactory learning associated with milk/ non-nutritive sucking and with stroking
Rabbit	Inhibition in non-pregnant females	Preference for young of own colony	Mammary pheromone: 2-methylbut-2-enal	Preference for lactating females and for own nest	Transfer of food aromas from the mother
Hamster	Inhibition in non-pregnant females	ND	ND	ND	ND
Mouse	Necessary in non-pregnant and in parturient females	Preference for own young	Amniotic fluid/saliva/ milk	Preference for own mother	ND
Precocial species					
Sheep/goat	Necessary in parturient inexperienced mothers	Necessary for individual offspring recognition	Complementary to other senses	Not necessary	In association with mother interactions
Australian sea lion	ND	Necessary for individual offspring recognition	ND	ND	ND
Intermediate species					
Pig	ND	Preference for own young	Amniotic fluid/milk	Recognition of sow's faeces, urine and udder area	In association with mother interactions

ND: Non-determined.

single compounds. On the other hand, odours coding for individuality are more likely to be composite. Indeed, recent studies investigating the chemical nature of lamb odours have revealed a very complex mixture of volatile compounds, which also undergo changes over the lactation period (Viviers *et al.*, 2014). More knowledge of these volatile compounds may help in the improvement of fostering methods for sheep farmers.

Odours of new-borns not only facilitate maternal responsiveness and offspring recognition, they also have a synergistic effect with other infantile cues. For instance, rat pup ultrasounds also stimulate the initiation of maternal anogenital licking, which is then maintained by pup preputial secretions (Brouette-Lahlou *et al.*, 1999). Investigating how and where the brain handles these multisensory cues that shape offspring identity offers a wealth of exciting future research opportunities.

References

Al Ain, S., Belin, L., Schaal, B. and Patris, B. (2013) How does a newly born mouse get to the nipple? Odor substrates eliciting first nipple grasping and sucking responses. *Developmental Psychobiology* 55, 888–901. DOI: 10.1002/dev.21082

Baldwin, B.A. and Shillito, E.E. (1974) Effects of ablation of olfactory bulbs on parturition and maternal-behavior in Soay sheep. *Animal Behaviour* 22, 220–223. DOI: 10.1016/S0003-3472(74)80072-2

Basiouni, G.F. and Gonyou, H.W. (1988) Use of birth fluids and cervical stimulation in lamb fostering. *Journal of Animal Science* 66, 872–879.

Bauer, J.H. (1983) Effects of maternal state on the responsiveness to nest odors of hooded rats. *Physiology and Behavior* 30, 229–232. DOI: 10.1016/0031-9384(83)90010-0

Beach, F.A. and Jaynes, J. (1956a) Studies of maternal retrieving in rats I: Recognition of young. *Journal of Mammalogy* 37, 177–180.

Beach, F.A. and Jaynes, J. (1956b) Studies of maternal retrieving in rats. III. Sensory cues involved in the lactating female's response to her young. *Behaviour* 10, 104–125.

Benuck, I. and Rowe, F.A. (1975) Centrally and peripherally induced anosmia: influences on maternal behavior in lactating female rats. *Physiology and Behavior* 14, 439–447. DOI: 10.1016/0031-9384(75)90009-8

Bilko, A., Altbacker, V. and Hudson, R. (1994) Transmission of food preference in the rabbit: the means of information transfer. *Physiology and Behavior* 56, 907–912.

Blass, E.M. and Teicher, M.H. (1980) Suckling. *Science* 210, 15–22. DOI: 10.1126/science.6997992

Brake, S.C. (1981) Suckling infant rats learn a preference for a novel olfactory stimulus paired with milk delivery. *Science* 211, 506–508. DOI: 10.1126/science.7192882

Breen, M.F. and Leshner, A.I. (1977) Maternal pheromone: a demonstration of its existence in mouse (*Mus musculus*). *Physiology and Behavior* 18, 527–529. DOI: 10.1016/0031-9384(77)90269-4

Brouette-Lahlou, I. (1989) Le ciblage et la régulation des comportements, une fonction nouvelle des phéromones, cas particulier du lèchement anogénital chez le rat. Thèse de doctorat en Neurosciences, Université C. Bernard, Lyon I, France, 138 pp.

Brouette-Lahlou, I., Amouroux, R., Chastrett, I., Cosnier, J., Stoffelsma, J. and Vernet-Maury, E. (1991a) Dodecyl propionate, the attractant from rat pups preputial gland, characterization and identification. *Journal of Chemical Ecology* 17, 1343–1354. DOI: 10.1007/BF00983767

Brouette-Lahlou, I., Vernet-Maury, E. and Chanel, J. (1991b) Is rat dam licking behavior regulated by pups' preputial gland secretion? *Animal Learning and Behavior* 19, 177–184. DOI: 10.3758/BF03197874

Brouette-Lahlou, I., Godinot, F. and Vernet-Maury, E. (1999) The mother rat's vomeronasal organ is involved in detection of dodecyl propionate, the pup's preputial gland pheromone. *Physiology and Behavior* 66, 427–436. DOI: 10.1016/S0031-9384(98)00334-5

Brunet, L.J., Gold, G.H. and Ngai, J. (1996) General anosmia caused by a targeted disruption of the mouse olfactory cyclic nucleotide-gated cation channel. *Neuron* 17, 681–693. DOI: 10.1016/S0896-6273(00)80200-7

Cannon, J.R., Bakker, H.R., Bradshaw, S.D. and McDonald, I.R. (1976) Gravity as sole navigational aid to newborn quokka. *Nature* 259, 42. DOI: 10.1038/259042a0

Carr, W.J., Marasco, E. and Landauer, M.R. (1979) Responses by rat pups to their own nest versus a strange conspecific nest. *Physiology and Behavior* 23, 1149–1151. DOI: 10.1016/0031-9384(79)90310-X

Carretero, M.I., Segovia, S., Gomez, F. and Del Cerro, M.C. (2003) Bicuculline infusion into the accessory olfactory bulb facilitates the induction of maternal behavior in rats. *Scandinavian Journal of Psychology* 44, 273–277. DOI: 10.1111/1467-9450.00345

Chirino, R., Beyer, C. and González-Mariscal, G. (2007) Lesion of the main olfactory epithelium facilitates maternal behavior in virgin rabbits. *Behavioral Brain Research* 180, 127–132. DOI: 10.1016/j.bbr.2007.02.039

Coureaud, G. and Schaal, B. (2000) Attraction of newborn rabbits to abdominal odors of adult conspecifics differing in sex and physiological state. *Developmental Psychobiology* 36, 271–281. DOI: 10.1002/(SICI)1098-2302(200005)36:4<271::AID-DEV2>3.0.CO;2-3

Coureaud, G., Schaal, B., Langlois, D. and Perrier, G. (2001) Orientation response of newborn rabbits to odours of lactating females: relative effectiveness of surface and milk cues. *Animal Behaviour* 61, 153–162. DOI: 10.1006/anbe.2000.1563

Coureaud, G., Schaal, B., Hudson, R., Orgeur, P. and Coudert, P. (2002) Transnatal olfactory continuity in the rabbit: behavioral evidence and short-term consequence of its disruption. *Developmental Psychobiology* 40, 372–390. DOI: 10.1002/dev.10038

Coureaud, G., Moncomble, A.S., Montigny, D., Dewas, M., Perrier, G. and Schaal, B. (2006) A pheromone that rapidly promotes learning in the newborn. *Current Biology* 16, 1956–1961. DOI: 10.1016/j.cub.2006.08.030

Coureaud, G., Fortun-Lamothe, L., Langlois, D. and Schaal, B. (2007) The reactivity of neonatal rabbits to the mammary pheromone as a probe for viability. *Animal* 1, 1026–1032. DOI: 10.1017/S1751731107000389

de Passillé, A.M., Rushen, J. and Hartsock, G. (1988) Ontogeny of teat fidelity in pigs and its relation to competition at suckling. *Canadian Journal of Animal Science* 68, 325–338.

Dickinson, C. and Keverne, E.B. (1988) Importance of noradrenergic mechanisms in the olfactory bulbs for the maternal behaviour of mice. *Physiology and Behavior* 43, 313–316. DOI: 10.1016/0031-9384(88)90193-X

Doane, H.M. and Porter, R.H. (1978) The role of diet in mother–infant reciprocity in the spiny mouse. *Developmental Psychobiology* 11, 271–277.

Elwood, R.W. (1977) Changes in the responses of male and female gerbils (*Meriones unguiculatus*) towards test pups during the pregnancy of the female. *Animal Behaviour* 25, 46–51.

Elwood, R.W. (1981) Post-parturitional establishment of pup cannibalism in female gerbils. *Developmental Psychobiology* 14, 209–212.

Fleming, A.S. and Rosenblatt, J. (1974a) Olfactory regulation of maternal behavior in rats. I. Effects of olfactory bulb removal in experienced and inexperienced lactating and cycling females. *Journal of Comparative and Physiological Psychology* 86, 221–232.

Fleming, A.S. and Rosenblatt, J.S. (1974b) Olfactory regulation of maternal behavior in rats. II. Effects of peripherally induced anosmia and lesions of the lateral olfactory tract in pup-induced virgins. *Journal of Comparative and Physiological Psychology* 86, 233–246.

Fleming, A.S., Vaccarino, F., Tambosso, L. and Chee, P. (1979) Vomeronasal and olfactory system modulation of maternal behavior in the rat. *Science* 203, 372–374. DOI: 10.1126/science.760196

Fleming, A.S., Cheung, U., Myhal, N. and Kessler, Z. (1989) Effects of maternal hormones on 'timidity' and attraction to pup-related odors in female rats. *Physiology and Behavior* 46, 440–453. DOI: 10.1016/0031-9384(89)90019-X

Fleming, A.S., Gavarth, K. and Sarker, J. (1992) Effects of transections to the vomeronasal nerves or to the main olfactory bulbs on the initiation and long-term retention of maternal behavior in primiparous rats. *Behavioral and Neural Biology* 57, 177–188. DOI: 10.1016/0163-1047(92)90122-K

Fraser, E.J. and Shah, N.M. (2014) Complex chemosensory control of female reproductive behaviors. *PLoS ONE* 9, e90368. DOI: 10.1371/journal.pone.0090368

Galef, B.G. and Clark, M.M. (1971) Parent–offspring interactions determine time and place of first ingestion of solid food by wild rat pups. *Psychonomic Science* 25, 15–16.

Galef, B.G. and Clark, M.M. (1972) Mother's milk and adult presence: Two factors determining initial dietary selection by weanling rats. *Journal of Comparative and Physiological Psychology* 78, 220–225.

Galef, B.G. and Henderson, P.W. (1972) Mother's milk: A determinant of the feeding preferences of weaning rat pups. *Journal of Comparative and Physiological Psychology* 78, 213–219.

Gandelman, R., Zarrow, M.X. and Denenberg, V.H. (1971) Stimulus control of cannibalism and maternal behavior in anosmic mice. *Physiology and Behavior* 7, 583–586. DOI: 10.1016/0031-9384(71)90112-0

Gaudin, S., Chaillou, E., Cornilleau, F., Moussu, C., Boivin, X. and Nowak, R. (2015) Daughters are more strongly attached to their mother than sons: a possible mechanism for early social segregation. *Animal Behaviour* 102, 33–43. DOI: 10.1016/j.anbehav.2015.01.003

Gerling, S. and Yahr, P. (1982) Maternal and paternal pheromones in gerbils. *Physiology and Behavior* 28, 667–673. DOI: 10.1016/0031-9384(82)90049-X

González-Mariscal, G. (2001) Neuroendocrinology of maternal behavior in the rabbit. *Hormones and Behavior* 40, 125–132. DOI: 10.1006/hbeh.2001.1692

González-Mariscal, G. and Poindron, P. (2002) Parental care in mammals: immediate internal and sensory factors of control. In: Pfaff, D.W., Arnold, A.P., Etgen, A.M., Fahrfbach, S.E. and Rubin, R.T. (eds) *Hormones, Brain and Behavior*, Vol. 1. Academic Press, New York, pp. 215–298.

González-Mariscal, G., Melo, A.I., Chirino, R., Jimenez, P., Beyer, C. and Rosenblatt, J.S. (1998) Importance of mother/young contact at parturition and across lactation for the expression of maternal behavior in rabbits. *Developmental Psychobiology* 32, 101–111. DOI: 10.1002/(SICI)1098-2302(199803)32:2<101::AID-DEV3>3.0.CO;2-T

González-Mariscal, G., Chirino, R., Beyer, C. and Rosenblatt, J.S. (2004) Removal of the accessory olfactory bulbs promotes maternal behavior in virgin rabbits. *Behavioral Brain Research* 152, 89–95. DOI: 10.1016/j.bbr.2003.09.036

Gubernick, D.J. (1981) Mechanisms of maternal 'labelling' in goats. *Animal Behaviour* 29, 305–306.

Herscher, L., Richmond, J.B. and Moore, A.U. (1963) Maternal behavior in sheep and goats. In: Rheingold, H.L.

(ed.) *Maternal Behavior in Mammals*. John Wiley and Sons Inc., New York, pp. 203–232.
Hofer, M.A., Shair, H. and Singh, P. (1976) Evidence that maternal ventral skin substances promote suckling in infant rats. *Physiology and Behavior* 17, 131–136. DOI: 10.1016/0031-9384(76)90279-1
Holmes, W.G. (1990) Parent–offspring recognition in mammals: a proximate and ultimate perspective. In: Krasnegor, N.A. and Bridges, R.S. (eds) *Mammalian Parenting Biochemichal, Neurobiological, and Behavioral Determinants*. Oxford University Press, New York, pp. 441–460.
Holmes, W.G. (1997) Temporal aspects in the development of Belding's ground squirrels' litter-mate preferences. *Animal Behaviour* 53, 1323–1336. DOI: 10.1006/anbe.1996.0367
Horrell, I. (1982) Immediate behavioural consequences of fostering 1-week-old piglets. *The Journal of Agricultural Science, Cambridge* 99, 329–336.
Horrell, I. and Bennett, J. (1981) Disruption of teat preferences and retardation of growth following cross-fostering of 1-week-old pigs. *Animal Production* 33, 99–106.
Horrell, I. and Hodgson, J. (1992) The bases of sow–piglet identification. 2. Cues used by piglets to identify their dam and home pen. *Applied Animal Behaviour Science* 33, 329–343. DOI: 10.1016/S0168-1591(05)80070-X
Hudson, R. (1985) Do newborn rabbits learn the odor stimuli releasing nipple-search behavior? *Developmental Psychobiology* 18, 575–585. DOI: 10.1002/dev.420180612
Hudson, R. and Distel, H. (1983) Nipple location by newborn rabbits: behavioural evidence for pheromonal guidance. *Behaviour* 85, 260–275.
Hudson, R. and Distel, H. (1986) Pheromonal release of suckling in rabbits does not depend on the vomeronasal organ. *Physiology and Behavior* 37, 123–128. DOI: 10.1016/0031-9384(86)90394-X
Hudson, R., Labra-Cardero, D. and Mendoza-Soylovna, A. (2002) Sucking, not milk, is important for the rapid learning of nipple-search odors in newborn rabbits. *Developmental Psychobiology* 41, 226–235. DOI: 10.1002/dev.10073
Hudson, S.J. and Mullord, M.M. (1977) Investigations on maternal bonding in dairy cattle. *Applied Animal Ethology* 3, 271–276.
Jeppesen, L.E. (1982) Teat-order in groups of piglets reared on an artificial sow. II. Maintenance of teat-order with some evidence for the use of odour cues. *Applied Animal Ethology* 8, 347–355.
Jirik-Babb, P., Manaker, S., Tucker, A.M. and Hofer, M.A. (1984) The role of the accessory and main olfactory systems in maternal behavior of the primiparous rat. *Behavioral and Neural Biology* 40, 170–178. DOI: 10.1016/S0163-1047(84)90267-X
Keil, W., von Stralendorff, F. and Hudson, R. (1990) A behavioral bioassay for analysis of rabbit nipple-search pheromone. *Physiology and Behavior* 47, 525–529. DOI: 10.1016/0031-9384(90)90120-S
Kindermann, U., Hudson, R. and Distel, H. (1994) Learning of suckling odors by newborn rabbits declines with age and suckling experience. *Developmental Psychobiology* 27, 111–122. DOI: 10.1002/dev.420270205
Kinsley, C.H. and Bridges, R.S. (1990) Morphine treatment and reproductive condition alter olfactory preferences for pup and adult male odors in female rats. *Developmental Psychobiology* 23, 331–347. DOI: 10.1002/dev.420230405
Klopfer, P.H., Adams, D.K. and Klopfer, M.S. (1964) Maternal imprinting in goats. *Proceedings of the National Academy of Sciences USA* 52, 911–914.
Kolunie, J.M. and Stern, J.M. (1995) Maternal aggression in rats: effects of olfactory bulbectomy, $ZnSO_4$-induced anosmia, and vomeronasal organ removal. *Hormones and Behavior* 29, 492–518. DOI: 10.1006/hbeh.1995.1285
Korhonen, H., Marnila, P. and Gill, H.S. (2000a) Bovine milk antibodies for health. *British Journal of Nutrition* 84 Suppl. 1, S135–S146.
Korhonen, H., Marnila, P. and Gill, H.S. (2000b) Milk immunoglobulins and complement factors. *British Journal of Nutrition* 84 Suppl. 1, S75–S80.
Kristal, M.B. (1980) Placentophagia: a biobehavioral enigma (or *De gustibus non disputandum est*). *Neuroscience and Biobehavioral Reviews* 4, 141–150. DOI: 10.1016/0149-7634(80)90012-3
Kristal, M.B. and Graber, G.C. (1976) Placentophagia in nonpregnant rats: influence of estrous cycle stage and birthplace. *Physiology and Behavior* 17, 599–605. DOI: 10.1016/0031-9384(76)90158-X
Leon, M. (1974) Maternal pheromone. *Physiology and Behavior* 13, 441–453. DOI: 10.1016/0031-9384(74)90098-5
Leon, M. (1983) Chemical communication in mother–young interactions. In: Vandenbergh, J. (ed.) *Pheromones and Reproduction in Mammals*. Academic Press, Orlando, Florida, pp. 39–69.
Leon, M. and Moltz, H. (1971) Maternal pheromone: discrimination by pre-weanling albino rats. *Physiology and Behavior* 7, 265–267. DOI: 10.1016/0031-9384(71)90296-4
Leon, M. and Moltz, H. (1972) The development of the pheromonal bond in the albino rat. *Physiology and Behavior* 8, 683–686. DOI: 10.1016/0031-9384(72)90095-9
Leonard, B.E. and Tuite, M. (1981) Anatomical, physiological, and behavioral aspects of olfactory bulbectomy in the rat. *International Review of Neurobiology* 22, 251–286.
Lepri, J., Wysocki, C. and Vandenbergh, J. (1985) Mouse vomeronasal organ: effects on chemosignal production and maternal behavior. *Physiology and Behavior* 35, 809–814. DOI: 10.1016/0031-9384(85)90416-0

Lévy, F. and Keller, M. (2009) Olfactory mediation of maternal behavior in selected mammalian species. *Behavioural Brain Research* 200, 336–345. DOI: 10.1016/j.bbr.2008.12.017

Lévy, F. and Poindron, P. (1984) Influence du liquide amniotique sur la manifestation du comportement maternel chez la brebis parturiente [The influence of amniotic fluid on maternal behavior in parturient sheep]. *Biology and Behavior* 9, 65–88.

Lévy, F. and Poindron, P. (1987) The importance of amniotic fluids for the establishment of maternal behavior in experienced and non-experienced ewes. *Animal Behaviour* 35, 1188–1192.

Lévy, F., Poindron, P. and Le Neindre, P. (1983) Attraction and repulsion by amniotic fluids and their olfactory control in the ewe around parturition. *Physiology and Behavior* 31, 687–692. DOI: 10.1016/S0031-9384(83)80004-3

Lévy, F., Gervais, R., Kindermann, U., Litterio, M., Poindron, P. and Porter, R. (1991) Effects of early post-partum separation on maintenance of maternal responsiveness and selectivity in parturient ewes. *Applied Animal Behaviour Science* 31, 101–110.

Lévy, F., Locatelli, A., Piketty, V., Tillet, Y. and Poindron, P. (1995) Involvement of the main but not the accessory olfactory system in maternal behavior of primiparous and multiparous ewes. *Physiology and Behavior* 57, 97–104. DOI: 10.1016/0031-9384(94)00200-O

Lévy, F., Kendrick, K., Keverne, E.B., Porter, R.H. and Romeyer, A. (1996) Physiological, sensory and experiential factors of parental care in sheep. *Advances in the Study of Behavior* 25, 385–473.

Lévy, F., Keller, M. and Poindron, P. (2004) Olfactory regulation of maternal behavior in mammals. *Hormones and Behavior* 46, 284–302. DOI: 10.1016/j.yhbeh.2004.02.005

Logan, D.W., Brunet, L.J., Webb, W.R., Cutforth,T., Ngai, J. and Stowers, L. (2012) Learned recognition of maternal signature odors mediates the first suckling episode in mice. *Current Biology* 22, 1998–2007. DOI: 10.1016/j.cub.2012.08.041

Luo, M., Fee, M.S. and Katz, L.C. (2003) Encoding pheromonal signals in the accessory olfactory bulb of behaving mice. *Science* 299, 1196–1201.

Maletinska, J., Spinka, M., Vichova, J. and Stehulova, I. (2002) Individual recognition of piglets by sows in the early post-partum period. *Behaviour* 139, 975–991. DOI: 10.1163/156853902320387927

Marques, D.M. (1979) Roles of the main olfactory and vomeronasal systems in the response of the female hamster to young. *Behavioral and Neural Biology* 26, 311–329. DOI: 10.1016/S0163-1047(79)91300-1

Marques, D.M. and Valenstein, E.S. (1976) Another hamster paradox: more males carry pups and fewer kill and cannibalize young than do females. *Journal of Comparative and Physiological Psychology* 90, 653–657.

Mateo, J.M. (2006) Development of individually distinct recognition cues. *Developmental Psychobiology* 48, 508–519. DOI: 10.1002/dev.20156

Meese, G.B. and Baldwin, B.A. (1975) Effects of olfactory bulb ablation on maternal behaviour in sows. *Applied Animal Ethology* 1, 379–386.

Melo, A.I. and González-Mariscal, G. (2003) Placentophagia in rabbits: incidence across the reproductive cycle. *Developmental Psychobiology* 43, 37–43. DOI: 10.1002/dev.10118

Mirza, S.N. and Provenza, F.D. (1990) Preference of the mother affects selection and avoidance of foods by lambs differing in age. *Applied Animal Behaviour Science* 28, 255–263. DOI: 10.1016/0168-1591(90)90104-L

Moore, C.L. (1981) An olfactory basis for maternal discrimination of sex of offspring in rats (*Rattus norvegicus*). *Animal Behaviour* 29, 383–386. DOI: 10.1016/S0003-3472(81)80097-8

Moore, C.L. (1985) Sex differences in urinary odors produced by young laboratory rats (*Rattus norvegicus*). *Journal of Comparative Psychology* 99, 336–341.

Morgan, M., Pack, R.J. and Howe, A. (1975) Nerve-endings in rat carotid-body. *Cell and Tissue Research* 157, 255–272.

Morrow-Tesch, J. and McGlone, J.J. (1990) Sensory systems and nipple attachment behavior in neonatal pigs. *Physiology and Behavior* 47, 1–4. DOI: 10.1016/0031-9384(90)90034-2

Mykytowycz, R. (1985) Odours signals in the life of the wild rabbit. *Australian Science Magazine* 1, 44–50.

Mykytowycz, R. and Dudzinski, M.L. (1972) Aggressive and protective behaviour of adult rabbits *Oryctolagus cuniculus* (L.) towards juveniles. *Behaviour* 43, 97–119.

Nowak, R. (1991) Senses involved in the discrimination of Merino ewes at close contact and from a distance by their newborn lambs. *Animal Behaviour* 42, 357–366. DOI: 10.1016/S0003-3472(05)80035-1

Oostindjer, M., Bolhuis, J.E., van den Brand, H., Roura, E. and Kemp, B. (2010) Prenatal flavor exposure affects growth, health and behavior of newly weaned piglets. *Physiology and Behavior* 99, 579–586. DOI: 10.1016/j.physbeh.2010.01.031

Oostindjer, M., Bolhuis, J.E., Mendl, M., Held, S., van den Brand, H. and Kempa, B. (2011) Learning how to eat like a pig: effectiveness of mechanisms for vertical social learning in piglets. *Animal Behaviour* 82, 503–511. DOI: 10.1016/j.anbehav.2011.05.031

Parfet, K.A.R. and Gonyou, H.W. (1991) Attraction of newborn piglets to auditory, visual, olfactory and tactile stimuli. *Journal of Animal Science* 69, 125–133.

Patris, B., Perrier, G., Schaal, B. and Coureaud, G. (2008) Early development of filial preferences in the rabbit: implications of nursing- and pheromone-induced odour learning? *Animal Behaviour* 76, 305–314. DOI: 10.1016/j.anbehav.2008.01.017

Pedersen, C.A. and Blass, E. (1981) Olfactory control over suckling in albino rats. In: Aslin, R.N., Alberts, J.R. and

Petersen, M.R. (eds) *The Development of Perception: Psychobiological Perspectives*. Academic Press, New York, pp. 359–381.:

Pitcher, B.J., Harcourt, R.G. and Charrier, I. (2010) Rapid onset of maternal vocal recognition in a colonially breeding mammal, the Australian sea lion. *PLoS ONE* 5, e12195. DOI: 10.1371/journal.pone.0012195

Poindron, P. and Le Neindre, P. (1980) Endocrine and sensory regulation of maternal behavior in the ewe. In: Rosenblatt, J.S., Hinde, R.A. and Beer, C. (eds) *Advances in the Study of Behavior*, Vol. 11. Academic Press, New York, pp. 75–119.

Poindron, P., Lévy, F. and Krehbiel, D. (1988) Genital, olfactory, and endocrine interactions in the development of maternal behaviour in the parturient ewe. *Psychoneuroendocrinology* 13, 99–125. DOI: 10.1016/0306-4530(88)90009-1

Poindron, P., Lévy, F. and Keller, M. (2007) Maternal responsiveness and maternal selectivity in domestic sheep and goats: the two facets of maternal attachment. *Developmental Psychobiology* 49, 54–70. DOI: 10.1002/dev.20192

Porter, R.H. and Doane, H.M. (1977) Dietary-dependent cross-species similarities in maternal chemical cues. *Physiology and Behavior* 19, 129–131. DOI: 10.1016/0031-9384(77)90169-X

Porter, R.H., Fullerton, C. and Berryman, J.C. (1973) Guinea-pig maternal-young attachment behaviour. *Zeitschrift Fur Tierpsychologie* 32, 489–495.

Porter, R.H., Lévy, F., Poindron, P., Litterio, M., Schaal, B. and Beyer, C. (1991) Individual olfactory signatures as major determinants of early maternal discrimination in sheep. *Developmental Psychobiology* 24, 151–158. DOI: 10.1002/dev.420240302

Price, E.O., Hutson, G.D., Price, M.I. and Borgwardt, R. (1994) Fostering in swine as affected by age of offspring. *Journal of Animal Science* 72, 1697–1701.

Richards, M.P. (1966) Maternal behaviour in the golden hamster: responsiveness to young in virgin, pregnant, and lactating females. *Animal Behaviour* 14, 310–313.

Risser, J.M. and Slotnick, B.M. (1987) Nipple attachment and survival in neonatal olfactory bulbectomized rats. *Physiology and Behavior* 40, 545–549. DOI: 10.1016/0031-9384(87)90042-4

Romeyer, A., Porter, R.H., Lévy, F., Nowak, R., Orgeur, P. and Poindron, P. (1993a) Maternal labelling is not necessary for the establishment of discrimination between kids by recently parturient goats. *Animal Behaviour* 46, 705–712. DOI: 10.1006/anbe.1993.1247

Romeyer, A., Porter, R.H., Poindron, P., Orgeur, P., Chesné, P. and Poulain, N. (1993b) Recognition of dizygotic and monozygotic twin lambs by ewes. *Behaviour* 127, 119–139. DOI: 10.1163/156853993X00452

Romeyer, A., Poindron, P. and Orgeur, P. (1994) Olfaction mediates the establishment of selective bonding in goats. *Physiology and Behavior* 56, 693–700. DOI: 10.1016/0031-9384(94)90229-1

Rosenblatt, J.S. and Lehrman, D.S. (1963) Maternal behavior in the laboratory rat. In: Rheingold, H.L. (ed.) *Maternal Behavior in Mammals*. John Wiley & Sons, Inc., New York, pp. 8–57.

Rosenblatt J.S., Siegel, H.I. and Mayer, A.D. (1979) Progress in the study of maternal behavior in the rat: hormonal, nonhormonal, sensory, and developmental aspects. *Advances in the Study of Behavior* 10, 225–311.

Rowell, T.E. (1960) On the retrieving of young and other behaviour in lactating golden hamsters. *Proceedings of the Zoological Society of London* 135, 265–282.

Saint-Dizier, H., Lévy, F. and Ferreira, G. (2007) Influence of the mother in the development of flavored-food preference in lambs. *Developmental Psychobiology* 49, 98–106. DOI: 10.1002/dev.20197

Schaal, B., Orgeur, P. and Arnould, C. (1995) Olfactory preferences in newborn lambs: possible influence of prenatal experience. *Behaviour* 132, 5–6. DOI: 10.1163/156853995X00603

Schaal, B., Coureaud, G., Langlois, D., Ginies, C., Semon, E. and Perrier, G. (2003) Chemical and behavioural characterization of the rabbit mammary pheromone. *Nature* 424, 68–72. DOI: 10.1038/nature01739

Schneider, N.Y., Fletcher, T.P., Shaw, G. and Renfree, M.B. (2009) The olfactory system of the tammar wallaby is developed at birth and directs the neonate to its mother's pouch odours. *Reproduction* 138, 849–857. DOI: 10.1530/REP-09-0145

Schneirla, T.C., Rosenblatt, J.S. and Tobach, E. (1963) Maternal behavior in the cat. In: Rheingold, H.L. (ed.) *Maternal Behavior in Mammals*. John Wiley & Sons, Inc., New York, pp. 122–168.

Schwartz, E. and Rowe, F.A. (1976) Olfactory bulbectomy: influences on maternal behavior in primiparous and multiparous rats. *Physiology and Behavior* 17, 876–883.

Seegal, R.F. and Denenberg, V.H. (1974) Maternal experience prevents pup-killing in mice induced by peripheral anosmia. *Physiology and Behavior* 13, 339–341.

Serra, J. and Nowak, R. (2008) Olfactory preference for own mother and litter in 1-day-old rabbits and its impairment by thermotaxis. *Developmental Psychobiology* 50, 542–553. DOI: 10.1002/dev.20327

Sevelinges, Y., Lévy, F., Mouly, A.M. and Ferreira, G. (2009) Rearing with artificially scented mothers attenuates conditioned odor aversion in adulthood but not its amygdala dependency. *Behavioural Brain Research* 198, 313–320. DOI: 10.1016/j.bbr.2008.11.003

Singh, P.J. and Tobach, E. (1975) Olfactory bulbectomy and nursing behavior in rat pups (Wistar DAB). *Developmental Psychobiology* 8, 151–164. DOI: 10.1002/dev.420080207

Singh, P.J., Tucker, A.M. and Hofer, M.A. (1976) Effects of nasal $ZnSO_4$ irrigation and olfactory bulbectomy on rat pups. *Physiology and Behavior* 17, 373–382. DOI: 10.1016/0031-9384(76)90094-9

Smith, F.V., Van-Toller, C. and Boyes, T. (1966) The 'critical period' in the attachment of lambs and ewes. *Animal Behaviour* 14, 120–125.

Swanson, L.J. and Campbell, C.S. (1979) Maternal behavior in the primiparous and multiparous golden hamster. *Zeitschrift für Tierpsychologie* 50, 96–104.

Teicher, M.H. and Blass, E.M. (1976) Suckling in newborn rats: eliminated by nipple lavage, reinstated by pup saliva. *Science* 193, 422–425. DOI: 10.1126/science.935878

Terrazas, A., Nowak, R., Serafin, N., Ferreira, A., Lévy, F. and Poindron, P. (2002) Twenty-four-hour-old lambs rely more on maternal behavior than on the learning of individual characteristics to discriminate between their own and an alien mother. *Developmental Psychobiology* 40, 408–418. DOI: 10.1002/dev.10041

Terry, L.M. and Johanson, I.B. (1987) Olfactory influences on the ingestive behavior of infant rats. *Developmental Psychobiology* 20, 313–331. DOI: 10.1002/dev.420200309

Thorhallsdottir, A.G., Provenza, F.D. and Balph, D.F. (1990) Ability of lambs to learn about novel foods while observing or participating with social models. *Applied Animal Behaviour Science* 25, 25–33. DOI: 10.1016/0168-1591(90)90066-M

Val-Laillet, D. and Nowak, R. (2008) Early discrimination of the mother by rabbit pups. *Applied Animal Behaviour Science* 111, 173–182. DOI: 10.1016/j.applanim.2007.06.003

Vince, M.A. (1993) Newborn lambs and their dams: the interaction that leads to sucking. *Advances in the Study of Behavior* 22, 239–268. DOI: 10.1016/S0065-3454(08)60408-8

Vince, M.A. and Ward, T.M. (1984) The responsiveness of newly-born Clun Forest lambs to odour sources in the ewe. *Behaviour* 89, 117–127. DOI: 10.1163/156853984X00065

Vince, M.A., Lynch, J.J., Green, G.C. and Elwin, R.L. (1987) Interactions between normal ewes and newly born lambs deprived of visual, olfactory and tactile sensory information. *Applied Animal Ethology* 19, 119–136. DOI: 10.1016/0168-1591(87)90209-7

Viviers, M.Z., Burger, B.V., le Roux, N.J., Morris, J. and le Roux, M. (2014) Temporal changes in the neonatal recognition cue of Dohne Merino lambs (*Ovis aries*). *Chemical Senses* 39, 249–262. DOI: 10.1093/chemse/bjt075

Wallace, P., Owen, K. and Thiessen, D.D. (1973) The control and function of maternal scent marking the Mongolian gerbil. *Physiology and Behavior* 10, 463–466.

Wang, Z. and Storm, D.R. (2011) Maternal behavior is impaired in female mice lacking type 3 adenylyl cyclase. *Neuropsychopharmacology* 36, 772–781. DOI: 10.1038/npp.2010.211

Weiss, J., Pyrski, M., Jacobi, E., Bufe, B., Willnecker, V., Schick, B., Zizzari, P., Gossage, S.J., Greer, C.A., Leinders-Zufall, T., Woods, C.G., Wood, J.N. and Zufall, F. (2011) Loss-of-function mutations in sodium channel $Na_V1.7$ cause anosmia. *Nature* 472, 186–190. DOI: 10.1038/nature09975

Wyatt, T.D. (2014) *Pheromones and Animal Behavior: Chemical Signals and Signatures*, 2nd edn. Cambridge University Press, Cambridge, UK.

Yamazaki, K., Beauchamp, G.K., Imai, Y., Bard, J. and Boyse, E.A. (1992) Expression of urinary H-2 odortypes by infant mice. *Proceedings of the National Academy of Sciences USA* 89, 2756–2758. DOI: 10.1073/pnas.89.7.2756

Yamazaki, K., Beauchamp, G.K., Curran, M., Bard, J. and Boyse, E.A. (2000) Parent–progeny recognition as a function of MHC odortype identity. *Proceedings of the National Academy of Sciences USA* 97, 10500–10502. DOI: 10.1073/pnas.180320997

9 The Role of Olfaction in Disease Detection and Prevention

Tadeusz Jezierski

Institute of Genetics and Animal Breeding, Polish Academy of Sciences, Department of Animal Behaviour, Jastrzebiec, Poland

For almost all diseases an early diagnosis increases the chance for successful treatment. There is therefore a constant demand for reliable and easily available detection and diagnostic methods. The rapid development and availability of diagnostic and therapeutic tools, employing advanced technologies, has led to contemporary medicine turning away from previously used simple methods based on practical experiences of generations of physicians and therapists. However, modern high-tech diagnostic equipment is often expensive, requires highly qualified specialist staff and is stationary, i.e. located at hospitals or laboratories, thus requiring patients to go to a clinic or to deliver biological material, such as urine samples for analyses. Many diagnostic methods and techniques are also time-consuming and not readily adapted to a high daily throughput. Perhaps more importantly, they are less likely to be available in rural environments or developing countries. Many diagnostic tools are often used only when symptoms are detectable and the patient is feeling ill. Additionally, for some diseases sampling of biological material for diagnostic purposes is often invasive (e.g. biopsy), with sampling or examination procedures involving pain, discomfort or undesirable side effects (e.g. colonoscopy or contrasts used for computerized tomography (CT) scanning or X-ray). This, in turn, may lead to some patients avoiding medical examinations. As a consequence, many modern diagnostic methods, although available, are not applicable for screening large human populations to pick up cases where a disease is suspected, in order to direct the client to a more detailed examination for an ultimate diagnosis. This is where the smell of certain diseases can play a role.

Detecting human disease by odour has a long history. The first published report of detecting human disease by odour appeared in the 3rd century BC in a Chinese medical text (McCulloch *et al.*, 2012). An earlier writing by Hippocrates (ca. 460 BC to ca. 370 BC) included a more general mention of changes in the odour of febrile patients (McCulloch *et al.*, 2012). In the past, mainly the human sense of smell was employed for the detection of disease-specific odours, and the odours were named according to how they were perceived by humans as for example *rotten apple-like* or *sweetish*. Fast-forward to modern times, where chemical analytical methods for identification of disease by detection of volatile organic compounds (VOCs) are used objectively and based on acknowledged scientific studies. In contrast, human olfaction can be subjective depending on individual skills and practical experience or training (see also Chapter 5). In order to use odours for diagnostic purposes, it is necessary not only to be able to smell a disease, but also to understand the pathophysiological mechanisms and stages of disease-specific VOCs. Some diseases are accompanied by unspecific symptoms that are often ignored or overlooked by a patient; specific odours accompanying a disease may therefore not only be used for early diagnosis, but also for better understanding of pathogenesis and for monitoring the therapy process.

The main advantages of employing olfaction for disease detection are non-invasiveness, simplicity and low cost. These advantages apply both when olfaction is used as a real-time procedure by directly smelling the patient, and when used remotely by smelling odour samples that can be stored without special conditions. The former may yield immediate availability of results in terms of olfactory impressions that could be easily interpreted. For the latter, the patient does not need to be present during olfactory investigation, which can be performed with some time delay. Human noses can be used to perform olfactory tests, based on spontaneous odour detection or after special training to enhance the detection ability. In the past few decades, different animal species, either behaving spontaneously or trained using operant conditioning, have been found capable of detecting diseases by odours. Olfaction as a disease screening or diagnostic tool is, however, not always possible, i.e. not all diseases have a specific odour, and for those that do, diagnoses often need to be supported and confirmed by other, more technologically and medically advanced methods. Also, odour detection has a degree of variability, sometimes with insufficient detection sensitivity and specificity. The human body emits hundreds of different VOCs via exhaled breath, sweat or urine, and these odours vary not only depending on conditions related to a disease, but also due to other endogenous factors such as diet and medication, as well as exogenous factors like cosmetics and body hygiene. In this chapter, I will give some illustrative examples of the scope and limitations in the use of odours as diagnostic biomarkers.

Why Do Some Diseases Smell?

Shirasu and Touhara (2011) in their review argue that diseases result in a change in body odour, and that detection of VOCs present in breath, blood, urine, sweat, skin, sputum, faeces, sebum (oily skin secretions) or vaginal discharges can be used as diagnostic olfactory biomarkers of infectious, metabolic, genetic and other kinds of diseases. Various pathological processes involve production of either new VOCs that were not present in healthy individuals, or changes in the proportion of particular VOCs. Such alterations in the VOC profile may be a consequence of oxidative stress in the cells of a body organ, of changes in hormonal or enzyme status, or a result of bacterial activity both of endogenous and exogenous origin. The most detailed hypotheses on how disease-specific odour may be produced concern the origin of hypothetical cancer odour markers. However, the metabolic origin of tumour-associated VOCs remains speculative (Sponring *et al.*, 2009). Williams and Pembroke (1989) suggested that aberrant protein synthesis associated with the growth of melanomas produced a melanoma-specific odour that the dog responded to in their study. Balseiro and Correia (2006) hypothesized that cancer odour is a product of alterations in the major histocompatibility complex (MHC). As to the cancer odour in breath, Phillips *et al.* (1999) proposed an explanation assuming that oxygen free-radical activity increases in cancerous cells. This would degradate cell membranes by lipid peroxidation and convert polyunsaturated fatty acids to volatile alkanes, which are considered to be the main odorous cancer marker in exhaled breath. However, it is clear that different diseases give rise to different odour profiles, and that their origins are still not fully understood.

Human Diseases Detectable by a Human Nose

Although the ability of humans to diagnose disease by smell only rarely has been the subject of comparative and quantitative studies, Bijland *et al.* (2013) published a literature survey of key scent detection studies using human, animal and electronic noses for detection of different types of cancer and infectious diseases. Comparing published studies they found that using different kinds of odour samples (breath, sputum, urine, faeces and tissue), different types of cancer and infectious diseases were detected by dogs and electronic noses with varying sensitivity, mostly around 60–85%, but ranging from 40–100%. In their review on the scent of disease Shirasu and Touhara (2011) indicate that quite different kinds of diseases can be related to changes in body odour (Table 9.1), including bacterial infections, inherited metabolic abnormalities, toxin ingestions, different kinds of cancer and even mental illnesses. Although for the diseases listed in Table 9.1 the characteristic odour in terms of human olfactory impression has been identified, the molecule(s) at the origin of this odour have been identified only for some diseases.

Although often possible, bacterial infections, inherited metabolic abnormalities, toxin ingestion and mental illnesses are not routinely identified using characteristic odours and olfaction as diagnostic tools. In contrast, odour markers of different types of cancer are now widely used (see section on 'Cancer in humans' below). Also, rapid changes in odours emanating from patients during unpredictable episodes of some metabolic diseases, e.g. hypoglycaemia in patients suffering from diabetes type 1, and in connection with some disorders of the central nervous system leading to seizures, are now being scrutinized (see section on 'Seizure-alerting dogs' below). In 2015, a woman from Perth in Scotland, Joy Milne, was found to be able to detect Parkinson's disease based on odours alone. It should, however, be

Table 9.1. Examples of odours emitted by humans (from the body or from faeces and breath) as detected by the human sense of smell, and which may indicate diseases. (Modified from Shirasu and Touhara, 2011.)

	Odour source		
	Faeces	**Breath**	**Body/Sweat**
Odour type/description:			
Sweetish	**Cholera** *Garner et al. (2009)* *Probert et al. (2009)*	**Leukaemia** *Blackburn (1957)*	**Diphteria** *Pavlou & Turner (2000)* **Smallpox** *Pavlou & Turner (2000)*
Foul/malodour		**Scarlet fever** *Honig et al. (2003)* **Pneumonia** *Liddell (1976)* **Tuberculosis** *Turck (1985)* **Liver disease** *Chen et al. (1970)*	**Scarlet fever** *Honig et al. (2003)*
Rotting/putrid **Rotten apple/acetone-like**		**Diabetes/diabetic ketoacidosis** *Manolis (1983)* *Liebich (1983)*	**Scurvy** **Ozena** *Liddell (1976)*
Butcher's shop			**Yellow fever** *Liddell (1976)*
Urine		**Uraemia/kidney failure** *Simenhoff et al. (1977)*	
Cheesy/acrid/sweaty feet			**Isovaleric acidemia** *Burke et al. (1983)* *Vockley & Ensenauer (2006)*

stressed that detection of any disease by characteristic odour alone should not be regarded as diagnosis. Like many other symptoms, odour can only provide a suspicion of a disease, whereas a full diagnosis can only be made using traditional medical methods.

Animal Diseases Associated with Odour

Changes in smells emanating from animals, in particular the presence of malodours, can, if associated with a disease, be a useful cue for a farmer to recognize first symptoms in a herd and for a veterinarian as guidance for a subsequent diagnosis. There are several diseases that are usually accompanied by malodours. For example, inflammation of the external ear canal (otitis externa) in dogs is characterized among other symptoms by a discharge with very bad-smelling odour. The nature of the odour and discharge depends to some extent on what the causative agent is, and experienced veterinarians have been known to distinguish bacterial from fungal otitis externa on the basis of their characteristic odour. In dogs, similar to humans, some diseases like kidney failure and diabetes can cause a change in breath odour. Very foul breath in dogs (and humans) may indicate dental disease or gum infection. Badly smelling odours around particular body parts of animals usually indicate infections, inflammation and/or purulent processes, such as anal gland/sac infection in dogs, endometritis in cows (Burfeind *et al.*, 2014) and hoof canker in horses (Neto *et al.*, 2001). Dobbelaar *et al.* (1996) showed in dairy cows that ketosis, which is a metabolic disorder and risk factor for other diseases, can be detected by measuring acetone concentrations in the exhaled breath of the cow. This could provide a simple method for detection of subclinical ketosis enabling preventative measures to be taken before the onset of the clinical stage.

Recognition of Disease in Conspecifics

Animals generally possess the ability to detect sick conspecifics, in part by means of odours emitted by these individuals, but possibly also by changes in their behaviour or body posture. It has been well documented in the literature that mice can distinguish between sick and healthy individuals of their own species by odour. This has been shown for parasite infections (Kavaliers and Colwell, 1995; Ehman and Scott, 2002; Kavaliers *et al.*, 2003), influenza (Penn *et al.*, 1996), mouse mammary tumour virus (Yamazaki *et al.*, 2002) and inflammatory processes (Arakawa *et al.*, 2010). Social communication between animals is influenced by the animals' ability to detect and recognize smells, including those carrying specific information on the health status of conspecifics. Odour from sick conspecifics usually evokes an avoidance response in odour-recipient individuals. For example, female meadow voles have been found to distinguish between bedding from parasitized and unparasitized conspecific males, and preferred to spend time with bedding from unparasitized males (Klein *et al.*, 1999). The ability of animals to recognize odour cues associated with infection and sickness has developed during the course of evolution as it reduces the risk of contagion and potential danger (Thomas *et al.*, 2005; Arakawa *et al.*, 2009). Trained mice have been shown capable of distinguishing urinary odours from conspecifics with and without experimentally induced lung cancer tumours, and the presence of certain volatile odorants is sufficient to identify tumour-bearing mice (Matsumura *et al.*, 2010).

Recognition of Human Diseases by Animals

Cancer in humans

In the past decade, case reports and experiments using canine olfaction to distinguish patients suffering from different forms of cancer from healthy controls have led to more focus on using odours as a diagnostic tool for cancer. The importance of using olfaction as an aid in cancer diagnosis is enhanced by some types of cancer, e.g. lung cancer, being accompanied by no early symptoms or by symptoms that are unspecific and easily overlooked or ignored. Many cases of cancer are diagnosed in an advanced stage, too late for a successful treatment. Thus, a crucial role for a successful cancer therapy is played by rapid and reliable screening methods, which would enable non-invasive examination of larger human populations on a regular basis in order to identify those who are not aware of having the early stages of cancer. Screening results based on odour detection could be used to indicate potential cancer suspicions and the patient could be directed to a more detailed and invasive medical examination, involving histopathology, to confirm or revoke the diagnosis.

Shirasu and Touhara (2011) indicate that a characteristic odour of breast and head-and-neck cancers can be identified only in advanced stages by human sense of smell; however, no data could be found in the scientific literature on practical sensitivity and specificity of cancer detection by human noses. According to Liddell (1976) and Shirasu *et al.* (2009) the affected area of advanced breast cancer emits a rotting odour of dimethyl trisulfide and fatty acids. However, breath samples of patients with breast cancer contain 2-propanol, 2,3-dihydro-1-phenyl-4 (1H)-quinazolinone, 1-phenyl-ethanone, heptanal and isopropyl myristate (Phillips *et al.*, 2006), but these odorants are not perceived by humans (Shirasu and Touhara, 2011). Also, alkanes and alkane derivatives found in the breath of patients suffering from lung cancer (Phillips *et al.*, 1999, 2007a) are according to Shirasu and Touhara (2011) not perceived by the human sense of smell. On the other hand, gynaecological tumours are detectable by the human sense of smell due to a rotting odour related with emission of volatile fatty acids (Dankert *et al.*, 1981).

Detection of cancer on the basis of characteristic odours has gained public attention mainly due to studies using dogs. The first case reports that triggered an interest from the mass media (and among canine scientists on the possibility of using dogs specifically for this purpose) were published in the medical journal *The Lancet* by Williams and Pembroke (1989) and by Church and Williams (2001). These reports described cases of untrained pet dogs that showed spontaneous peculiar behaviours consisting of constant sniffing and trying to bite off a mole on their owner's leg. This made the owner aware of a lesion that was subsequently diagnosed as malignant melanoma. A hypothesis was that dogs are able to detect cancer in humans through an odour signature of the disease. More systematic experimental studies started as a result of a conference initiated in 2003 by Dr John Church, who was one of the authors of the first case reports

Table 9.2. Summary of studies on detection of cancer by trained dogs.

Cancer type	Relative successfulness of detection	Odour source	Detection accuracy		References
			Sensitivity[a] (%)	Specificity[b] (%)	
Melanoma	Most	Tissue scraps	100	Not indicated	Pickel *et al.* (2004)
	Least	Breath	66–80	77–87	Walczak *et al.* (2012)
Lung cancer	Most	Breath	99	99	McCulloch *et al.* (2006)
	Least	Breath	56–64	8–33	Amundsen *et al.* (2014)
Breast cancer	Most	Breath	88	98	McCulloch *et al.* (2006)
	Least	Urine	22	(no better than chance)	Gordon *et al.* (2008)
Prostate cancer	Most	Urine	99	97	Taverna *et al.* (2014, 2015)
	Least	Urine	13–25 (no better than chance)	71–75	Elliker *et al.* (2014)
Bladder cancer		Urine	41		Willis *et al.* (2004)
Ovarian cancer		Tumour	100	95	Horvath *et al.* (2010)
		Blood	100	98	Horvath *et al.* (2010)
Colorectal cancer		Breath	91	99	Sonoda *et al.* (2011)
		Faeces	97	99	Sonoda *et al.* (2011)

[a]Sensitivity indicates the proportion of positives that are correctly identified as such.
[b]Specificity indicates the proportion of negatives that are correctly identified as such.

and an advocate of so-called bio-detection methods in medicine. In the past decade not only melanoma but six other types of cancer have been investigated with respect to detection by trained dogs on the basis of odour signatures (Table 9.2). However, it is difficult to compare directly the results of experimental studies conducted so far on canine detection of cancer, as these studies differ methodically in many aspects; for example, different kinds of odour samples were used for the same kind of cancer (e.g. urine, Gordon *et al.*, 2008; breath, McCulloch *et al.*, 2006).

In contrast to case reports where the dogs behaved spontaneously, the use of specially trained dogs to detect cancer odour in biological samples has to be statistically tested to assess the validity of the method. This can be done either by the dog selecting the target odour sample out of several samples in a line-up, or as a yes/no response to each of the samples sniffed. The probability of the dog correctly indicating the target sample by chance alone should always be considered. There is a great variability in the detection accuracy in published experimental studies on cancer detection by trained canines, ranging from no better than chance to 100% (see Table 9.2). This variability can be attributed at least in part to a lack of methodological standardization; some authors use methodologies that, although based on similar principles, differ in details that could have impact on the results (Jezierski *et al.*, 2015). On the basis of results published so far it is acknowledged that trained dogs can discriminate breath samples of patients with diagnosed cancer disease from those of healthy people at a rate that is better than by chance. Out of several cancer-related odour sources, exhaled breath appears to be the most investigated using chemical analytical methods (e.g. Phillips *et al.*, 1999, 2007a, 2013; Deng *et al.*, 2004; Buszewski *et al.*, 2007; Amann *et al.*, 2014). The first attempts to correlate the results of analyses using gas chromatography coupled with mass spectrometry (GC–MS) of breath samples from cancer patients and healthy donors with those obtained by parallel testing using trained dogs were undertaken by Buszewski *et al.* (2012a,b). According to these authors, out of 18 selected compounds present in human breath the highest positive correlations with dogs' indications were found for 2-pentanone and ethyl acetate ($r = 0.97$ and $r = 0.85$, respectively). In other words, breath samples containing higher amounts of these two compounds were better detected by dogs. In contrast, the content of 1-propanol and propanal in breath samples was negatively correlated with the dogs' indications ($r = -0.98$ and $r = -0.87$, respectively), i.e. the percentage of correct indications by dogs decreased when the content of these two compounds in breath samples was high.

Experimental design is pivotal when trying to assess the capability of dogs to detect cancer and other diseases based on odour. This is neatly demonstrated in the study carried out by Elliker *et al.* (2014), who trained ten dogs to indicate (by placing their nose on) urine samples from patients with prostate cancer. The dogs were trained in a large (60 m^2) arena, and urine samples were presented four at a time in open-top flasks (9 cm deep) underneath four 20-mm scent-holes spaced 75 cm apart. A positive reinforcement schedule was used (clicker training, food reward and praise) when the dogs sat next to the correct scent-hole – initially baited with food rewards. These were subsequently replaced by a urine sample from a prostate cancer patient (CaP) placed at random in one of the scent-holes, whilst the three remaining scent-holes stayed empty. The next training stage involved one CaP and three different non-CaP urine samples. It should be noted that the person present in the room was blinded to the position of the CaP sample, and only rewarded the dog if told to by a person not visible to the dog. This was to prevent any micro-movements of the handler from somehow indicating the correct response to the dog. Due to a limited supply, it was sometimes necessary to present urine samples from the same donors several times during training. Two dogs showed promising ability to discriminate following 5 months of training. They were tested in the same set-up using 15–16 sets of four urine samples, and where one sample in each set was a CaP sample. However, during two test-runs where all of the samples were new and unknown to the dogs, one of the dogs identified only 2/15 and 2/16 of the CaP samples. The other dog, tested once, identified only 4/16 samples. These results indicate that the dogs had not learned to identify specific cancer odours during training, but rather had learned which specific urine samples would be rewarded. This is in itself impressive, as the number of training samples was large, and the authors expected the dogs' olfactory memory to be exceeded (Elliker *et al.*, 2014). The authors do not exclude the possibility that dogs could be trained to detect prostate cancer, but they call for caution and the use of carefully controlled experimental procedures.

Walczak *et al.* (2012) used breath samples (Fig. 9.1), placed in a line-up of five samples, with one of the samples originating from a patient with a diagnosed lung or breast cancer. Significant individual differences were found in the results achieved by the dogs, and the authors concluded that it is difficult to predict the performance of fully trained dogs (Fig. 9.2) on the basis of the results achieved during the training (Walczak *et al.*, 2012).

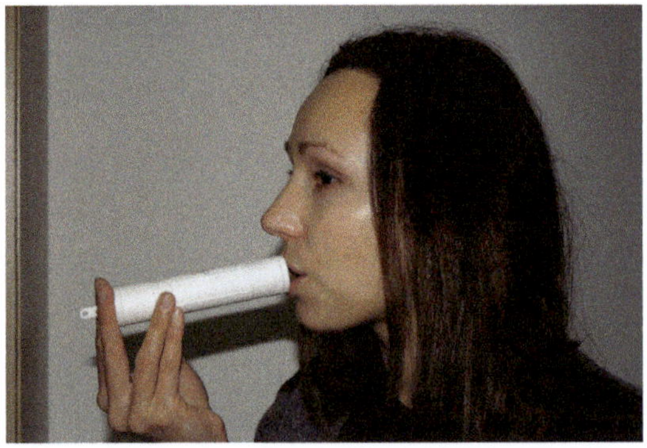

Fig. 9.1. Method for taking breath samples.

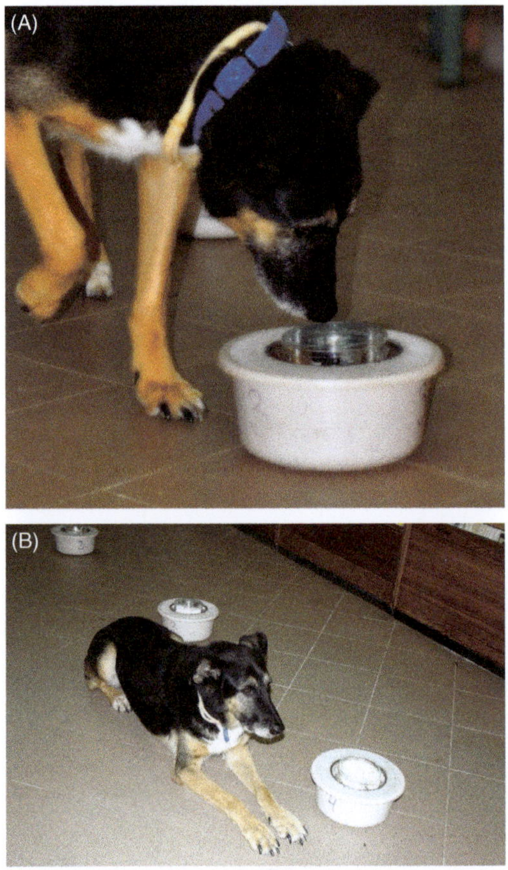

Fig. 9.2. Dog working in a scent line-up: (A) sniffing a sample, and (B) indicating a sample.

It is not only dogs that are used to detect human cancer odour. Nematodes of wild type *Caenorhabditis elegans* display attractive chemotaxis towards human cancer cell secretions, cancer tissues and urine from cancer patients but avoid control urine (Hirotsu *et al.*, 2015). With regard to distinguishing between odours from human cancerous versus healthy cells, the first step in using biological sensors, integrating artificial and biological olfaction was made by Strauch *et al.* (2014). These authors have shown that an array of odorant receptors on the fruit fly's antenna can be recorded by calcium imaging. The response of these receptors when presented with odours from cancerous and healthy cell samples gives rise to distinct odour response patterns on the antenna, and these were comparable across flies.

In addition to the issues raised by Elliker *et al.* (2014), it has still not been shown that dogs or other animals can sniff out cancer odour markers in the *early stages* of the disease, before confirmed medical diagnosis. All the detection parameters achieved so far concern pattern odour samples collected from patients with already diagnosed cancers. Thus, the usefulness of trained dogs for practical cancer screening is still being questioned. The identity of the chemical compound(s) and their qualitative or quantitative combination, which constitute the cancer odour signature to which the sniffer dogs react, are still unknown and the use of canines for cancer screening is still considered to be a black-box technology.

Tuberculosis

Human tuberculosis has in recent years become a recurring health problem in some countries, thus rapid, sensitive, specific, inexpensive and non-invasive detection methods are urgently needed. It has been reported that tuberculosis in humans can be detected by trained rats on the base of sputum odour (Fig. 9.3). The detection sensitivity was found to vary between 72% and 100% and the rate of false alarms varied between 0.7 and 8.1% (Weetjens *et al.*, 2009). Using electronic sensors that employ fuzzy logic and pattern recognition, sputum samples from the patients with tuberculosis can be distinguished from control samples with 89% detection sensitivity and 91% specificity (Pavlou *et al.*, 2004; Fend *et al.*, 2006; Phillips *et al.*, 2007b). A theoretical support for this method was provided by Syhre and Chambers (2008), who identified four specific compounds (methyl phenylacetate, methyl p-anisate, methyl nicotinate and p-phenilanisole) to be distinctive volatile markers of tuberculosis.

Hypoglycaemia

Hypoglycaemia is associated with numerous symptoms triggered by falling glucose levels in the blood, such as weakness, fatigue, apathy, lethargy, ataxia, loss of consciousness, sleep and, in severe cases, coma and death. Occurrence of frequent hypoglycaemia episodes concerns primarily patients with type 1 diabetes who use medications such as insulin to treat the condition. Specific symptoms may vary and are often vague as they can also be produced by other conditions. Patients with recurrently low glucose levels may not be aware of their threshold symptoms without much warning and many measurement methods (especially glucose meters) – as well as being impractical – are imprecise at low levels. Anecdotal evidence that pet dogs are able to detect hypoglycaemia episodes during or before the patient is aware of his/her falling blood sugar levels was confirmed in at least 14 publications published in

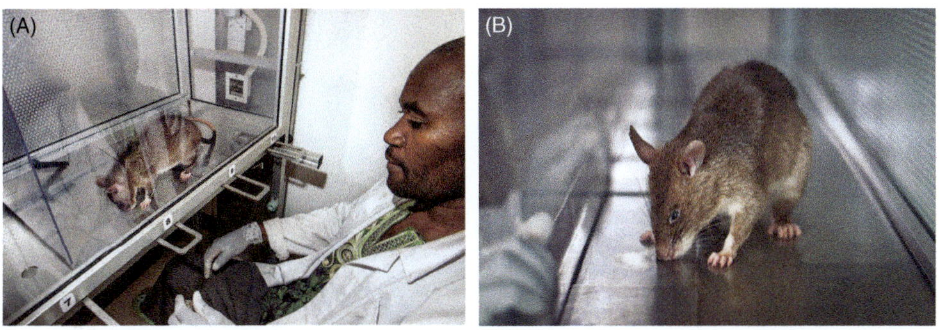

Fig. 9.3. (A) Automated tuberculosis detection cage for rats (© APOPO), with (B) trained rat sniffing a sputum sample (© Maria Anna Caneva Saccardo).

recent years (Weber *et al.*, 2015). In a questionnaire study by Wells *et al.* (2008) including 212 dog owners with medically diagnosed type 1 diabetes, 138 (65%) respondents indicated that their dogs had spontaneously demonstrated a behavioural reaction (vocalizing, 61% of dogs; licking the owner, 49%; starring at owner's face, 41%; nuzzling, 41%; jumping on the owner, 30%) to at least one of their episodes of hypoglycaemia; and 32% of animals reacted to more than 11 events. Moreover, 71 (34%) of respondents claimed that their dogs reacted before the owner was aware of having low blood sugar and 41% of dogs showed behavioural responses approximately at the same time as the owners themselves were aware they were hypoglycaemic. Wells *et al.* (2008) also mention incidents where cat owners suffering from diabetes report that their pets have woken them up from nocturnal episodes of hypoglycaemia. Rooney *et al.* (2013) showed that specially trained hypoglycaemia alert dogs that are placed with clients with diabetes alerted the clients to hypoglycaemia episodes with significant – though variable – accuracy. Hardin *et al.* (2012, 2013, cited by Weber *et al.*, 2015) assessed the sensitivity and specificity of trained alerting dogs to distinguish between euglycaemic and hypoglycaemic odour samples of patients with type 1 diabetes in two randomized, controlled trials, comprising four dogs each. The sensitivity ranged from 22 to 100% in one study and from 22 to 71% in another study, both with a specificity between 71 and 90%.

Most authors of publications on hypoglycaemia alerting dogs (e.g. Wells *et al.*, 2008; Rooney *et al.*, 2013; Weber *et al.*, 2015) emphasize that the physiological basis of how dogs can sense hypoglycaemia is still unclear, but odour cues resulting from changes in the chemical composition of the sweat and/or breath during hypoglycaemia are regarded as the main signal. This assumption was supported by observations that dogs sometimes alerted without visual cues, e.g. being in another room, or during patient's sleep (Chen *et al.*, 2000; O'Connor *et al.*, 2008; Rooney *et al.*, 2013). Additional cues could be used by dogs when alerting to a hypoglycaemic state, for example, slight behavioural alterations in the patient, such as muscle tremor, changes in breath frequency or even energy wave changes in a person's electrical and/or magnetic fields (Chen *et al.*, 2000; O'Connor *et al.*, 2008). The hypothesis that odour cues may be a signal for hypoglycaemia alerting dogs was supported in a paper by Novak *et al.* (2007), who proposed exhaled methyl nitrate as a non-invasive odour marker of hyperglycaemia in type 1 diabetes. More recently it was found that levels of isoprene in the breath rose significantly during hypoglycaemia and, although isoprene is one of the most common molecules in human breath, dogs may be sensitive to its rise (Neupane *et al.*, 2016).

Seizure-alerting dogs

Some dogs are able to alert spontaneously seizures in adults and children (e.g. Strong *et al.*, 1999, 2002; Brown and Strong, 2001; Dalziel *et al.*, 2003; Kirton *et al.*, 2004). According to Kirton *et al.* (2004) about 40% of pet dogs living with epileptic patients demonstrate seizure-specific behaviour and 15% of these show anticipatory ability. Another study showed that out of 29 pet dogs owned by epilepsy patients, only nine dogs responded to seizures and three were reported to alert to seizure onset (Dalziel *et al.*, 2003). Moreover, trained dogs not only provide an overt signal that acts as a useful warning to the human about upcoming seizures, there are also indications that seizure frequency may be reduced by as much as 43% as a consequence (Strong *et al.*, 2002). The mechanisms of reducing seizure frequency by the presence of seizure-alerting dogs remain speculative. One possible explanation is that dog's companionship is reassuring and relaxing, and increases the ability of the patients to predict their seizures, thus reducing some of the fear of the condition, which is beneficial in reducing seizure frequency (Strong *et al.*, 2002). As seen with hypoglycaemia, the seizure-alerting dogs may alert not only to olfactory cues but most likely also to subtle human behaviour changes and even to heart rate changes (Brown and Strong, 2001). No false positive alerts were found by Kirton *et al.* (2004), suggesting that it cannot be excluded that dogs not only alert but may to some extent induce seizure.

Schizophrenia

A 'peculiar smell' in psychiatric hospitals has been reported already in 19th century literature (Shirasu and Touhara, 2011). In 1960, Smith and Sines demonstrated that trained rats can distinguish sweat from patients with schizophrenia from that of healthy controls. Several attempts have been made to identify the compound(s) giving rise to this particular smell in people diagnosed with mental illnesses. Gordon *et al.* (1973) could not confirm an earlier finding by Smith *et al.* (1969) that *trans*-3-methyl-2-hexenoic acid was the only component that differentiated sweat of patients with schizophrenia from healthy controls. Although Phillips *et al.* (1993) found an increased concentration of pentane and carbon disulfide in the breath of patients with schizophrenia, Shirasu and Touhara (2011) maintain that the peculiar smell of psychiatric wards has not yet been elucidated.

Modern Technologies to Detect Disease-related Volatiles

In the past decades, increasingly advanced technologies have been developed and employed to analyse volatiles produced or altered during disease. Since 1990, many modern sampling methods, pre-concentration techniques and types of analytical equipment have been developed and introduced. Breath analysis in particular is thought to have a great potential for medical diagnosis (Amann *et al.*, 2014). Breath samples can be collected non-invasively, as often as required, and analyses can be done in real time, e.g. during physical exertion, sleep, surgery or in an intensive care unit (Amann *et al.*, 2014). Modern chemical analytical methods enable breath-to-breath resolution, which may have great potential not only for disease detection but also for therapeutic monitoring and determination of the phenotype of enzyme activity for personalized medicine.

Since the 1980s, so-called 'electronic noses', devices that electronically mimic the human olfactory system, have been developed and repeatedly improved (Turner and Magan, 2004). Electronic noses consist of several gas sensors that are combined with a pattern recognition system to analyse and characterize sample-derived complex VOCs without separation of the mixture into individual components. Recently, an artificial multiplexed superbioelectronic (MSB) nose has been described that mimics the human olfactory sensory system, showing a high-performance odorant discriminatory ability in mixtures (Kwon *et al.*, 2015). This portable MSB-nose was constructed using highly uniform graphene micropatterns that were conjugated with two different human olfactory receptors.

At present, it is necessary to improve the accuracy of the sensors and their sensitivity to marker VOCs. Once these improvements are made, disease-specific volatile biomarkers can be used more regularly in clinical practice (Pavlou and Turner, 2000; Wilson, 2015). According to Bijland *et al.* (2013) the overall sensitivity and specificity of e-noses are high in the published studies, but e-noses are not widely implemented in daily practice: they are relatively expensive and different types of e-noses with a variety of underlying techniques mean that results from one type of e-nose are difficult to generalize to another.

More recently Van Alstine (2016) analysed how artificial and animal olfactory diagnostics are developed and made to work within the existing framework of health innovations. One of the barriers to successful implementation of olfactory diagnostics is the standardization of methods. In her conclusion, Van Alstine (2016) states that the current paradigm of standardization of health innovations seems to be more accommodating to artificial olfaction development than to animal olfaction, which faces a number of tricky paradoxes and major barriers as a result of a number of contextual limitations, including the involvement of animal learning.

Impairment of Olfaction in Disease

When discussing the relationships between olfaction and disease, it is worth mentioning that diseases not only can change odours emanating from an ill organism, but some diseases may also impact the sense of smell. It is well known that olfactory capacity of rodents declines with age (e.g. Kondo *et al.*, 2010). In humans, the sense of smell is considered to play a minor role compared with other modalities, and olfactory screening is therefore rarely conducted as part of a routine medical evaluation. It is well established that the two main human neurodegenerative disorders, Parkinson's disease and Alzheimer's disease, compromise the olfactory function (e.g. Devanand *et al.*, 2000). Olfactory dysfunction is one of the earliest symptoms of neurodegenerative diseases and includes higher olfactory thresholds, diminished olfactory identification and discrimination as well as odour memory (for a review, see Rahayel *et al.*, 2012; Doty and Kamath, 2014; Ottaviano *et al.*, 2016; White *et al.*, 2016). Other animal models of Alzheimer's disease have been developed in pigs, and one of the symptoms to be used is a reduction in odour sensitivity (Søndergaard *et al.*, 2010; see also Chapter 4).

Conclusions

Recognizing characteristic odours is an ancient diagnostic tool in traditional human and veterinary medicine. For a long time it has been to some extent forgotten or ignored by modern medicine, however, in recent years more and more scientific publications deliver evidence that odours may be important disease symptoms that in some cases may be crucial for early detection of disease. Moreover, studies have shown how VOCs that give rise to the characteristic odour of a disease can be produced by changes in metabolism, oxidative stress or decomposition of cells that accompany a disease. Changes in the odour emanating from an organism may be the first signs of infectious, inherited or acquired metabolic disorders, cancerous diseases and even mental disorders. The advantages of odours as disease indicators are related to the simplicity of detection by the sense of smell, either by humans or by trained or untrained animals. In the case of trained animals, it is important to use stringent and controlled training schedules to ensure reliable detection. The recent development of chemical analysis technologies, coupled with bio-electronic methods, has the potential to deliver new tools that will enable us to identify disease-specific VOCs not only in tiny concentrations but also in real time. This would create new prospects for rapid disease detection, allowing patients to be submitted to more advanced diagnostic tools much earlier. As such, and in agreement with Bijland *et al.* (2013), scent detection used for medical purposes holds great promise for the future.

References

Amann, A., Miekisch, W., Schubert, J., Buszewski, B., Ligor, T., Jezierski, T., Pleil, J. and Risby, T. (2014) Analysis of exhaled breath for disease detection. *Annual Review of Analytical Chemistry* 7, 455–482. DOI: 10.1146/annurev-anchem-071213-020043

Amundsen, T., Sundstrøm, S., Buvik, T., Gederaas, O.A. and Haarverstad, R. (2014) Can dogs smell lung cancer? First study using exhaled breath and urine screening in unselected patients with suspected lung cancer. *Acta Oncologica* 53, 307–315. DOI: 10.3109/0284186X.2013.819996

Arakawa, H., Arakawa, K. and Deak, T. (2009) Acute illness induces the release of aversive odor cues from adult, but not prepubertal, male rats and suppresses social investigation by conspecifics. *Behavioural Neuroscience* 123, 964–978. DOI: 10.1037/a0017114

Arakawa, H., Arakawa, K. and Deak, T. (2010) Sickness-related odor communication signals as determinants of social behavior in rat: a role for inflammatory processes. *Hormones and Behavior* 57, 330–341. DOI: 10.1016/j.yhbeh.2010.01.002

Balseiro, S.C. and Correia, H.R. (2006) Is olfactory detection of human cancer by dogs based on major histocompatibility complex-dependent odour component? – A possible cure and a precocious diagnosis of cancer. *Medical Hypotheses* 66, 270–272.

Bijland, E.K., Bomers, M.K. and Smulders, Y.M. (2013) Smelling the diagnosis: a review on the use of scent in diagnosing disease. *The Netherlands Journal of Medicine* 71, 300–307.

Blackburn, E.K. (1957) The breath in acute leukaemia. *British Medical Journal* 1, 146–147.

Brown, S.W. and Strong, V. (2001) The use of seizure-alert dogs. *Seizure* 10, 39–41. DOI:10.1053/seiz.2000.0481

Burfeind, O., Bruins, M., Bos, A., Sannmann, I., Voigtsberger, R. and Heuwieser, W. (2014) Diagnosis of acute puerperal metritis by electronic nose device analysis of vaginal discharge in dairy cows. *Theriogenology* 82, 64–70. DOI: 10.1016/j.theriogenology.2014.03.001

Burke, D.G., Halpern, B., Malegan, D., McCairns, E., Danks, D., Schlesinger, P. and Wilken, B. (1983) Profiles of urinary volatiles from metabolic disorders characterized by unusual odors. *Clinical Chemistry* 29, 1834–1838.

Buszewski, B., Kęsy, M., Ligor, T. and Amann, A. (2007) Human exhaled air analytics: biomarkers of diseases. *Biomedical Chromatography* 21, 553–566. DOI: 10.1002/bmc.835

Buszewski, B., Ligor, T., Jezierski, T., Wenda-Piesik, A., Walczak, M. and Rudnicka, J. (2012a) Identification of volatile lung cancer markers by gas chromatography-mass spectrometry: comparison with discrimination by canines. *Analytical and Bioanalytical Chemistry* 404, 141–146. DOI: 10.1007/s00216-012-6102-8

Buszewski, B., Rudnicka, J., Ligor, T., Walczak, M. and Jezierski, T. (2012b) Analytical and unconventional methods of cancer detection using odor. *Trends in Analytical Chemistry* 38, 1–12. DOI: 10.1016/j.trac.2012.03.019

Chen, M., Daly, M., Williams, N., Williams, S., Williams, C. and Williams, G. (2000) Non-invasive detection of hypoglycaemia using a novel, fully biocompatible and patient friendly alarm system. *British Medical Journal* 321, 1565–1566. DOI: 10.1136/bmj.321.7276.1565

Chen, S., Zieve, L. and Mahadevan, V. (1970) Mercaptans and dimethyl sulfide in the breath of patients with cirrhosis of the liver. Effect of feeding methionine. *Journal of Laboratory and Clinical Medicine* 75, 628–635.

Church, J. and Williams, H. (2001) Another sniffer dog for the clinic? *The Lancet* 358, 930.

Dalziel, D.J., Uthman, B.M., McGorray, S.P. and Reep, R.L. (2003) Seizure-alert dogs: a review and preliminary study. *Seizure* 12, 115–120. DOI: 10.1016/S1059-1311(02)00225-X

Dankert, J., Holloway, Y., Bouma, J., van der Werf, J. and Wolters, B.G. (1981) Metronidazaole in smelly gynaecological tumors. *Lancet* 2, 1295.

Deng, C., Zhang, J., Xiaofeng, Y., Zhang, W. and Zhang, X. (2004) Determination of acetone in human breath by gas chromatography–mass spectrometry and solid-phase microextraction with on-fiber derivatization. *Journal of Chromatography B* 810, 269–275. DOI: 10.1016/j.jchromb.2004.08.013

Devanand, D.P., Michaels-Marston, K.S., Liu, X., Pelton, G.H., Padilla, M., Marder, K., Bell, K., Stern, Y. and Mayeux, R. (2000) Olfactory deficits in patients with mild cognitive impairment predict Alzheimer's disease at follow-up. *American Journal of Psychiatry* 157, 1399–1405. DOI: 10.1176/appi.ajp.157.9.1399

Dobbelaar, P., Mottram, T.T.F., Nyabadza, C., Hobbs, P., Elliott-Martin, R.J. and Schukken, Y.H. (1996) Detection of ketosis in dairy cows by analysis of exhaled breath. *Veterinary Quarterly* 18, 151–152, DOI: 10.1080/01652176.1996.9694638

Doty, R.L. and Kamath, V. (2014) The influences of age on olfaction: a review. *Frontiers in Psychology* 5, 20. DOI: 10.3389/fpsyg.2014.00020

Ehman, K.D. and Scott, M.E. (2002) Female mice mate preferentially with non-parasitized males. *Parasitology* 125, 461–466. DOI: 10.1017/S003118200200224X

Elliker, K., Sommerville, B.A., Broom, D.M., Neal, D.E., Armstrong, S. and Williams, H.C. (2014) Key considerations for the experimental training and evaluation of cancer odour detection dogs: lessons learnt from a double-blind, controlled trial of prostate cancer detection. *BMC Urology* 14, 22–30. DOI: 10.1186/1471-2490-14-22

Fend, R., Kolk, A.H., Bessant, C., Buijtels, P., Klatser, P.R. and Woodman, A.C. (2006) Prospects for clinical application of electronic-nose technology to early detection of *Mycobacterium tuberculosis* in culture and sputum. *Journal of Clinical Microbiology* 44, 2039–2045. DOI: 10.1128/JCM.01591-05

Garner, C.E., Smith, S., Bardhan, P.K., Ratcliffe, N.M. and Probert, C.S. (2009) A pilot study of faecal volatile organic compounds in faeces from cholera patients in Bangladesh to determine their utility in disease diagnosis. *Transactions of the Royal Society of Tropical Medicine and Hygiene* 103, 1171–1173. DOI: 10.1016/j.trstmh.2009.02.004

Gordon, R.T., Schatz, C.B., Myers, L.J., Kosty, M., Gonczy, C., Kroener, J., Tran, M. *et al*. (2008) The use of canines in the detection of human cancers. *Journal of Alternative and Complementary Medicine* 14, 61–67. DOI: 10.1089/acm.2006.6408

Gordon, S.G., Smith, K., Rabinowitz, J.L. and Vagelos, P.R. (1973) Studies of *trans*-3-methyl-2-hexenoic acid in normal and schizophrenic humans. *Journal of Lipid Research* 14, 495–503.

Hirotsu, T., Sonoda, H., Uozumi, T., Shinden, Y., Mimori, K., Maehara, Y., Ueda, N. and Hamakawa, M. (2015) A highly accurate inclusive cancer screening test using *Caenorhabditis elegans* scent detection. *PLoS ONE* 10, e0118699. DOI: 10.1371/journal.pone.0118699

Honig, P.J., Frieden, I.J., Kim, H.J. and Yan, A.C. (2003) Streptococcal intertrigo: an underrecognized condition in children. *Pediatrics* 112, 1427–1429. DOI: 10.1542/peds.112.6.1427

Horvath, G., Andersson, H. and Paulsson, G. (2010) Characteristic odor in the blood reveals ovarian carcinoma. *BMC Cancer* 10, 643–649. DOI: 10.1186/1471-2407-10-643

Jezierski, T., Walczak, M., Ligor, T., Rudnicka, J. and Buszewski, B. (2015) Study of the art: canine olfaction used for cancer detection on the basis of breath odour. Perspectives and limitations. *Journal of Breath Research* 9, 027001. DOI: 10.1088/ 1752-7155/9/2/027001

Kavaliers, M. and Colwell, D.D. (1995) Discrimination by female mice between the odours of parasitized and non-parasitized males. *Proceedings of the Royal Society B: Biological Sciences* 282, 31–35. DOI: 10.1098/rspb.1995.0113

Kavaliers, M., Colwell, D.D., Braun, W.J. and Choleris, E. (2003) Brief exposure to the odour of a parasitized male alters the subsequent mate odour responses of female mice. *Animal Behaviour* 65, 59–68. DOI: 10.1006/anbe.2002.2043

Kirton, A., Wirrell, E., Zhang, J. and Hamiwka, L. (2004) Seizure-alerting and -response behaviors in dogs living with epileptic children. *Neurology* 62, 2303–2305.

Klein, S.L., Gamble, H.R. and Nelson, R.J. (1999) *Trichinella spiralis* infection in voles alters female odor preference but not partner preference. *Behavioral Ecology and Sociobiology* 45, 323–329. DOI: 10.1007/s002650050567

Kondo, K., Suzukawa, K., Sakamoto, T., Watanabe, K., Kanaya, K., Ushio, M., Yamaguchi, T., Nibu, K.-I., Kaga, K. and Yamasoba, T. (2010) Age-related changes in cell dynamics of the postnatal mouse olfactory neuroepithelium: cell proliferation, neuronal differentiation, and cell death. *The Journal of Comparative Neurology – Research in Systems Neuroscience* 518, 1962–1975. DOI: 10.1002/cne.22316

Kwon, O.S., Song, H.S., Park, S.J., Lee, S.H., An, J.H., Park, J.W., Yang, H., Yoon, H., Bae, J., Park, T.H. and Jang, J. (2015) An ultrasensitive, selective, multiplexed superbioelectronic nose that mimics the human sense of smell. *Nano Letters* 15, 6559–6567. DOI: 10.1021/acs.nanolett.5b02286

Liddell, K. (1976) Smell as a diagnostic marker. *Postgraduate Medical Journal* 52, 136–138.

Liebich, H.M. (1983) Analysis of acidic metabolites by capillary column GC and GC/MS. *Journal of High Resolution Chromatography and Chromatography Communications* 6, 640–650. DOI: 10.1002/jhrc.1240061202

Manolis, A. (1983) The diagnostic potential of breath analysis. *Clinical Chemistry* 29, 5–15.

Matsumura, K., Opiekun, M., Oka, H., Vachani, A., Albelda, S.M., Yamazaki, K. and Beauchamp, G.K. (2010) Urinary volatile compounds as biomarkers for lung cancer: a proof of principle study using odor signatures in mouse models of lung cancer. *PLoS ONE* 5, e8819. DOI: 10.1371/journal.pone.0008819

McCulloch, M., Jezierski, T., Broffman, M., Hubbard, A., Turner, K. and Janecki, T. (2006) Diagnostic accuracy of canine scent detection in early- and late-stage lung and breast cancers. *Integrative Cancer Therapies* 5, 1–9. DOI: 10.1177/1534735405285096

McCulloch, M., Turner, K. and Broffman, M. (2012) Lung cancer detection by canine scent: will there be a lab in the lab? *European Respiratory Journal* 39, 511–512. DOI: 10.1183/09031936.00215511

Neto, J.C.L., Martins, L.P., Alessi, A.C., Ferraz, G.C., Marques, J.A. and Valadao, C.A.A. (2001) Clinical aspects, anatomopathologic observations and treatment of six cases of chronic hypertrophic pododermatitis (equine canker). *Arquivo Brasileiro de Medicina Veterinaria e Zootecnia* 53, 303–309.

Neupane, S., Peverall, R., Richmond, G., Blaikie, T.P.J., Taylor, D., Hancock, G. and Evans, M.L. (2016) Exhaled breath isoprene rises during hypoglycaemia in type 1 diabetes. *Diabetes Care* 39, e97–e98, DOI: 10.2337/dc16-0461

Novak, B.J., Blake, D.R., Meinardi, S., Rowland, F.S., Pontello, A., Cooper, D.M. and Galassetti, P.R. (2007) Exhaled methyl nitrate as a noninvasive marker of hyperglycemia in type 1 diabetes. *Proceedings of the National Academy of Sciences USA* 104, 15613–15618. DOI: 10.1073/pnas.0706533104

O'Connor, M.B., O'Connor, C. and Walsh, C.H. (2008) A dog's detection of low blood sugar: a case report. *Irish Journal of Medical Science* 177, 155–157. DOI: 10.1007/s11845-008-0128-0

Ottaviano, G., Frasson, G., Nardello, E. and Martini, A. (2016) Olfaction deterioration in cognitive disorders in the elderly. *Aging Clinical and Experimental Research* 28, 37–45. DOI 10.1007/s40520-015-0380-x

Pavlou, A.K. and Turner, A.P. (2000) Sniffing out the truth: clinical diagnosis using the electronic nose. *Clinical Chemistry and Laboratory Medicine* 38, 99–112. DOI: 10.1515/CCLM.2000.016

Pavlou, A.K., Magan, N., Jones, J.M., Brown, J., Klaster, P. and Turner, A.P.F. (2004) Detection of *Mycobacterium tuberculosis* (TB) in vitro and in situ using an electronic nose in combination with a neural network system. *Biosensor and Bioelectronic* 20, 538–544. DOI: 10.1016/j.bios.2004.03.002

Penn, D., Schneider, G., White, K., Slev, P. and Potts, W. (1996) Influenza infection neutralizes the attractiveness of male odor to female mice (*Mus musculus*). *Ethology* 104, 685–694.

Phillips, M., Sabas, M. and Geenberg, J. (1993) Increased pentane and carbon disulfide in the breath of patients with schizophrenia. *Journal of Clinical Pathology* 46, 861–864. DOI: 10.1136/jcp.46.9.861

Phillips, M., Gleeson, K., Hughes, J.M.B., Greenberg, J., Cataneo, R.N., Baker, L. and McVay, W.P. (1999) Volatile organic compounds in breath as markers of lung cancer: a cross-sectional study. *The Lancet* 353, 1930–1933. DOI: 10.1016/S0140-6736(98)07552-7

Phillips, M., Cataneo, R.N., Ditkoff, B.A., Fisher, P., Greenberg, J., Gunawardena, R., Kwon, C.S., Tietje, O. and Wong, C. (2006) Prediction of breast cancer using volatile biomarkers in the breath. *Breast Cancer Research and Treatment* 99, 19–21. DOI: 10.1007/s10549-006-9176-1

Phillips, M., Altorki, N., Austin, J.H., Cameron, R.B., Cataneo, R.N., Greenberg, J., Kloss, R. *et al.* (2007a) Prediction of lung cancer using volatile biomarkers in breath. *Cancer Biomarkers* 3, 95–109.

Phillips, M., Cataneo, R.N., Condos, R., Ring Erckson, G.A., Greenberg, J., La Bombardi, V., Munawar, M.I. and Tietje, O. (2007b) Volatile biomarkers of pulmonary tuberculosis in the breath. *Tuberculosis* 87, 44–52. DOI: 10.1016/j.tube.2006.03.004

Phillips, M., Cataneo, R.N., Chaturvedi, A., Kaplan, P.D., Libardoni, M., Mundada, M., Patel, U. and Zhang, X. (2013) Detection of an extended human volatome with two-dimensional gas chromatography time-of-flight mass spectrometry. *PLoS ONE* 8, e75274. DOI: 10.1371/journal.pone.0075274

Pickel, D.P., Manucy, G.P., Walker, D.B., Hall, S.B. and Walker, J.C. (2004) Evidence of canine olfactory detection of melanoma. *Applied Animal Behaviour Science* 89, 107–116. DOI: 10.1016/j.applanim.2004.04.008

Probert, C.S., Ahmed, I., Khalid, T., Johnson, E., Smith, S. and Ratcliffe, N. (2009) Volatile organic compounds as diagnostic biomarkers in gastrointestinal and liver diseases. *Journal of Gastrointestinal and Liver Diseases* 18, 337–343.

Rahayel, S., Frasnelli, J. and Joubert, S. (2012) The effect of Alzheimer's disease and Parkinson's disease on olfaction: a meta-analysis. *Behavioural Brain Research* 231, 60–74. DOI: 10.1016/j.bbr.2012.02.047

Rooney, N.J., Morant, S. and Guest, C. (2013) Investigation into the value of trained glycaemia alert dogs to clients with type 1 diabetes. *PLoS ONE* 8, e369921. DOI: 10.1371/journal.pone.0069921

Shirasu, M. and Touhara, K. (2011) The scent of disease: volatile organic compounds of the human body related to disease and disorder. *The Journal of Biochemistry* 150, 257–266. DOI: 10.1093/jb/mvr090

Shirasu, M., Nagai, S., Hayashi, R., Ochiai, A. and Touhara, K. (2009) Dimethyl trisulfide as characteristic odor associated with fungating cancer wounds. *Bioscience Biotechnology and Biochemistry* 73, 2117–2120. DOI: 10.1271/bbb.90229

Simenhoff, M.L., Burke, J.F., Saukkonen, J.J., Ordinario, A.T. and Doty, R. (1977) Biochemical profile or uremic breath. *New England Journal of Medicine* 297, 132–135.

Smith, K. and Sines, J.O. (1960) Demonstration of a peculiar odor in the sweat of schizophrenic patients. *Archives of General Psychiatry* 2, 184–188.

Smith, K., Thompson, G.F. and Koster, H.D. (1969) Sweat of schizophrenic patients of the odorous substance. *Seizure* 166, 398–399.

Søndergaard, L.V., Holm, I.E., Herskin, M.S., Dagnaes-Hansen, F., Johansen, M.G., Jørgensen, A.L. and Ladewig, J. (2010) Determination of odor detection threshold in the Göttingen minipig. *Chemical Senses* 35, 727–734. DOI: 10.1093/chemse/bjq076

Sonoda, H., Kohnoe, S., Yazamato, T., Satoh, Y., Morizono, G., Shokata, K., Morita, M. *et al.* (2011) Colorectal cancer screening with odour material by canine scent detection. *Gut (BMJ Gastroenterology)* 60, 814–819.

Sponring, A., Filipiak, W., Mikoviny, T., Ager, C., Schubert, J., Miekisch, W. *et al.* (2009) Release of volatile organic compounds from the lung cancer cell line NCI-H2087 *in vitro*. *Anicancer Research* 29, 419–426.

Strauch, M., Lüdke, A., Münch, D., Laudes, T., Galizia, G., Martinelli, E., Lavra, L., Paolesse, R., Ulivieri, A., Catini, A., Capuano, R. and Di Natale, C. (2014) More than apples and oranges – detecting cancer with a fruit fly's antenna. *Scientific Reports* 4, 3576. DOI: 10.1038/srep03576

Strong, V., Brown, S.W. and Walker, R. (1999) Seizure-alert dogs: fact or fiction? *Seizure* 8, 62–65. DOI: 10.1053/seiz.1998.0250

Strong, V., Brown, S., Huyton, M. and Coyle, H. (2002) Effect of trained Seizure Alert Dogs® on frequency of tonic–clonic seizures. *Seizure* 11, 402–405. DOI: 10.1053/seiz.2001.0656

Syhre, M. and Chambers, S.T. (2008) The scent of *Mycobacterium tuberculosis*. *Tuberculosis* 88, 317–323. DOI: 10.1016/j.tube.2008.01.002

Taverna, G., Tidu, L., Grizzi, F., Giuisti, G., Seveso, M., Benetti, A., Hurle, R. *et al.* (2014) Prostate cancer urine detection through highly-trained dogs' olfactory system: a real clinical opportunity. *The Journal of Urology* 191(Supplement), e546.

Taverna, G., Tidu, L., Grizzi, F., Torri, V., Mandressi, A., Sardella, P., La Torre, G. *et al.* (2015) Olfactory system of highly trained dogs detects prostate cancer in urine samples. *The Journal of Urology* 193, 1382–1387. DOI: 10.1016/j.juro.2014.09.099

Thomas, F., Adamo, S. and Moore, J. (2005) Parasitic manipulation: where are we and where should we go? *Behavioural Processes* 68, 185–199. DOI: 10.1016/j.beproc.2004.06.010

Turck, M. (1985) Foul breath and a productive cough. *Hospital Practice* 20, 50.

Turner, A.P. and Magan, N. (2004) Electronic noses and disease diagnostics. *Nature Reviews Microbiology* 2, 161–166. DOI: 10.1038/nrmicro823

Van Alstine, L. (2016) How olfactory diagnostics are developed and made to work: a qualitative study. MSc thesis, i6119468 in Global Health FHML/Health Science, Maastricht University, The Netherlands, 31 August 2016.

Vockley, J. and Ensenauer, R. (2006) Isovaleric acidemia: new aspects of genetic and phenotypic heterogeneity. *American Journal of Medical Genetics Part C: Seminars in Medical Genetics* 142C, 95–103. DOI: 10.1002/ajmg.c.30089

Walczak, M., Jezierski, T., Górecka-Bruzda, A., Sobczyńska, M. and Ensminger, J. (2012) Impact of individual training parameters and manner of taking breath odor samples on the reliability of canines as cancer screeners. *Journal of Veterinary Behavior – Clinical Applications and Research* 7, 283–294. DOI: 10.1016/j.jveb.2012.01.001

Weber, K.S., Roden, M. and Muessig, K. (2015) Short Report: Treatment, do dogs sense hypoglycaemia? *Diabetic Medicine* 33, 934–938. DOI: 10.1111/dme.12975

Weetjens, B.J., Mgode, G.F., Machang'u, R.S., Kazwala, R., Mfi nanga, G. *et al.*(2009) African pouched rats for the detection of pulmonary tuberculosis in sputum samples. *The International Journal of Tuberculosis and Lung Disease* 13, 737–743.

Wells, D.L., Lawson, S.W. and Siriwardena, A.N. (2008) Canine responses to hypoglycemia in patients with type 1 diabetes. *The Journal of Alternative and Complementary Medicine* 14, 1235–1241. DOI: 10.1089/acm.2008.0288

White, T., Sadikot, A.F. and Djordjevic, J. (2016) Metacognitive knowledge of olfactory dysfunction in Parkinson's disease. *Brain and Cognition* 104, 1–6. DOI: 10.1016/j.bandc.2016.01.004

Williams, H. and Pembroke, A. (1989) Sniffer dogs in the melanoma clinic? *The Lancet* 333, 734.

Willis, C.M., Church, S.M., Guest, C.M., Cook, A.W., McCarthy, N., Bransbury, A.J., Church, M.R.T. and Church, J.C.T. (2004) Olfactory detection of human bladder cancer by dogs: proof of principle study. *British Medical Journal* 329, 712–715. DOI: 10.1136/bmj.329.7468.712

Wilson, A.D. (2015) Advances in electronic-nose technologies for the detection of volatile biomarker metabolites in the human breath. *Metabolites* 5, 140–163. DOI: 10.3390/metabo5010140

Yamazaki, K., Boyse, E.A., Bard, J., Curran, M., Kim, D., Ross, S.R. and Beauchamp, G.K. (2002) Presence of mouse mammary tumor virus specifically alters the body odor of mice. *Proceedings of the National Academy of Sciences USA* 99, 5512–5515. DOI: 10.1073/pnas.082093099

10 The Role of Olfaction in Relation to Stress and Fear

VINCENT BOMBAIL

Neurobiology of Olfaction, INRA, Université Paris-Saclay, Jouy-en-Josas, France

Like other sensory functions in animals, olfaction contributes to an understanding of the environment; odours include signals that may be considered as threatening and may induce fear and a stress response. These olfactory cues can originate from conspecifics or predators and constitute a wealth of information that is used to promote survival. Chemical signals can be used to assess the presence of unknown conspecifics, possibly competitors for precious resources (food or sexual partners), or predators. Interestingly, olfactory perception and its influence on behaviour are affected by the inner state of the animal via hormones. Here, after defining the stress response and fear, I will review evidence that stressed animals might perceive odours differently. Next, we will see how olfactory signals can generate a stress response and induce fear, and look into experimental support for the claim that odours can be used for stress relief.

Definitions: The Stress Response and Fear

Bickering about the exact meaning of words is not the exclusive remit of linguists and angry lovers; in order to understand the biology of 'stress' and 'fear', a few definitions can go a long way. My intention here is to give a brief outline of the stress response in order to understand how it relates to fear and how a stress response can be induced by odours and affect animal behaviour. Research is a dynamic and ongoing process, and I invite readers to refer to more eminent colleagues for good generalist reviews of the literature on stress biology (e.g. Marian Joëls, Stafford Lightman, Bruce McEwen and Robert Sapolsky, to name but a few). It might also be relevant to compare and contrast laboratory findings to those from studies in wild animals; ecological physiology can enlighten us about the significance of stress (Boonstra, 2013). The word and concept of 'stress' is used in material sciences to describe the application of a force onto an object, a strain leading to bending or torsion and eventually breakage. In our everyday life, stress is commonly used to describe the psychological strain resulting from adverse events (Fig. 10.1). Therefore, stress is sometimes also defined as a lack of 'well-being' (physical and psychological), but an insight about how this concept was developed will enlighten us further.

Hans Selye: the strange case of the clumsy experimentalist

The application of the word stress to animal physiology is generally (but perhaps wrongly, see Jackson, 2014) attributed to Hans Selye, an Austro-Hungarian scientist working in Montreal around the mid-20th century. At that time in the field of endocrinology (the study of hormones), gland extracts were prepared and injected into laboratory animals in order to observe and understand the effects of hormones. According to the legend, Selye was not very skilled at handling rats and perhaps over the course of the injection treatments, a few were dropped or jumped away from the syringe and had to be chased around the lab. Any proper experiment includes a comparison with a control group in order to allow the identification of the specific treatment effects, in this case rats injected with a neutral substance such as physiological saline. Intriguingly, Selye observed symptoms in both groups: enlarged adrenal glands, smaller immune tissue (e.g. thymus) and the presence of ulcerations in the digestive tract (Selye, 1950). This triad of symptoms turned out to be caused by the chronic experimental disturbance to the rats,

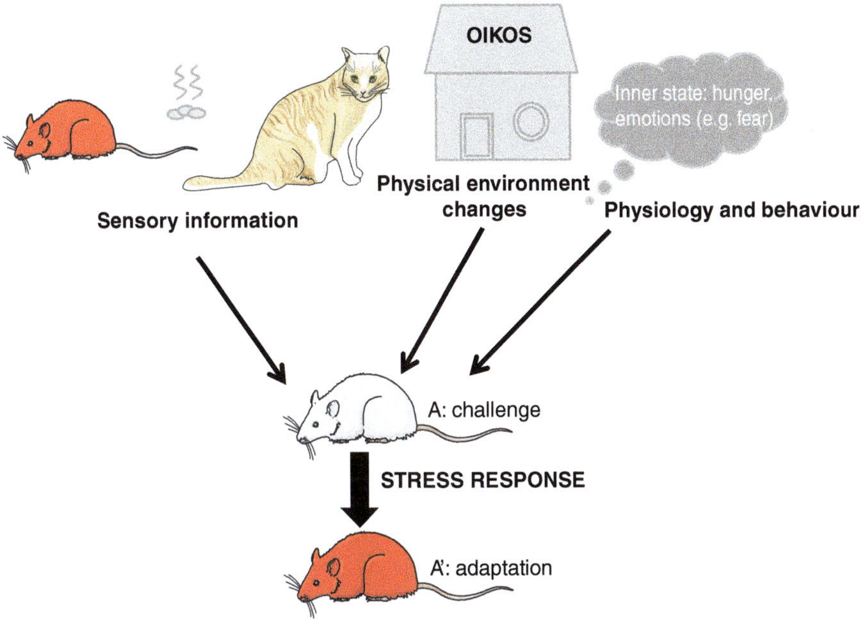

Fig. 10.1. If animal A experiences a challenging situation, its stress response allows integrated body and mind adaptation (A'). Stressors can be diverse in their nature (e.g. sight, sound and odours) and origin (social signals including competitors, predators), and past experience can influence the animal's reaction.

which seemingly induced an adaptive physiological response, rather than the specific effect of the gland extracts. Selye popularized the notion of stress, following his work on this 'general adaptation syndrome', which is now better known as stress response. Acquisition of scientific knowledge always builds upon previous findings. It is worth mentioning that this model was built on advances from earlier physiologists, such as Claude Bernard who had developed the notion of '*milieu intérieur*' (that the composition of internal fluids are tightly regulated), and Walter Cannon who worked on the regulation of homeostasis through the famous 'fight or flight' response. In experiments that would nowadays be seen as cruel and require much argumentation in terms of research ethics, Selye later showed that noxious physical stimuli (injection of toxins, exposure to cold, haemorrhage, etc.) or stressors that disturb homeostasis induce an adaptive response in animals, and that sustained activation of this initial stress response can lead to pathological states (Selye, 1950). Now we know that stressors also include psychological/emotional insults, such as social isolation, overcrowding, and exposure to dominant individuals in what is known as social defeat paradigms. This is a concrete example of the mind–body relationship, which is easy to manipulate and relatively easy to induce/observe experimentally. Selye's 'general adaptation syndrome' can be divided into three phases: an acute alarm phase involving the sympathetic nervous system (with action of the adrenaline and noradrenaline hormones), a resistance phase where body resources are mobilized to allow adaptation to the stressor (with action of the glucocorticoid hormones) and finally a recovery phase, where resources are replenished. In the case of repeated stimulation or insufficient resources for recovery to occur, an exhaustion phase can follow. This is where the adaptive effects of the *acute stress* (single stimulation) can lead to the pathological effects of *chronic stress* (repeated and inappropriate stimulation of the stress response). In addition to the famous triad of symptoms induced by chronic stress, further consequences have since been reported and will be described below. Those pathological effects of chronic stress often require physiological alterations in order to achieve homeostasis, a process named *allostasis*. Authors therefore refer to the *allostatic load* as the 'wear and tear', the damage caused by chronic stress (McEwen, 2007). In addition to being a scientist, Selye became a prolific popular science writer, spreading his work and his views on stress. It is noteworthy that a distinction was made between what he named distress (negative stress) and eustress (positive stress), the notion that not all stress is detrimental. Things are not this simple indeed: the arousal caused by confronting a pleasant or positive yet new situation can lead to physiological manifestations consistent with the stress response caused by negative stimuli. Finally, the physiological impact of stress depends on the context in which it is encountered and the state of mind of the individual. When rats are chronically treated with electric shocks, they develop stomach ulcers. By modulating the shock delivery protocol, it is possible to alter stressor predictability or induce helplessness and this affects the development of those ulcers (reviewed in Weiss, 1972). Chronic stressors have a more negative impact if they are unpredictable and if the animal lacks (perceived) control. Now that we appreciate what stress is, we can ask: what are the manifestations of the stress response and how does it work?

The stress response: mechanisms

The stress response is a coordinated physiological reaction to threats and environmental change. It is a complex endocrine signalling system. It involves what is known as the hypothalamic–pituitary–adrenal (HPA) axis. In order to help us understand in detail some of the interactions between stress/fear and olfaction, and assess how it can relate to animal behaviour, it is necessary to give a short introduction to the more mechanistic aspects of the stress response. I have tried to keep it as simple as possible; Fig. 10.2 provides brief graphical support for explanations that could otherwise induce fear or stress in you, the reader.

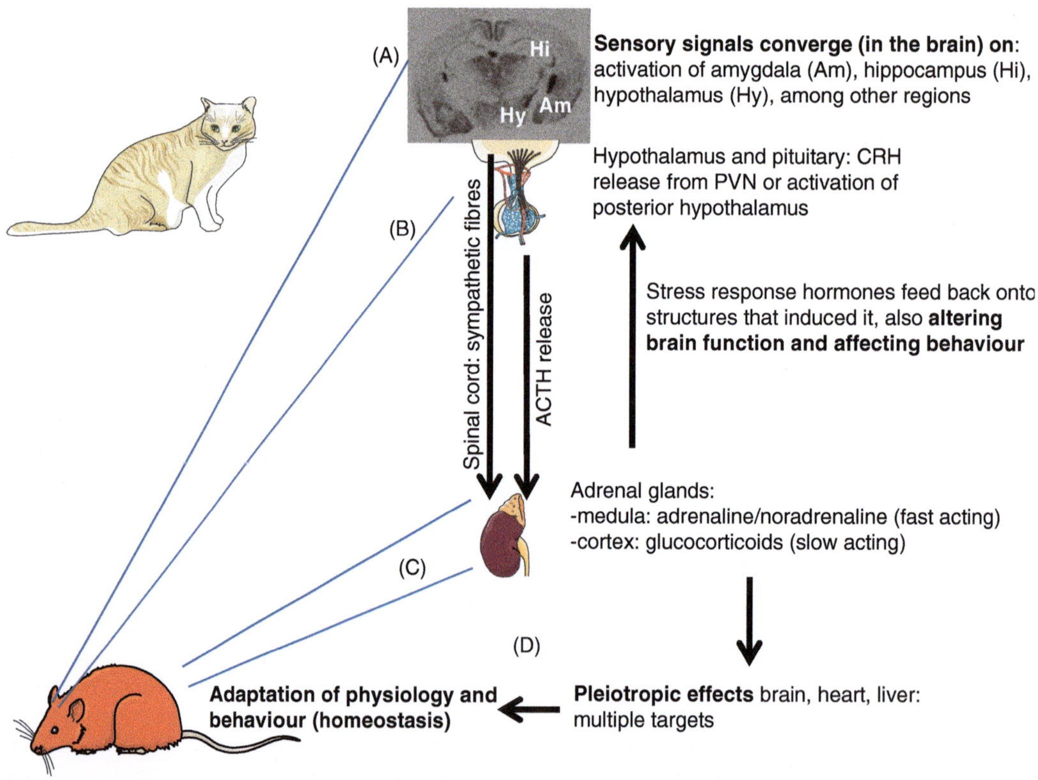

Fig. 10.2. Experiencing adversity, a physiological challenge or certain emotions (fear) leads to activation of various brain centres (A), that converge on the hypothalamus (B) whereby signals are sent, in order to activate the adrenal glands (C), to trigger the release of stress hormones. As the hormones act on various targets, multiple body systems are recruited (D) in order for the physiology and behaviour of the organism to be adaptive. Sustained activation of this stress axis can lead to pathologies. ACTH, adrenocorticotropic hormone; CRH, corticotropin-releasing hormone; PVN, paraventricular nucleus.

When an animal is placed into a stressful situation, a series of events takes place: signals from several brain regions, gathering information about the environment (including the amygdala, see below), converge and act on the hypothalamus. The hypothalamus is a communication centre, a brain region involved in sending and receiving signals between the brain and the rest of the body. Stimulation of the posterior region of the hypothalamus leads to signals activating the adrenal gland medulla via the spinal cord and the splanchnic nerve, causing the rapid release of adrenaline and noradrenaline. The effect of these hormones is called pleiotropic: they act on multiple body systems in order to generate an adaptive response. For instance, the effects of adrenaline on cardiopulmonary function and perspiration are well described (remember this next time you have to speak in front of strangers, with sweaty palms and your heart thumping in your chest). In parallel, the paraventricular nucleus of the hypothalamus (PVN) also receives inputs that lead to secretion of corticotropin-releasing hormone (CRH). This peptide has effects on various brain regions to coordinate the stress response, but mostly signals to the pituitary gland to release adrenocorticotropic hormone (ACTH), which travels through the bloodstream and in turn stimulates the cortex of the adrenal glands to release glucocorticoid hormones (predominantly cortisol in large mammals and fish, and corticosterone in rodents, birds and reptiles). Glucocorticoids also affect the brain to inhibit in turn the signals that lead to their secretion. This negative feedback ensures a tight control over adrenal hormone secretion.

These steroid stress hormones are also pleiotropic and affect virtually all aspects of physiology with a vast range of consequences. From their name (gluco-) we can recognize they will affect the metabolic tissues (liver, fat) to mobilize energy resources (e.g. induce release of glucose for muscles to work), but also the nervous system to provide an adaptive behavioural response. It is the sustained secretion of glucocorticoids that caused the triad of symptoms in Selye's rats and affected both the thymus (lymphatic immune system) and the gastrointestinal tract. Since the nervous system is also a target system affected by this hormonal response, behaviour will be impacted (moods/emotions, decision-making and perhaps – as we will see – perception). Although secreted as part of the same stress response, catecholamine and glucocorticoid hormones have different effects on physiology. If we were to use a running metaphor, the catecholamine response can be compared to what is required to run a 100 m race, whereas the steroid response is more like what is required to run a marathon. As described above, stressors can be physical factors from the environment (temperature, noise, pollution) or social signals (predators, dominant conspecific). Odours can act as stressors (see below). Captivity can be a stressor for wild and domesticated animals because it reduces their ability to express the full spectrum of naturally occurring behaviours. Certain production practices used on farms can induce stress in the animals concerned, such as tail docking in pigs and beak trimming in poultry, and the effects of these procedures on the animals' stress response are well known (e.g. Prunier *et al.*, 2005; Xie *et al.*, 2013).

Who is feeling stressed? How to detect a stress response, and its behavioural consequences

Even if animals could talk, would they be able to tell us whether they are stressed? Would a human always truly report this? It is sometimes challenging to identify and quantify stress. One difficulty in assessing stress levels is determining and interpreting whether those measures relate to arousal in a positive experience, or a negative stress stimulus that results in less well-being. Physiological markers can be useful but we shall see that behavioural assessments may be more informative as they integrate physiological effects and life experiences.

Most of the aforementioned mediators of the endocrine stress response can be detected and measured in biological fluids. For instance, glucocorticoid hormones (corticol, corticosterone) are assayed in tissue or body fluids such as blood and saliva. Glucocorticoids are frequently used as stress biomarkers but since they also have circadian and ultradian secretion patterns, they vary during the day in the absence of stressors. Faeces and urine can be alternative sources as they are hormone excretion routes (depending on the species) and integrate secretion over time, smoothing out this variability. Such measures can be used to monitor baseline hormone levels in control animals or investigate longer-term secretion of stress hormones, for instance in animals subjected to chronic stress. Adrenaline and ACTH are less frequently used, as they are quickly secreted and therefore require special sampling techniques. However, physiological manifestations of the early stress response caused by adrenaline (sweating, heart rate or blood pressure, body temperature) can be more non-invasively quantified.

In a world where resources are limited, there is a strong evolutionary pressure to maintain an ability to respond to variations in an ever-changing environment, whether you are a predator (find food) or prey (avoid being someone's dinner). Behavioural effects of stress exposure can be species specific, depending on ecology and evolution, but there are striking similarities across vertebrates. Most importantly there are individual variations, some due to genetics and some due to life experience, which can remain fairly consistent over the animal's life. For instance, several studies report the numerous consequences of early life adversity (social or nutritional stressors, early weaning). Some authors refer to those traits consistent across situations as the personality of an animal. In other words, some animals are more prone than others to fearful reactions (see below) or to the effect of the stress response. A consequence of this variability of stress effects and sensitivity is the difficulty to extrapolate from published data to our own experimental paradigm. The impact of the same stressor can differ enormously, depending on the animals concerned and the environment they live in, and this must be characterized accordingly.

The induction of a stress response will result in various behaviours that can be understood as adaptations to the presence of a real or perceived threat (Greenberg *et al.*, 2002). But as stressor exposure is prolonged, other behaviours – perhaps more pathological – can be expressed. Stress will increase arousal and attention, perhaps leading to changes in stimuli salience. There is a body of evidence suggesting that sensory perception can be modified, e.g. increased for sounds and reduced for pain. If free movement is possible, and depending on whether the location is familiar or not, locomotor and exploratory activities can be induced or repressed. Stress will affect mood, resulting in increased signs of anxiety – these will depend on the ecology and history of the species considered, e.g. in rodents they are manifested by fear of open spaces and freezing (alert immobilization) and more generally by certain types of vocalizations, increased defecation/micturition and fleeing. The intraspecific social relationships (e.g. dominance) might be perceived differently, perhaps causing aggression and fighting between individuals. Besides the many behavioural responses that can be observed following stressor exposure, there are clear effects of stress hormones on memory, including fear-memory, thereby potentially impacting future behavioural patterns. The consequences for the stress response of altering a set of (stress) hormones will result in alterations in other hormone levels, including reproductive hormones and hormones related to metabolism. In other words, reproductive physiology (fertility, sexual or parenting behaviour) and metabolic physiology (metabolism, appetite and ingestive behaviour) can also be impacted by the induction of the stress response (for an in-depth review, see Greenberg *et al.*, 2002). The behavioural effects of the acute stress response can thus be extensive.

The behavioural effects of chronic stress can appear from as early as 2 weeks of exposure in animal models, depending on the stressors applied. Chronic stress can cause alterations in behaviours, sometimes similar to acute stress as discussed above, and the repeated exposure to stress hormones affects numerous brain systems. This will impact motivation, affective behaviour and mood. The appearance of depressive-like symptoms and altered mood-related behaviours (anxiety disorders) are among the more recognized consequences of chronic stress (McEwen, 2007). In humans, depression is a persistent feeling of sadness and loss of interest, an ill-defined nebula of symptoms, diagnosed via the reported persistence of some of those symptoms for a certain time. In the most severe cases of major depressive disorders (MDDs), symptoms can sometimes be alleviated with a number of drugs. However, MDDs are only one extreme of the full range of mood disorders and these are possibly harder to cure than to define and identify, but advances have been made using animal models. In laboratory animal models of depression, two behavioural traits, corresponding to the human depressive symptoms, are usually studied: despair and anhedonia. Despair, or learned helplessness, is a behaviour pattern that develops in animals that have been exposed to aversive stimuli or noxious situations they could not escape or overcome, and therefore were unable to express the stress response's fight or flight. This is often assessed in laboratory rodents using the tail suspension test (Steru *et al.*, 1985) or the forced swim test (Porsolt *et al.*, 1977), whereby immobility (hanging still or floating, respectively) is considered a sign of resignation to adversity (behavioural despair). Anhedonia, loss of pleasure seeking behaviour, can be studied in tests where a choice is given between a pleasurable stimulus and a more neutral one, e.g. water or sweetened water, where loss of preference for the sweet water is considered to be anhedonia-like (Katz, 1982). More recent and more refined paradigms involve testing choices in the animal, with a view to assess cognitive bias along the lines of the famous image: is this glass half-empty or half-full (Harding *et al.*, 2004; Brydges *et al.*, 2012)? Depressive-like and anxious disorders are often associated with other behavioural changes such as inhibited social behaviour (Beery and Kaufer, 2015), alterations in self grooming (Kalueff *et al.*, 2016) and increased irritability or aggressiveness (Neumann *et al.*, 2010). Finally, at least in humans, chronic stress leading to allostasis is not solely a risk factor for the development of affective disorders, it also acts as a trigger that promotes the development of other psychiatric conditions such as bipolar disorders or addiction (Pettorruso *et al.*, 2014). Clearly there is progress to be made on the understanding and assessment of the pathological effects of long-term stress on animals, but interesting patterns are emerging.

Fear

Emotions are cognitive and physiological alterations in an individual that result from exposure to certain environmental or social signals. Fear is an emotion that is generally induced by detection or even suspicion of a danger in the environment. Fear behaviour will manifest differently according to the ecology of the species concerned. For rodents it is the smell of predators and being in open spaces where they might feel vulnerable, whereas equids might seek open spaces where they are more likely to see predators. Fear responses include behaviours that promote danger avoidance (e.g. increased vigilance or anxiety-related behaviours) and physiological readiness for the 'fight or flight' response (Nettle and Bateson, 2012). Fear is transient by nature, often associated with activation of the amygdala region of the brain, which will result in stress responses (see above); sustained exposure to stimuli that induce fear can sometimes affect physiology and behaviour in ways consistent with chronic stress exposure.

Odours are powerful sensory cues when it comes to fear: they can generate fear reactions in an acquired/learnt and innate fashion. In a classical (Pavlovian) negative conditioning paradigm, odours can be used as the conditioned stimuli: a neutral olfactory cue is administered with a noxious stimulus. Following several repetitions, learning occurs and a pairing is made between the cue and the noxious stimulus. Finally, the neutral olfactory cue alone can elicit a fear/stress reaction (expectation of noxious stimulus). Interestingly, fear learning generally occurs with high glucocorticoid levels (which are known to affect memory), indicating an interesting area of interaction between those phenomena.

Certain odours can create an innate fear reaction: without prior exposure to the olfactory stimulus, spontaneous fear and stress reactions can be induced, such as seen in some prey species exposed to predator odours (see below). This trait would be highly favoured through the course of evolution, and avoidance of potential danger could be seen as homologous to the ancestral fear some of us have of the sight of snakes or spiders (though it is difficult to unravel the innate from the cultural in humans). It should be noted that numerous odorant compounds become aversive at high concentration and can induce avoidance behaviours due to their unpleasantness.

Olfaction is Affected by Stress

Our perception is affected by our internal state. I despise the bakery shop next to the train station, fanning out the tempting smell of fresh croissants every morning, because it is a daily reminder that I cannot eat *ad libitum*. Hunger or satiety make us react differently to an appetizing food smell, since feeding-related hormones (e.g. insulin) affect olfactory function (see Chapter 6). The same goes for the effects of our stress hormones (catecholamines, CRH, ACTH, glucocorticoids such as cortisol or corticosterone) whose secretion has profound impacts on physiology and behaviour. Hormones act on receptors, with the presence of receptors in a cell type or a tissue reflecting the potential for responsiveness to a given hormone. Several brain regions associated with olfactory responses (e.g. the olfactory epithelium and bulb) express such receptors for adrenaline, CRH or glucocorticoids, suggesting olfaction can be impacted by exposure to stressors and that altered olfaction could indeed be part of the coordinated body response. Given that sensory inputs can be affected by the inner state of the animal, one can hypothesize this might have evolved in order to optimize adequate perception, maximize decision-making and enhance survival. There are conflictual early reports in rodent models, where animals were adrenalectomized and olfactory performance was either reduced (Sakellaris, 1972) or unaltered (Doty *et al.*, 1991). More recent research brings some elements of response to the question of how stress might impact olfaction.

Acute stress has been reported to impact aspects of olfactory perception. It has been shown that acute injection of a synthetic glucocorticoid (dexamethasone) acts on the olfactory mucosa, where the first step of olfactory detection takes place (Raynaud *et al.*, 2015b). This pharmacological treatment mimics the glucocorticoid stress response and appears to induce a more acute sense of smell in a rat model. In another study, noradrenaline released during a mild stressor exposure has been shown to act on the olfactory bulb to repress the formation of olfactory memories in an odour recognition task (Manella *et al.*, 2013). This suggests that a mildly stressed animal may not remember odours so well, because stress hormones could directly affect the olfactory bulb. Other studies show quite the opposite for strong stressors and trauma-associated odours in mice (Jones *et al.*, 2008). When an electric foot shock was associated with an odorant molecule (acetophenone) for which the olfactory receptor is known (M71), this led to increased freezing in the presence of acetophenone alone after 3 weeks of training (ten pairings per week), which is characteristic of Pavlovian conditioning. Intriguingly, this was accompanied by an increased number of M71 neurons expressing the cognate receptor to this odorant, suggesting increased sensitivity to acetophenone. This morphological change is an example of experience-dependent plasticity in the olfactory system, induced by associative learning (McGann, 2015). It seems that acute stressors or aversive learning can impact olfactory function in animals, sometimes rapidly but also more slowly (tissue reorganization); those findings require further validation in other models in order to be generalized to species other than rodents.

Repeated exposure to stressful situations has been shown to affect olfactory perception. Chronic stress has profound effects on metabolism or the immune system, which can shape and affect the substratum for animal behaviour (the brain) and any epithelia of the body (including the olfactory epithelium). Classically, stress is known to affect brain structures involved in emotion processing, memory formation and fear: the prefrontal cortex, the hippocampus and amygdala (fear-associated region). But new work shows olfactory regions can be affected. We recently demonstrated that the first step of olfactory detection was reduced in the olfactory epithelium following a regimen of chronic variable stress (unpredictable stressors), where rats developed depressive-like symptoms (Raynaud *et al.*, 2015a). These findings were later extended to the olfactory bulb, where mice chronically treated with corticosterone in their drinking water developed anxious and depressive-like disorders (Siopi *et al.*, 2016). This was accompanied by changes in cell survival in the olfactory bulb and a loss of olfactory discrimination, acuity and memory at the behavioural level. Those two studies on the negative impact of chronic stress on the olfactory system in rodent models mirror a wealth of data from human patient studies. A range of effects have been reported in depressive patients, from loss of olfactory sensitivity (Buron and Bulbena, 2013) to alterations in the pleasant sensation brought by odours (olfactory anhedonia; Naudin *et al.*, 2012). Qualitative and quantitative effects on olfactory perception are interesting to study in humans, because we can accurately describe how we feel and smell, something that we can only infer from animal studies. In another patient population (people suffering from panic disorders; Wintermann *et al.*, 2013), brain imaging studies showed that brain regions were differently activated in response to odours, suggesting changes in the way the information is processed. Interestingly, Krusemark *et al.* (2013) showed this also to be the case in healthy subjects, in which a transient anxious state is induced: different brain regions were activated and the interactions between the regions were also different. Those last studies confirm that although odour perception was not found to be affected, the brain processes in response to odour presentation differed. This supports the idea that emotional and behavioural responses to odours can be affected by stress. Finally, it is worth mentioning that in animal models where olfactory sensitivity is reduced through genetic modification of the olfactory detection machinery, anxiety and depressive-like effects can be detected in rodents (Glinka *et al.*, 2012; Chen *et al.*, 2014) and fish (Abreu *et al.*, 2016).

From a hormonal point of view, it is not just the stress hormones described above that are released in response to stress exposure, a number of other hormonal systems can be stimulated or repressed. Chronic stress can affect the reproductive axis, for instance by decreasing testosterone production in males, disrupting the ovulation/oestrous cycle in females and even altering sex-specific behaviour in many species (Toufexis *et al.*, 2014). These are examples of some of the indirect effects of the pervasive stress response. Olfactory loss in animals appears to be comorbid with symptoms of chronic stress; this may manifest in reduced feeding, social abilities or reproductive function. Animal management practices should take into account these effects, in species and situations where odours represent important sensory information.

In summary, olfaction is affected by hormones of the stress response. Acute stress, both through general arousal and specific olfactory mechanisms, modifies the salience of olfactory stimuli and alters their perception and memorability. Chronic stress, often in association with depressive-like symptoms, is associated with losses in olfactory perception.

Odours Can Induce a Stress Reaction

Stressed animals might perceive odours differently, and odours themselves can induce a stress response. Although olfactory cues are potent conditioned stimuli in associative learning, I will only focus here on how fear and stress reactions might be induced in an innate fashion.

Fear from predator odours

Strong unpleasant smells can assault our senses and make us feel uncomfortable. Butyric acid is the product of fatty acid degradation by bacteria and is well known for smelling like rancid butter or sweaty armpits. Whether you or your neighbours on the train suffer from a lack of fridge or deodorant, you might avert your nostrils from the source or even breathe through your mouth (and risk retronasal olfaction). When rats are allowed to explore a test arena containing an odorant source in a corner, they appear to avoid spending time near butyric acid and also avoid 2,4,5-trimethyl-3-thiazoline (TMT), a component of fox faeces odour (Endres and Fendt, 2009). TMT has an unpleasant smell, but unlike butyric acid it also causes an increase in the amount of time the rats spend freezing. Freezing is a transient immobilization whilst the rat appears alert, thought to help reduce harm in threatening situations and classically recognized as a fear-related behaviour. This shows that it is not only unpleasantness that drives seemingly innate fear behaviours, but that certain compounds specifically induce a fear reaction (see Chapter 2).

Historically, laboratory scientists have induced fear in rodents using a variety of odours sourced from diverse predator species (cat, puma, wolf, fox): cloth rubbed on fur, used bedding, anal gland secretions, faeces and urine (Apfelbach *et al.*, 2005). Further field studies of prey reaction to scary predator odours include a large bestiary of farm and wild animals from all over the world (Apfelbach *et al.*, 2005). Generally, these innate fear reactions can also be induced by purified compounds identified in secretions from predators; for instance, TMT (as above) and various pyridine analogues (Brechbuhl *et al.*, 2015) including 2-phenylethylamine (PEA; Ferrero *et al.*, 2011). There is evidence that the fear-inducing properties of predator smell are dependent on their meat-based diet (Berton *et al.*, 1998; Osada *et al.*, 2015). Cats are carnivorous and yet some French scientists still managed to feed one a vegetarian diet. The same animal was subsequently fed freshly killed mice and in both cases the faeces were collected (Berton *et al.*, 1998). Next, mice were tested in an exploratory task, and the meat faeces caused a reduction in exploration and induced more anxious-like behaviours. Smelling a digested fellow mouse in faeces form might be scary. Likewise urine from wolves fed melon did not cause the same avoidance reaction as urine from predators fed meat in four prey species: the mountain beaver (*Aplodontia rufa*), the house mouse (*Mus musculus*), the deer mouse (*Peromyscus maniculatus*) and the guinea pig (*Cavia porcellus*). Is it because feeding melons to meat-eating predators must be somewhat disturbing for them? Or most likely, because the odorant compounds are meat-derived metabolites? Urine odours are a complex mixture of signals indeed. Nevertheless, the captive vegan wolf is not so mighty in the eye of the rodent prey.

Behavioural responses of rats to TMT (Endres *et al.*, 2005) and other predator odours (Apfelbach *et al.*, 2005) have been extensively reviewed elsewhere; these responses include freezing, avoidance behaviour, defensive burying, reduced eating and changes in the acoustic startle response (reaction to a sudden loud noise). TMT odour exposure also leads to a stress response, as evidenced by increases in blood pressure and plasmatic corticosterone levels. This stress response will in turn affect various biological functions and behaviours. The neural pathway of TMT-induced fear and stress responses has been recently described in detail (Kondoh *et al.*, 2016). Upon odour detection, the olfactory bulb is directly connected to a brain nucleus related to the amygdala, the amygdalo-piriform transition area, which in turn activates the PVN of the hypothalamus and induces the endocrine stress response (corticosterone production). Several structures dedicated to odour recognition are involved in fear-related molecule detection (Takahashi, 2014): the main olfactory epithelium, the vomeronasal organ or the Grueneberg ganglia (Brechbuhl *et al.*, 2013, 2015). Different brain regions will also be activated by various fear-inducing odorants and this may translate to slightly different physiological and behavioural responses (Takahashi, 2014).

Responses to frightening predator odours are modulated in health and disease. Several studies point to the role of past experience. For instance in cows, the response to wolf urine odour (together with auditory stimulation of recorded wolf howling) was increased in animals bred in areas inhabited by wolves, in comparison to naïve animals that had not experienced wolves (Cooke *et al.*, 2013). In other ungulates, the heart rate of elk (*Cervus elaphus canadensis*) exposed to various predator odours was highly variable between individuals (Chabot *et al.*, 1996). One could hypothesize that these variations are the product of genetic variability (e.g. in emotional or endocrine responsiveness) and life experience (e.g. previous stress exposure). The effect of life experience also applies to attempts to repel rodents using odour sources: they can become less effective, as there is evidence of habituation following repeated exposure to commercial prey repellent (Apfelbach *et al.*, 2005). As seen in Chapter 2, it has been reported that infection with the parasite *Toxoplasma gondii* can lead to alterations in olfactory perception of fear-related odours. Infected rats might become less fearful of cat odours (Vyas *et al.*, 2007), thereby leading to higher chances of an unfortunate final meeting with the predator and increased transmission of the parasite. Science is exciting because some results may force the reassessment of what we think we know: the whole *Toxoplasma* mind-control story might not stand up to scrutiny according to some authors (Worth *et al.*, 2013) and might be debatable.

Interestingly (and perhaps frustratingly for those concerned), there is evidence that TMT or cat urine does not always induce freezing or avoidance in laboratory and field conditions (Apfelbach *et al.*, 2005). Failure to observe the expected result can sometimes be explained. There are species differences in fear responses to TMT, for instance between Sprague-Dawley, Long-Evans and Wistar rats (all laboratory rat strains), the latter being less responsive in terms of freezing (Rosen *et al.*, 2015). Could it be that, after generations of captive breeding in small plastic boxes, no selective pressure for predator avoidance has led to a reduction in the innate fear trait? Odours are also notoriously difficult to experiment with, as delivery is difficult to assess (has the animal already smelled the test odour?), the odorant concentrations are difficult to control (depending on room temperature or air movements), and the experimental paradigm used will strongly affect the results (size of the experimental arena in which odours are presented, distance to the odour source, illumination); thus, experimentation on fear-related odours can be a scarily complex affair for scientists.

Stress communicated from conspecifics

Empathy can be seen as a complex process only encountered in humans (possibly not even all of us), but it can be homologous to the emotional contagion observed in many animal species. Farmed pigs have been reported to show signs of emotional contagion, both for positive and negative emotions (Reimert *et al.*, 2013). Several laboratories have reported that a stress response can be induced by exposure to odours from stressed conspecifics. This olfactory based 'empathy', a form of chemical communication, can be seen as advantageous, in order to alert individuals to possible threats experienced by others.

In an experiment with pigs, when a food dispenser was sprayed with urine from a restrained animal, the latency to feed from it was longer than when sprayed with control urine (Vieuille-Thomas and Signoret, 1992). This suggests pig urine might contain an alarm pheromone, transmitting negative experiences to naïve sows. Studies in cows show they also appear to detect stress in a conspecific. Using feeding suppression as a measure of stress, naïve individuals exhibited a longer latency to feed in the presence of urine from a heifer subjected to electric shocks (Boissy *et al.*, 1998). The smell of stressed heifer urine also induced a rise in cortisol levels in the exposed animals. Faeces may also contain such stress signals. It has been shown that rats can distinguish between faeces from stressed and non-stressed conspecifics (Valenta and Rigby, 1968; Mackay-Sim and Laing, 1980). However, chemical signals are not all contained in urine and faeces. Exposure to various stressors has been shown to cause the production of an airborne signal inducing a number of physiological and behavioural stress responses in naïve rats (Kiyokawa *et al.*, 2013). Even in anaesthetized rats, electric stimulation of the perianal gland led to the production of this alarm pheromone (Kiyokawa *et al.*, 2013; Inagaki *et al.*, 2014). This social signalling of arousal in animal groups is perhaps an evolutionary strategy in order to rapidly propagate information that can be crucial to survival and those odorant molecules are the volatile homologues of sonic alarm calls.

What is the nature of such odorant alarm molecules? Sometimes they can be compounds with pheromone properties (Chapter 3). In the case of the rat alarm pheromone, a mixture of two compounds (4-methylpentanal and hexanal) is reported to be responsible for the information transfer (Inagaki *et al.*, 2014). In other cases those compounds can be evolutionarily 'borrowed' from other species. For instance, stressed mice secrete a sulfur-containing volatile compound that shares chemical features with compounds from meat-eating predator urine (2-*sec*-butyl-4,5-dihydrothiazoles). Mice use it to signal non-specific dangers to others, as this alarm signal is detected by the Grueneberg ganglia (Brechbuhl *et al.*, 2013).

Where do these chemical signals come from? As seen above they can be produced by various glands. Additionally, adrenaline might induce sweating, causing body odour (skin/hair bacteria), thereby leaving chemical cues about our past (and perhaps present) emotional state. Faeces and urine are frequently sources of such odorants, which might be the reason why increased defecation/micturition are observed in cases of stress, perhaps under the influence of stress mediators (and perhaps the reason for the expression encountered in several languages that links defecation and fear). Finally, chronic stress alters immunity at the level of epithelial tissue (e.g. digestive tract or skin), thereby altering the qualitative and quantitative properties of the microbial communities living there and therefore the odorant metabolites they release. It is conceivable that chronic stress might be signalled or detected through olfactory signals.

Stress odours may have other uses than communicating distress to conspecifics. In an extensive survey of frog secretions, green tree frogs (*Litoria caerulea*) are reported to emit a nutty odour from their parotid gland when stressed (Smith *et al.*, 2004). It is not known whether conspecifics are impacted; it is thought to serve as an aposematic deterrent, a strong signal to deter and scare predators away. One can wonder whether this odour causes stress in their predators. Finally, I mentioned earlier the notion that not all stress is detrimental, and a rise in cortisol levels may simply reflect arousal caused by a novel situation. Studies of African wild dogs (*Lycaon pictus*) in a zoo show that an endocrine stress response can be induced by exposure to prey odour such as gazelle and that this could be used for enrichment purposes (Rafacz and Santymire, 2014), although see Chapter 13 for potential pitfalls in the use of odours in zoos. From a practical perspective, this knowledge should be incorporated into animal husbandry and care practices. Perhaps potential stressful signals should be washed away, in order to avoid communication of distress states that would generate a stress response (e.g. see Chapter 14 on the effects of odour removal on the behaviour of laboratory rodents). There might be a huge animal welfare potential to harvest from an improved knowledge of the odorant signal semantics, which could potentially allow us to show more empathy to animals in our care.

Odours that Alleviate the Effects of the Stress Response?

If odours can induce a stress response, could there be odours that reduce it? An open mind allows you to embrace ideas you'd never considered before, but – to paraphrase the comedian Tim Minchin – never be so open-minded that your brain falls out. Nowadays, we believe we can treat the pathological manifestations and consequences of inappropriate or exacerbated stress responses. Several pharmacological agents can act with an alleged high specificity on targets in the nervous system, affecting adrenalin effects or serotonin levels, for instance, to reduce anxiety or depression. Beta blockers are known to reduce the effects of adrenal catecholamines on physiological and psychological responses to acute stress. Although they are not always effective at the human population level, selective serotonin reuptake inhibitors (SSRI, aka Prozac and other trademarks) are reported to alleviate in some cases the depressive symptoms often precipitated by chronic stress. In the past and present, other cultures with differing views of the medical process have used inhalation aromatherapy for diverse ailments. Several scientific studies point towards a role for odours in the amelioration of stress in various experimental models. Many plant-derived products have remarkable pharmacological effects when used topically or internally. Here, I will solely discuss inhalation of odorant compounds. Is there a way to manipulate the stress response and its effects using odours?

Oils extracted from citrus fruits have been a popular source of odorants in stress-relief research. Essential oils from bergamot, *Citrus bergamia*, have positive effects on anxiety and depression symptoms (Navarra *et al.*, 2015). Patients waiting for their dentist appointment are an interesting population in which to study the potential beneficial effects of odours (Lehrner *et al.*, 2000). Men and women were asked to fill out questionnaires about their inner state and emotions while sitting in the dentist waiting room, in the unannounced presence or absence of orange oil (*Citrus sinensis*). Men were not affected by the presence of odours, but female patients reported less anxiety, a more positive mood and a higher level of calmness. Such effects of odours in humans could be attributed to the aesthetic element of odour exposure, perhaps helping humans dealing with adversity and awaited loss of control of sitting in a dentist chair. There is anecdotal evidence that odour exposure during therapy sessions can help patients suffering from post-traumatic stress disorder (PTSD; Daniels and Vermetten, 2016). This might have grounding and calming effects, a distraction from the negative experience of reliving a traumatic experience. Animals cannot fill out questionnaires about their feelings, but can be tested in the lab with anxiety and depression tests in response to odorant exposure. Rats were subjected to tests (Porsolt *et al.*, 1977) in which they were forced to swim in a beaker of water for 15 min on day 1, then re-immersed in a beaker of water on day 2 and their swimming activity quantified. Duration of immobility, i.e. the time spent floating by the animal, was measured during the day 2 swim episode. This reflects learned helplessness, as the floating is assumed to indicate the rat has given up attempting to escape. Learned helplessness is a symptom of depressive disorders, and it is widely used to detect anti-depressant activity of pharmaceuticals. Rats that were exposed to lemon oil vapours between days 1 and 2 showed reduced immobility time, a marker of anti-depressant activity (Komori *et al.*, 1995). This reduction in immobility time was confirmed independently in mice (Komiya *et al.*, 2006). Lemon oil also had anxiolytic effects in an elevated plus maze task (Komiya *et al.*, 2006). In this latter study, lavender and rose oils did not produce any anti-depressant or anxiolytic effects. However, rose oil had anxiolytic effects in rats in another report (de Almeida *et al.*, 2004) and was also shown to reduce the endocrine corticosterone induction response (Fukada *et al.*, 2012).

While their colleagues were busy testing essential oils from all sorts of fruits, others worked on the effect of another odour source: a mixture of *trans*-2-hexenal and *cis*-3-hexenol called green odour (GO), which can evoke in humans the smell of fresh cut grass. GO has been reported to have anti-stress properties in laboratory rodents for a wide range of experimental paradigms. In models of prenatal stress, application of stressors to gravid dams sometimes leads to depressive-like behaviours in offspring (Fujita *et al.*, 2010). Following 10 consecutive days of forced immobilization for an hour in a restraint tube during the second half of gestation, dams gave birth to offspring that developed abnormal stress responses (as measured by plasma markers) and depressive-like symptoms. Offspring from a parallel group of stressed females treated with GO during the stress exposure regimen did not develop such stress-related traits (Fujita *et al.*, 2010), suggesting GO countered the effect of restraint stress. In an experiment where rats were subjected to a daily 3-min forced swim test for 10 days, leading to increased immobility times, GO exposure re-established locomotor behaviour, suggesting a loss in learned helplessness (Watanabe *et al.*, 2011).

As seen above, exposure to the component of fox odour TMT or an electric foot shock can induce an endocrine stress response and increased anxiety in behavioural tests. Rats pre-exposed to GO before TMT or electric foot shock treatment showed a lower stress response than non-GO-exposed conspecifics (Nikaido *et al.*, 2011). Finally, to resonate with the human example mentioned earlier, rats subjected to a fear conditioning paradigm aimed at modelling PTSD showed ameliorated symptoms following GO exposure (Nikaido *et al.*, 2016).

It seems surprising that the claim of an odorant-based therapeutic strategy for acute and chronic stress relief has not been taken up by a larger population of the scientific and medical community. Most of the scientific work published on this topic in the reputable peer-reviewed scientific literature is suitable by scientific standards. The devil is in the detail. There are inconsistencies in terms of protocols used: diversity of control odours, e.g. triethyl citrate (Fujita *et al.*, 2010) and butyric acid (Watanabe *et al.*, 2011) or distilled water (Nikaido *et al.*, 2011), diversity of behavioural effects and endocrine endpoints. It is conceivable that in certain situations, the presence of an odour could act as an environmental enrichment during a stressful situation, in an otherwise relatively barren experimental cage. There are inconsistencies as to what odours have what effect (e.g. rose oil has anxiolytic effects in some reports but not others). The Cochrane collaboration concentrates on reviewing medical data in order to critically assess the efficacy of treatments and promote evidence-based medical care. A quick search through the Cochrane reviews (comprehensive large-scale meta-analysis studies) indicate the evidence for any effect of inhalation aromatherapy is at best 'equivocal' (if not ineffective) for alleviating suffering in dementia and in labour pain management (Smith *et al.*, 2011; Forrester *et al.*, 2014). Absence of proof is not proof of absence, but these analyses are not supporting measurable effects of inhalation aromatherapy. Overall, in naïve laboratory animals that have never experienced a peaceful walk in the fresh-cut grass, some odorants are reported to have appeasing properties. The lack of solid reproducible evidence warrants caution in generalizing this effect.

There are other possible avenues to explore. If animals can learn a negative association with an aversive stimulus (e.g. foot shock and odour) that induces a stress response, it is also possible to learn a positive association with a pleasant stimulus. Would a manifestation of this phenomenon be that animals that can be trained with this positive conditioning would exhibit an attenuated stress response to mild stressors? This could be used for instance to improve welfare of captive animals undergoing veterinary procedures, limiting their stress response. This positive conditioning has been shown in certain cases for pigs (Oostindjer *et al.*, 2011). Young animals whose mothers had been fed flavoured food during late gestation and lactation showed lower behavioural and hormonal signs of stress response upon weaning when the same odour was presented (see Fig. 6.4). It would be interesting to study whether other challenging situations can be made less stressful through this seemingly comforting role of odour application.

In line with what proponents of alternative medicines (such as inhalation aromatherapy) support, there might be a case for the use of odours in management of adverse and pathological effects of the stress response. A challenge in contemporary science is about overcoming positive publication bias whereby negative data cannot be easily published. More research is needed into what could be a fascinating therapeutic avenue. If real, this potential should be studied, understood and harvested, and if fictitious, the myth should be debunked. Open your mind, but be sure to apply your critical scientific thinking.

Conclusion

A spontaneous adaptive response takes place in animals subjected to changes in their environment, whether a perceived or a real threat. This stress response can affect olfactory perception and sometimes leads to the generation of odorant signals via bodily secretion, a form of chemical communication. Olfactory cues from other species can elicit a fear or stress response in a naïve animal. Perhaps as a species that does not often use olfaction as our primary sensory modality, we humans might lack some empathy towards animals that do. In order to improve the welfare of captive animals, we should take these aspects of sensory perception into account.

References

Abreu, M.S., Giacomini, A.C., Kalueff, A.V. and Barcellos, L.J. (2016) The smell of 'anxiety': behavioral modulation by experimental anosmia in zebrafish. *Physiology and Behavior* 157: 67–71 DOI: 10.1016/j.physbeh.2016.01.030

Apfelbach, R., Blanchard, C.D., Blanchard, R.J., Hayes, R.A. and McGregor, I.S. (2005) The effects of predator odors in mammalian prey species: a review of field and laboratory studies. *Neuroscience and Biobehavioral Reviews* 29, 1123–1144. DOI: 10.1016/j.neubiorev.2005.05.005

Beery, A.K. and Kaufer, D. (2015) Stress, social behavior, and resilience: insights from rodents. *Neurobiology of Stress* 1, 116–127. DOI: 10.1016/j.ynstr.2014.10.004

Berton, F., Vogel, E. and Belzung, C. (1998) Modulation of mice anxiety in response to cat odor as a consequence of predators diet. *Physiology and Behavior* 65(2): 247–254. DOI: 10.1016/S0031-9384(98)00126-7

Boissy, A., Terlouw, C. and Le Neindre, P. (1998) Presence of cues from stressed conspecifics increases reactivity to aversive events in cattle: evidence for the existence of alarm substances in urine. *Physiology and Behavior* 63, 489–495. DOI: 10.1016/S0031-9384(97)00466-6

Boonstra, R. (2013) Reality as the leading cause of stress: rethinking the impact of chronic stress in nature. *Functional Ecology* 27, 11–23. DOI: 10.1111/1365-2435.12008

Brechbuhl, J., Moine, F., Klaey, M., Nenniger-Tosato, M., Hurni, N., Sporkert, F., Giroud, C. and Broillet, M.C. (2013) Mouse alarm pheromone shares structural similarity with predator scents. *Proceedings of the National Academy of Sciences USA* 110, 4762–4767. DOI: 10.1073/pnas.1214249110

Brechbuhl, J., Moine, F., Tosato, M.N., Sporkert, F. and Broillet, M.C. (2015) Identification of pyridine analogs as new predator-derived kairomones. *Frontiers in Neuroscience* 9, 253. DOI: 10.3389/fnins.2015.00253

Brydges, N.M., Hall, L., Nicolson, R., Holmes, M.C. and Hall, J. (2012) The effects of juvenile stress on anxiety, cognitive bias and decision making in adulthood: a rat model. *PLoS ONE* 7, e48143. DOI: 10.1371/journal.pone.0048143

Buron, E. and Bulbena, A. (2013) Olfaction in affective and anxiety disorders: a review of the literature. *Psychopathology* 46, 63–74. DOI: 10.1159/000338717

Chabot, D., Gagnon, P. and Dixon, E.A. (1996) Effect of predator odors on heart rate and metabolic rate of wapiti (*Cervus elaphus canadensis*). *Journal of Chemical Ecology* 22, 839–868. DOI: 10.1007/BF02033590

Chen, Y., Liu, X., Jia, X., Zong, W., Ma, Y., Xu, F. and Wang, J. (2014) Anxiety- and depressive-like behaviors in olfactory deficient Cnga2 knockout mice. *Behavioural Brain Research* 275, 219–224. DOI: 10.1016/j.bbr.2014.08.042

Cooke, R.F., Bohnert, D.W., Reis, M.M. and Cappellozza, B.I. (2013) Wolf presence in the ranch of origin: impacts on temperament and physiological responses of beef cattle following a simulated wolf encounter. *Journal of Animal Science* 91, 5905–5911. DOI: 10.2527/jas.2013-6777

Daniels, J.K. and Vermetten, E. (2016) Odor-induced recall of emotional memories in PTSD – review and new paradigm for research. *Experimental Neurology* 284(B), 168–180. DOI: 10.1016/j.expneurol.2016.08.001

de Almeida, R.N., Motta, S.C., de Brito Faturi, C., Catallani, B. and Leite, J.R. (2004) Anxiolytic-like effects of rose oil inhalation on the elevated plus-maze test in rats. *Pharmacology Biochemistry and Behavior* 77, 361–364. DOI: 10.1016/j.pbb.2003.11.004

Doty, R.L., Risser, J.M. and Brosvic, G.M. (1991) Influence of adrenalectomy on the odor detection performance of rats. *Physiology and Behavior* 49, 1273–1277. DOI: 10.1016/0031-9384(91)90362-R

Endres, T. and Fendt, M. (2009) Aversion- vs fear-inducing properties of 2,4,5-trimethyl-3-thiazoline, a component of fox odor, in comparison with those of butyric acid. *Journal of Experimental Biology* 212, 2324–2327. DOI: 10.1242/jeb.028498

Endres, T., Apfelbach, R. and Fendt, M. (2005) Behavioral changes induced in rats by exposure to trimethylthiazoline, a component of fox odor. *Behavioral Neuroscience* 119, 1004–1010. DOI: 10.1037/0735-7044.119.4.1004

Ferrero, D.M., Lemon, J.K., Fluegge, D., Pashkovski, S.L., Korzan, W.J., Datta, S.R., Spehr, M., Fendt, M. and Liberles, S.D. (2011) Detection and avoidance of a carnivore odor by prey. *Proceedings of the National Academy of Sciences USA* 108, 11235–11240. DOI: 10.1073/pnas.1103317108

Forrester, L.T., Maayan, N., Orrell, M., Spector, A.E., Buchan L.D. and Soares-Weiser, K. (2014) Aromatherapy for dementia. *Cochrane Database of Systematic Reviews* 2, CD003150. DOI: 10.1002/14651858.CD003150.pub2

Fujita, S., Ueki, S., Miyoshi, M. and Watanabe, T. (2010) 'Green odor' inhalation by stressed rat dams reduces behavioral and neuroendocrine signs of prenatal stress in the offspring. *Hormones and Behavior* 58, 264–272. DOI: 10.1016/j.yhbeh.2010.03.007

Fukada, M., Kano, E., Miyoshi, M., Komaki, R. and Watanabe, T. (2012) Effect of 'rose essential oil' inhalation on stress-induced skin-barrier disruption in rats and humans. *Chemical Senses* 37, 347–356. DOI: 10.1093/chemse/bjr108

Glinka, M.E., Samuels, B.A., Diodato, A., Teillon, J., Feng Mei, D., Shykind, B.M., Hen, R. and Fleischmann, A. (2012) Olfactory deficits cause anxiety-like behaviors in mice. *Journal of Neuroscience* 32, 6718–6725. DOI: 10.1523/jneurosci.4287-11.2012

Greenberg, N., Carr, J.A. and Summers, C.H. (2002) Causes and consequences of stress. *Integrative and Comparative Biology* 42, 508–516. DOI: 10.1093/icb/42.3.508

Harding, E.J., Paul, E.S. and Mendl, M. (2004) Cognitive bias and affective state. *Nature* 427, 312. DOI: 10.1038/427312a

Inagaki, H., Kiyokawa, Y., Tamogami, S., Watanabe, H., Takeuchi, Y. and Mori, Y. (2014) Identification of a pheromone that increases anxiety in rats. *Proceedings of the National Academy of Sciences USA* 111, 18751–18756. DOI: 10.1073/pnas.1414710112

Jackson, M. (2014) Evaluating the role of Hans Selye in the modern history of stress. In: Cantor, D. and Ramsden, E. (eds) *Stress, Shock, and Adaptation in the Twentieth Century*. University of Rochester Press, Rochester, New York,. Available at: https://www.ncbi.nlm.nih.gov/books/NBK189532/ (accessed 1 February 2017).

Jones, S.V., Choi, D.C., Davis, M. and Ressler, K.J. (2008) Learning-dependent structural plasticity in the adult olfactory pathway. *Journal of Neuroscience* 28, 13106–13111. DOI: 10.1523/jneurosci.4465-08.2008

Kalueff, A.V., Stewart, A.M., Song, C., Berridge, K.C., Graybiel, A.M. and Fentress, J.C. (2016) Neurobiology of rodent self-grooming and its value for translational neuroscience. *Nature Reviews Neuroscience* 17, 45–59. DOI: 10.1038/nrn.2015.8

Katz, R.J. (1982) Animal model of depression: pharmacological sensitivity of a hedonic deficit. *Pharmacology Biochemistry and Behavior* 16, 965–968.

Kiyokawa, Y., Kodama, Y., Kubota, T., Takeuchi, Y. and Mori, Y. (2013) Alarm pheromone is detected by the vomeronasal organ in male rats. *Chemical Senses* 38, 661–668. DOI: 10.1093/chemse/bjt030

Komiya, M., Takeuchi, T. and Harada, E. (2006) Lemon oil vapor causes an anti-stress effect via modulating the 5-HT and DA activities in mice. *Behavioural Brain Research* 172, 240–249. DOI: 10.1016/j.bbr.2006.05.006

Komori, T., Fujiwara, R., Tanida, M. and Nomura, J. (1995) Potential antidepressant effects of lemon odor in rats. *European Neuropsychopharmacology* 5, 477–480. DOI: 10.1016/0924-977X(95)80007-O

Kondoh, K., Lu, Z., Ye, X., Olson, D.P., Lowell, B.B and Buck, L.B. (2016) A specific area of olfactory cortex involved in stress hormone responses to predator odours. *Nature* 532, 103–106. DOI: 10.1038/nature17156

Krusemark, E.A., Novak, L.R., Gitelman D.R. and Li, W. (2013) When the sense of smell meets emotion: anxiety-state-dependent olfactory processing and neural circuitry adaptation. *Journal of Neuroscience* 33, 15324–15332. DOI: 10.1523/jneurosci.1835-13.2013

Lehrner, J., Eckersberger, C., Walla, P., Potsch, G. and Deecke, L. (2000) Ambient odor of orange in a dental office reduces anxiety and improves mood in female patients. *Physiology and Behavior* 71, 83–86. DOI: 10.1016/S0031-9384(00)00308-5

Mackay-Sim, A. and Laing, D.G. (1980) Discrimination of odors from stressed rats by non-stressed rats. *Physiology and Behavior* 24, 699–704. DOI: 10.1016/0031-9384(80)90400-X

Manella, L.C., Alperin, S. and Linster, C. (2013) Stressors impair odor recognition memory via an olfactory bulb-dependent noradrenergic mechanism. *Frontiers in Integrative Neuroscience* 7, 97. DOI: 10.3389/fnint.2013.00097

McEwen, B.S. (2007) Physiology and neurobiology of stress and adaptation: central role of the brain. *Physiological Reviews* 87, 873–904. DOI: 10.1152/physrev.00041.2006

McGann, J.P. (2015) Associative learning and sensory neuroplasticity: how does it happen and what is it good for? *Learning and Memory* 22, 567–576. DOI: 10.1101/lm.039636.115

Naudin, M., El-Hage, W., Gomes, M., Gaillard, P., Belzung, C. and Atanasova, B. (2012) State and trait olfactory markers of major depression. *PLoS ONE* 7, e46938. DOI: 10.1371/journal.pone.0046938

Navarra, M., Mannucci, C., Delbo, M. and Calapai, G. (2015) Citrus bergamia essential oil: from basic research to clinical application. *Frontiers in Pharmacology* 6, 36. DOI: 10.3389/fphar.2015.00036

Nettle, D. and Bateson, M. (2012) The evolutionary origins of mood and its disorders. *Current Biology* 22, R712–R721. DOI: 10.1016/j.cub.2012.06.020

Neumann, I.D., Veenema, A.H. and Beiderbeck, D.I. (2010) Aggression and anxiety: social context and neurobiological links. *Frontiers in Behavioral Neuroscience* 4, 12. DOI: 10.3389/fnbeh.2010.00012

Nikaido, Y., Miyata, S. and Nakashima, T. (2011) Mixture of *cis*-3-hexenol and *trans*-2-hexenal attenuates behavioral and stress responses induced by 2,5-dihydro-2,4,5-trimethylthiazoline and electric footshock stress in rats. *Physiology and Behavior* 103, 547–556. DOI: 10.1016/j.physbeh.2011.04.011

Nikaido, Y., Yamada, J., Migita, K., Shiba, Y., Furukawa, T., Nakashima, T. and Ueno, S. (2016) *Cis*-3-hexenol and *trans*-2-hexenal mixture prevents development of PTSD-like phenotype in rats. *Behavioural Brain Research* 297, 251–258. DOI: 10.1016/j.bbr.2015.10.023

Oostindjer, M., Bolhuis, J.E., Simon, K., van den Brand, H. and Kemp, B. (2011) Perinatal flavour learning and adaptation to being weaned: all the pig needs is smell. *PLoS ONE* 6, e25318. DOI: 10.1371/journal.pone.0025318

Osada, K., Miyazono, S. and Kashiwayanagi, M. (2015) The scent of wolves: pyrazine analogs induce avoidance and vigilance behaviors in prey. *Frontiers in Neuroscience* 9, 363. DOI: 10.3389/fnins.2015.00363

Pettorruso, M., De Risio, L., Di Nicola, M., Martinotti, G., Conte, G. and Janiri, L. (2014) Allostasis as a conceptual framework linking bipolar disorder and addiction. *Frontiers in Psychiatry* 5, 173. DOI: 10.3389/fpsyt.2014.00173

Porsolt, R.D., Le Pichon, M. and Jalfre, M. (1977) Depression: a new animal model sensitive to antidepressant treatments. *Nature* 266, 730–732. DOI: 10.1038/266730a0

Prunier, A., Mounier, A.M. and Hay, M. (2005) Effects of castration, tooth resection, or tail docking on plasma metabolites and stress hormones in young pigs. *Journal of Animal Science* 83, 216–222. DOI: 10.2527/2005.831216x

Rafacz, M.L. and Santymire, R.M. (2014) Using odor cues to elicit a behavioral and hormonal response in zoo-housed African wild dogs. *Zoo Biology* 33, 144–149. DOI: 10.1002/zoo.21107

Raynaud, A., Meunier, N., Acquistapace, A. and Bombail, V. (2015a) Chronic variable stress exposure in male Wistar rats affects the first step of olfactory detection. *Behavioural Brain Research* 291, 36–45. DOI: 10.1016/j.bbr.2015.05.013

Raynaud, A., Meunier, N., Dewaele, A., Grebert, D., Durieux, D., Le Bourhis, M. and Bombail, V. (2015b) Stress and olfaction: glucocorticoids affect the first step of olfactory detection in male Wistar rat. *Psychoneuroendocrinology* 61, 73–74. DOI: 10.1016/j.psyneuen.2015.07.593

Reimert, I., Bolhuis, J.E., Kemp, B. and Rodenburg, T.B. (2013) Indicators of positive and negative emotions and emotional contagion in pigs. *Physiology and Behavior* 109, 42–50. DOI: 10.1016/j.physbeh.2012.11.002

Rosen, J.B., Asok, A. and Chakraborty, T. (2015) The smell of fear: innate threat of 2,5-dihydro-2,4,

5-trimethylthiazoline, a single molecule component of a predator odor. *Frontiers in Neuroscience* 9, 292. DOI: 10.3389/fnins.2015.00292

Sakellaris, P.C. (1972) Olfactory thresholds in normal and adrenalectomized rats. *Physiology and Behavior* 9, 495–500. DOI: 10.1016/0031-9384(72)90003-0

Selye, H. (1950) Stress and the general adaptation syndrome. *British Medical Journal* 1, 1383–1392. DOI: 10.1136/bmj.1.4667.1383

Siopi, E., Denizet, M., Gabellec, M.M., de Chaumont, F., Olivo-Marin, J.C., Guilloux, J.P., Lledo, P.M. and Lazarini, F. (2016) Anxiety- and depression-like states lead to pronounced olfactory deficits and impaired adult neurogenesis in mice. *Journal of Neuroscience* 36, 518–531. DOI: 10.1523/jneurosci.2817-15.201636/2/518

Smith, B.P.C., Williams, C.R., Tyler, M.J. and Williams, B.D. (2004) A survey of frog odorous secretions, their possible functions and phylogenetic significance. *Applied Herpetology* 2, 47–82. DOI: 10.1163/1570754041231587

Smith, C.A., Collins, C.T. and Crowther, C.A. (2011) Aromatherapy for pain management in labour. *Cochrane Database Systematic Reviews* 7, CD009215. DOI: 10.1002/14651858.CD009215

Steru, L., Chermat, R., Thierry, B. and Simon, P. (1985) The tail suspension test: a new method for screening antidepressants in mice. *Psychopharmacology (Berl)* 85, 367–370.

Takahashi, L.K. (2014) Olfactory systems and neural circuits that modulate predator odor fear. *Frontiers in Behavioral Neuroscience* 8, 72. DOI: 10.3389/fnbeh.2014.00072

Toufexis, D., Rivarola, M.A., Lara, H. and Viau, V. (2014) Stress and the reproductive axis. *Journal of Neuroendocrinology* 26, 573–586. DOI: 10.1111/jne.12179

Valenta, J.G. and Rigby, M.K. (1968) Discrimination of the odor of stressed rats. *Science* 161, 599–601. DOI: 10.1016/0031-9384(80)90400-x

Vieuille-Thomas, C. and Signoret, J.P. (1992) Pheromonal transmission of an aversive experience in domestic pig. *Journal of Chemical Ecology* 18, 1551–1557. DOI: 10.1007/bf00993228

Vyas, A., Kim, S.K., Giacomini, N., Boothroyd, J.C. and Sapolsky, R.M. (2007) Behavioral changes induced by *Toxoplasma* infection of rodents are highly specific to aversion of cat odors. *Proceedings of the National Academy of Sciences USA* 104, 6442–6447. DOI: 10.1073/pnas.0608310104

Watanabe, T., Fujihara, M., Murakami, E., Miyoshi, M., Tanaka, Y., Koba, S. and Tachibana, H. (2011) Green odor and depressive-like state in rats: toward an evidence-based alternative medicine? *Behavioural Brain Research* 224, 290–296. DOI: 10.1016/j.bbr.2011.06.001

Weiss, J.M. (1972) Psychological factors in stress and disease. *Scientific American* 226, 104–113.

Wintermann, G.B., Donix, M., Joraschky, P., Gerber, J. and Petrowski, K. (2013) Altered olfactory processing of stress-related body odors and artificial odors in patients with panic disorder. *PLoS ONE* 8, e74655. DOI: 10.1371/journal.pone.0074655

Worth, A.R., Lymbery, A.J. and Thompson, R.C. (2013) Adaptive host manipulation by *Toxoplasma gondii*: fact or fiction? *Trends in Parasitology* 29, 150–155. DOI: 10.1016/j.pt.2013.01.004

Xie, W.Y., Hou, X.Y., Yan, F.B., Sun, G.R., Han, R.L. and Kang, X.T. (2013) Effect of gamma-aminobutyric acid on growth performance and immune function in chicks under beak trimming stress. *Animal Science Journal* 84, 121–129. DOI: 10.1111/j.1740-0929.2012.01051.x

11 The Role of Olfaction in Animal Housing and as Enrichment

DEBORAH L. WELLS AND PETER G. HEPPER

Animal Behaviour Centre, School of Psychology, Queen's University Belfast, Northern Ireland, UK

Concern over the welfare of animals housed in captivity has prompted a considerable amount of attention on ways of improving the physical and/or social environment. The most common term for improvements of this nature is 'environmental enrichment'. Although a rather vague concept, and one that is used interchangeably between different studies and authors, enrichment can be broadly defined as any technique designed to improve the biological functioning of a captive animal via modifications to its environment (Newberry, 1995).

The goals of environmental enrichment are still relatively open to debate, but it is generally agreed that enrichment strategies should encourage more in the way of species-typical patterns of behaviour, enhance behavioural repertoire and/or reduce the propensity for animals to engage in stereotypic, or other types of aberrant behaviour such as self-mutilation or vacuum activities, i.e. innate fixed action patterns of behaviour that are performed in the absence of the appropriate external trigger (e.g. Young, 2003). Over the years, studies have explored the effects of enrichments including cage size, social contacts and the introduction of furniture, toys and other manipulanda on the welfare of a wide variety of species, with many achieving one or more of the suggested goals of environmental enrichment (for reviews, see Shepherdson *et al.*, 1998; Wells, 2004; Lutz and Novak, 2005).

Lately, some attention has been directed towards exploring the impact of sensory stimulation (i.e. stimuli designed to trigger one or more of an animal's senses) as a method of environmental enrichment for captive animals (for a review, see Wells, 2009). This chapter discusses some of the research exploring the effect of one type of sensory stimulation, namely olfactory enrichment, on the welfare of captive animals. In her discussion of enrichment, Newberry (1995) highlighted the paucity of research in relation to the influence of odours in animal housing. Recent studies have attempted to address this, and the current chapter describes some of the pertinent research in this area. Shortcomings in the research are discussed together with factors to consider when implementing enrichment of this nature.

The Rationale for Using Olfactory Stimulation

Numerous species of animal utilize olfactory signals in their natural environment, whether to communicate their intentions with inter- and conspecifics, locate prey, lure mates and/or find food (for a review, see Doty, 2015). Unfortunately, the captive environment can be surprisingly devoid of meaningful olfactory information (Clark and King, 2008). The use of disinfectants and other sanitizing chemicals can serve to mask or remove biologically meaningful, and potentially enriching, odour cues. Recently, attempts have been made to enhance the environment of captive animals via the introduction of olfactory stimuli, both those that could be considered 'animal-based' in nature, and those that are 'plant-based'.

Animal-based Olfactory Enrichment

A wide range of animal-based odours could be considered enriching in nature, but those that have received the most scientific attention are urine and faecal material from prey or predators found in the animals' natural habitat.

The introduction of odours from natural prey has been shown to have largely enriching effects, in many cases facilitating more species-typical patterns of behaviour, e.g. increased behavioural repertoire. For example, studies have reported increased activity and social affiliations in zoo-housed felids (e.g. African lions, black-footed cats) exposed to the scents of dung or body odour from prey species including zebra, gazelle, antelope and kudu (Baker *et al.*, 1997; Schuett and Frase, 2001; Wells and Egli, 2004).

Although perhaps at apparent odds with what might be perceived as 'good' for an animal's welfare, some studies point to an enriching effect of odours arising from predatory species. For example, Christensen and Rundgren (2008) found that horses were not frightened *per se* by the scent of wolf odour, as evidenced by only minor behavioural reactions (e.g. an increased number of eating bouts) and no increase in heart rate, following exposure to the olfactory stimulus. That said, the animals did exhibit a higher level of vigilance (e.g. increased sniffing) upon detection of this odour. Boon (2003) found no difference in the behavioural reaction of two female captive-born, zoo-housed Goldei's monkeys to predatory (cheetah and ocelot faeces) and non-predatory (peppermint oil) scents. One might have expected the animals in these two studies to have shown fear reactions to the predatory odours, in light of the belief that the olfactory recognition of predators may be innate (Buchanan-Smith *et al.*, 1993). These experiments raise the question of whether odours that an animal has never been exposed to in the past, even if considered to be inherently relevant, are really meaningful to all species, or individuals (see also Chapter 2 on innateness and learning).

Some studies do admittedly point to a detrimental effect of predatory odours in the captive environment. For example, increased intragroup aggression has been observed in male mice exposed to the smell of cat urine (Zhang *et al.*, 2008), and cotton-top tamarins have been shown to exhibit high anxiety responses to the faecal odour of potential predators, including wild cats and weasels (Buchanan-Smith *et al.*, 1993). Sheep and cattle have also been shown to display behavioural changes suggestive of anxiety (Terlouw *et al.*, 1998), including reduced feeding (Pfister *et al.*, 1990; Arnould and Signoret, 1993; Arnould *et al.*, 1993), when exposed to the faeces of domestic dogs, while tapirs have been shown to display 'stress' and 'jumpiness' in response to the scent of jaguar urine (Calderisi, 1997).

Plant-based Olfactory Enrichment

Essential oils and other aromatic compounds from plants have been used for decades in the treatment of human ailments, with effects dependent upon the type of stimulus employed. For example, peppermint, jasmine and rosemary have been shown to improve alertness and enhance cognitive performance, e.g. increase the speed of mental arithmetic challenges, enhance vigilance and speed of reaction on computer-based tasks (e.g. Kovar *et al.*, 1987; Diego *et al.*, 1998; Warm and Dember, 1990), while lavender, chamomile and sandalwood have been reported to reduce anxiety and encourage a more positive state of mind (e.g. Schwartz *et al.*, 1986; Roberts and Williams, 1992; Moss *et al.*, 2003).

Recent work has shown that essential oils and other plant-derived odours might be able to improve the welfare of some animals. As is the case with humans, some odours appear to encourage relaxation and alleviate stress. Studies on mice and rats, for example, have shown anxiolytic effects of inhaled valerian (Komori *et al.*, 2006), cedarwood (Kagawa *et al.*, 2003), lemon oil (Komiya *et al.*, 2006) and chamomile (Yamada *et al.*, 1996). The ambient odour of lavender has been shown to reduce activity and vocalizations in dogs housed in rescue shelters. Graham and colleagues (2005) exposed 55 dogs of mixed breed housed in an animal rescue shelter to four types of olfactory stimulation: the diffused essential oils of lavender, chamomile, rosemary and peppermint. Animals were also studied in their normal kennel environment minus the introduction of any artificial odours (control). The dogs received each condition of stimulation for 4 h a day for 5 days, with an intervening period of 2 days between conditions. Certain aspects of the dogs' behaviour were found to be influenced by the odours (Fig. 11.1). Specifically, dogs spent significantly more time resting and less time moving upon exposure to lavender and chamomile than any of the other olfactory stimuli. These odours also encouraged less vocalization than other types of scent. The diffusion of rosemary and peppermint into the dogs' environment encouraged significantly more standing, moving and vocalizing than other types of odour. It was suggested by the authors that the welfare of sheltered dogs may be enhanced through exposure to appropriate forms of olfactory stimulation. Lavender and chamomile, which are well renowned for their relaxant properties, appeared particularly beneficial, resulting in activities suggestive of relaxation. Although still only speculative, the authors proposed that these odours may also appeal to visitors, resulting in enhanced perceptions of the rescue

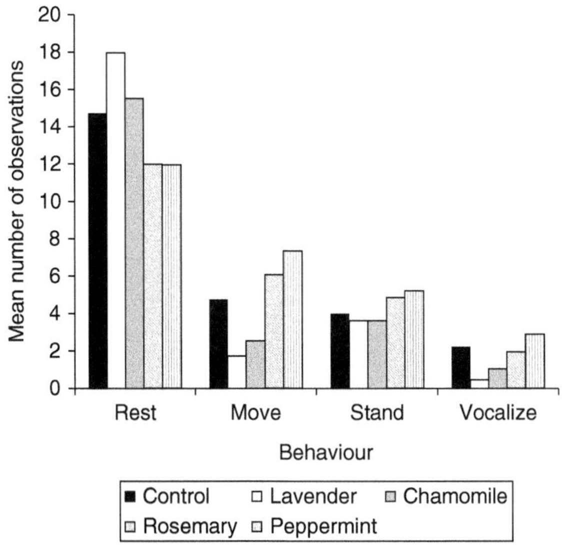

Fig. 11.1. The mean number of times dogs were observed performing specific behaviours according to condition of olfactory stimulation. (Data from Graham *et al.*, 2005.)

shelter and an increased desire to adopt a dog from such an environment.

Lavender may not only be of use in the housing of captive animals, but may also offer welfare advantages for individuals in transit. For example, both the incidence and severity of travel sickness in pigs have been found to be reduced significantly by the provision of lavender-scented straw (Bradshaw *et al.*, 1998). Wells (2006) also found that behavioural symptoms of travel-induced excitement (e.g. barking, hyperactivity) in dogs could be alleviated through the exposure to lavender-impregnated cloths. Thirty-two pet dogs were studied whilst travelling in their owners' car to their normal walking site during two conditions of olfactory stimulation: (i) a control condition, during which the animals were exposed to no odour other than that arising naturally from their environment, and (ii) an experimental condition, during which animals were exposed to the ambient odour of lavender impregnated onto flannel cloths. The behaviour of the dogs was recorded for the duration of their car journeys (between 20–30 min per trip) for 3 consecutive days of the control condition and 3 consecutive days of the experimental condition. Analysis showed that the dogs spent significantly more of their time resting and sitting, and less of their time moving and vocalizing in the experimental, compared with the control, condition (see Fig. 11.2). It was concluded that aromatherapy in the form of diffused lavender may offer a more appealing and practical method of therapy for travel-induced excitement in pet dogs than traditional methods of behavioural modification for this behavioural problem, e.g. sedatives, punishment, restraint. Further research, however, is still needed to explore the longer-term value of this therapy.

Some plant-based odours appear to have a more stimulating, than relaxing, effect on animals. For example, captive felids have been shown to gain arousing benefits from the introduction of spices such as chilli, cinnamon, cumin, nutmeg and ginger (Schuett and Frase, 2001; Pearson, 2002; Wells and Egli, 2004; Skibiel *et al.*, 2007), while zoo-housed Rothschild giraffes have been shown to spend less of their time inactive (standing and resting) following the introduction of scents including rose, banana and orange, although the authors noted significant individual differences in behaviour between the five animals under scrutiny (Fay and Miller, 2015). The essential oils of peppermint and/or rosemary have been found to increase the activity level of captive mice (Kovar *et al.*, 1987; Umezu *et al.*, 2001), dogs (Graham *et al.*, 2005), chimpanzees (Struthers and Campbell, 1996) and even lions. For example, Powell (1995) evaluated the influence of olfactory stimulation (in addition to the enrichments of frozen balls of ice and hanging logs) on the behaviour of four pair-housed (two adults, two cubs) African lions held in Zoo Atlanta. Animals were exposed to the scents of musk cologne, peppermint extract, allspice and almond extract. The various scents were chosen both for safety reasons (with some of the odours having already been used in other zoos) and biological relevance (peppermint, for example, is botanically

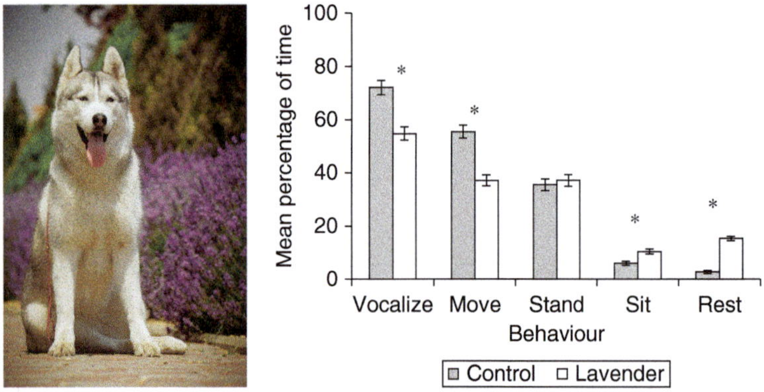

Fig. 11.2. The mean (± SE) percentage of travel time that dogs were observed vocalizing, moving, standing, sitting and resting according to condition of olfactory stimulation (* denotes significant effect, $P < 0.05$). (Data from Wells, 2006.)

related to catnip, see later). The lions were exposed to each of the odours separately following their application to furnishings within their enclosure. Behavioural observations were carried out for each pair of animals for two 30-min intervals for 24 days. During observations, the lions' enclosure was scanned every 20 s and the animals' activity was recorded according to an ethogram developed for the study. The occurrence of rare behaviours was also recorded, specifically, the licking or gnawing of an object, paw manipulation, sniffing or flehmen (where the animal curls back its upper lips whilst inhaling), face rubbing and back rolling. Analysis revealed significantly higher occurrences of sniffing and flehmen in both adults and cubs exposed to the scents, with the extract of peppermint encouraging stronger behavioural reactions than the other odours. The introduction of scents also encouraged more species-typical behaviour in the form of back rolling.

One plant-based odour that has attracted attention from a welfare perspective, specifically in cats, is catnip, *Nepeta cataria*. This herb, which is a member of the mint family, harbours particular welfare benefits for domestic cats, stimulating, in those animals genetically predisposed, a response similar to that elicited by certain pheromones. Ellis and Wells (2010) explored the influence of exposure to catnip, along with other odours, on the behaviour and welfare of cats housed in a rescue shelter. One hundred and fifty single-housed cats of mixed breed were randomly assigned to one of four conditions of olfactory stimulation: (i) a control (an odourless cloth); (ii) a 'biologically relevant' odour (a cloth impregnated with the scent of rabbit); and a range of 'biologically non-relevant' odours, specifically cloths impregnated with (iii) lavender, chosen for its relaxant properties (see earlier) and (iv) catnip. Animals were also studied in their normal kennel environment without the provision of cloths. Cats were exposed to the relevant olfactory stimuli for 3 h/day for 5 consecutive days. Each cat's behaviour was recorded every 5 min on days 1, 3 and 5 of olfactory exposure. Overall, the cats showed relatively little interest in the cloths, spending just over 6% of the total observation time interacting with these stimuli. However, animals exposed to the catnip-impregnated cloths exhibited significantly more interest in the stimulus than animals exposed to the other cloths, spending an average of 11% of the observation time interacting with the objects. Across all experimental conditions, interest in the cloths was significantly lower in the second and third hours of stimulus presentation compared to the first, suggesting habituation. Certain components of the cats' behavioural repertoire were influenced by olfactory stimulation (see Fig. 11.3). Catnip and prey scent encouraged a significantly higher frequency of behaviours indicative of reduced activity (e.g. more time sleeping, less time standing and actively exploring the environment) in comparison to the control condition. Catnip also encouraged play-like behaviour characterized as the 'catnip response'. Overall, the results suggested that certain odours, notably catnip, may hold potential as environmental enrichment for captive domestic cats. The welfare merits of this herb for other species of felids is less obvious. For instance, black-footed cats have been reported to show an increase in behavioural repertoire following exposure to cloths

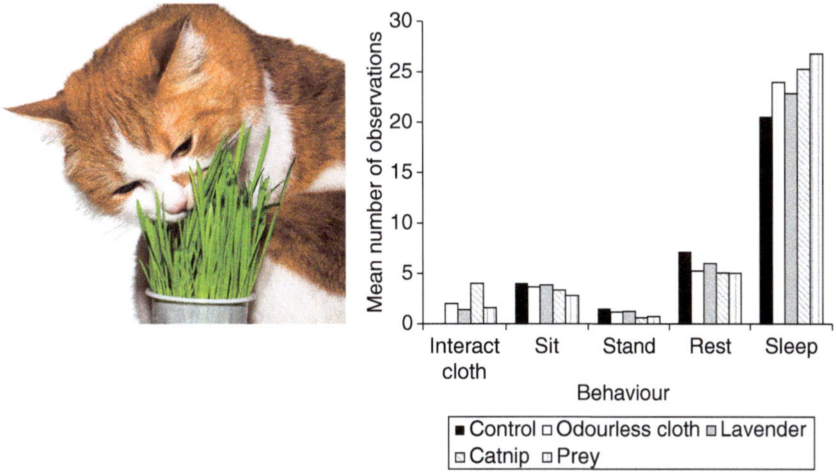

Fig. 11.3. The mean number of times cats were observed performing specific behaviours according to condition of olfactory stimulation. (Data from Ellis and Wells, 2010.)

impregnated with catnip (Wells and Egli, 2004). However, zoo-housed oncilla cats appear to gain no welfare benefits from catnip, displaying a similar occurrence of stereotypic pacing before, during and after exposure to this herb; cinnamon, by contrast, reduced the incidence of this behaviour pattern (Resende *et al.*, 2011).

Pheromones

An increasing amount of attention has been directed towards the use of pheromones (odours that elicit a specific reaction in a member of the same species; Wyatt, 2010; see also Chapter 3) as both a method of environmental enrichment for captive animals and a therapeutic treatment for domestic pets displaying behaviour patterns suggestive of reduced well-being.

Cats

Cats have been shown to gain various welfare benefits from exposure to feline pheromones. The most commonly used product to this end is Feliway (*Ceva Santé Animale*, France), a synthetic analogue of domestic cat pheromone. The product, which is available in plug-in and spray forms, has been shown to significantly reduce urine spraying (Mills and White, 2000) and aggressive interactions (Gaultier *et al.*, 2005) in pet cats and is now routinely used by rescue shelters and catteries around the world in an effort to make cats feel more confident in their captive surroundings (Beck, 2013).

The product also appears to harbour welfare benefits for larger felids. Spielman (2000) found that Feliway resulted in decreased levels of spraying and head-rubbing in zoo-housed tigers, while Gaultier and associates (2005) discovered that the same product could facilitate the introduction of a new tiger into an existing zoo group. More recently, Martínez-Macipe and colleagues (2015) showed that both cat facial pheromone and cat appeasing pheromone increased play behaviour in zoo-housed lions and highlighted the potential welfare benefits of semiochemical stimulation for such animals.

Dogs

The canine equivalent of Feliway, Dog Appeasing Pheromone (DAP; *Ceva Santé Animale*, France), also has potential as a method of enrichment for captive animals. In one of the first studies of its kind, Tod *et al.* (2005) explored the behavioural effects of this product on dogs housed in a Scottish animal rescue shelter. Thirty-seven dogs were exposed to the scent of DAP diffused into the air via electrical sockets, while 17 dogs were studied in their normal kennel environment. All animals were studied for 7 days. The authors reported a significantly lower frequency of barking in the dogs exposed to DAP in response to the presence of a person walking by their kennels, although maximum barking amplitude was not significantly altered. It was concluded that DAP may offer potential as a palliative tool for reducing behavioural signs of psychological stress in dogs. Interestingly, a more recent study (Grigg and Piehler, 2015) reported no effect of a DAP collar on the proportion of time that long-term kennelled dogs devoted to stress-related behaviours, although the small sample size (six dogs in the experimental condition vs two in the control), renders the conclusions difficult to generalize.

Canine pheromones may harbour benefits to animals with physical or psychological ailments. For example, DAP has been shown to be an effective therapeutic intervention for firework phobia and travel-induced anxiety in pet dogs (e.g. Sheppard and Mills, 2003; Estelles and Mills, 2006), although the longer-term efficacy of the treatment for behavioural problems of this type is unknown. More recently, Siracusa and colleagues (2010) studied the effects of DAP on stress responses in kennelled dogs subjected to orchiectomy or ovariohysterectomy. Dogs exposed to cages sprayed with DAP displayed greater alertness and visual exploration after their surgery than control animals exposed to pheromone-free cages sham treated with the carrier. The authors concluded that a synthetic appeasing pheromone may improve the recovery and welfare of dogs undergoing surgery.

Pigs

Livestock, notably pigs, have also been shown to gain benefits from semiochemical stimulation. Exposure to sexual pheromones, for instance, has been shown to reduce aggressive behaviour during the regrouping of such animals (McGlone *et al.*, 1987; Petherick and Blackshaw, 1987). Nursing behaviour has also been shown to be regulated by maternal pheromones in pigs. Morrow-Tesch and McGlone (1990a,b) found that piglets would not nurse when odours thought to contain maternal pheromones were removed from the skin of their lactating mothers. In a similar vein, piglets have been shown to engage in less aggressive behaviour and gain more weight during post-weaning mixing when exposed to synthetic compounds containing elements of a maternal pheromone (Pageat and Teissier, 1998; McGlone and Anderson, 2008), as detailed in Chapter 12 (see Fig. 12.4). Pigs subject to the vibrations of a transport simulation, a known stressor, have also shown significantly lower heart rate in response to the scent of a synthetic compound containing maternal pheromones, compared with the conditions of a non-odour control or a non-relevant unfamiliar odour, i.e. hartshorn oil (Driessen *et al.*, 2008).

Horses

Horses have been subject to little attention with regard to the value of pheromonal therapy, but the small amount of research conducted in this area points to potentially beneficial effects. Falewee and colleagues (2006) examined the effects of a synthetic equine appeasing pheromone (EAP) on a group of mixed-breed horses in a double-blinded controlled study. An experimental group of animals (n = 20) was administered two nasal sprays per nostril 18 min prior to a fear-eliciting test in which each horse was walked through a fringed curtain to gain access to a stable; 20 control animals undertook the same test having been sprayed with an aqueous solution devoid of pheromonal content. The behaviour of both groups of animals was recorded over the trial and behavioural patterns associated with fear (avoidance and emotional reaction) were assessed. Each animal's heart rate was also recorded using a Polar Horse Trainer Advanced heart rate monitoring system. Analysis revealed fewer behavioural signs of anxiety in the experimental group of animals compared with the control. Animals sprayed with EAP stopped less frequently and showed shorter hesitations when they encountered the fringed curtain, needing less time to enter the stable than horses in the placebo group (Fig. 11.4). Experimental horses also displayed a lower mean heart rate and maximum heart rate over the trial than their control counterparts. The authors concluded that EAP may be a useful method of curbing behavioural and physiological expressions of fear and anxiety, which can often prevent the management of such animals in a safe manner. The slight time lag (~20 min) between application of the pheromonal treatment and observable effects on behaviour, however, suggests that EAP may be of most use in foreseeable fear-eliciting situations.

Practical Applications, Problem Issues and Further Considerations

This chapter highlights the potential for certain odours to be used as a tool for improving the welfare of animals, captive or otherwise. Many of the studies undertaken have shown olfactory stimulation to result in changes in the biological functioning of animals in a manner suggestive of enhanced

Parameter	Control group	Experimental group
Mean heart rate (bpm)	87 (±22)	69 (±14)
Maximum heart rate (bpm)	175 (±23)	145 (±23)
Time taken to enter stall (s)	203 (±66)	115 (±76)

Fig. 11.4. Mean (± SD) performance measures for a control group of horses and for horses exposed to equine appeasing pheromone (EAP); bpm = beats per minute. (Adapted from Falewee *et al.*, 2006.)

physical and/or psychological welfare, and in this respect olfactory stimulation could be considered to meet many of the suggested goals of environmental enrichment, e.g. increased behavioural repertoire (see Young, 2003).

The success of olfactory stimulation as a method of enhancing animal well-being, however, may depend, to some degree, upon the species in question. Non-human primates, which rely more heavily upon their sense of sight, seem to gain fewer welfare advantages than more olfactory-driven species. Ostrower and Brent (2000), for instance, found that the smell of 'pleasant' (e.g. vanilla, orange, peach) and 'unpleasant' (e.g. moth balls, cigar smoke, limburger cheese) odours on fleece cloths elicited no more attention from captive chimpanzees than cloths devoid of olfactory stimuli. Wells and colleagues (2007) similarly found that olfactory stimulation in the form of odour-impregnated (orange, almond, vanilla, peppermint) cloths had no significant effect on the behaviour of zoo-housed gorillas. It is difficult to determine from these studies whether olfactory enrichment harbours no welfare advantages for great apes, or whether the odours employed in these studies simply held no emotional valence for the animals concerned (see later).

Some caution needs to be exerted with regard to the health and safely implications of certain odours. As highlighted above, some of the animal-based odours, particularly those arising from potential predatory species, may induce stress and should be used sparingly in the captive environment, if at all. Many of the other studies carried out in this area have utilized essential oils, which come with their own concerns. For example, whilst moderate doses of lavender can have an anxiolytic effect, higher doses can cause sedation (Shaw *et al.*, 2007), a potentially undesirable outcome in the captive environment. The issue of toxicity also needs to be addressed in relation to the use of essential oils and plant derivatives. Some authors have reported unwanted physical side effects (e.g. vomiting, skin irritation) in pet cats exposed to potpourri (Richardson, 1999; Foss, 2002), and deaths arising from nightshade plants have been reported in some primates (Engel, 2002). Since lavender and certain other essential oils are not normally recommended for use in pregnant woman, there may also be issues over the use of these stimuli in certain cohorts of animals.

As with the introduction of any enrichment programme, a careful cost–benefit analysis, weighing up the advantages and disadvantages of the olfactory enrichment, should be carried out before any scheme is put into practice. Importantly, enrichment strategies should be adopted that ultimately aim to improve the biological functioning of the species under scrutiny. Some of the least successful studies have utilized stimuli that, arguably, could be considered biologically 'meaningless' to the animals concerned. Stimuli specific to a species' natural habitat should not necessarily be considered meaningful, or advantageous, to the animal under scrutiny, and in many cases stimuli that do not occur naturally in the wild may offer more in the way of welfare advantages.

Although this chapter has focused on the value of odour *introduction* for captive animal welfare, the avoidance of natural odour *removal* also needs to be mentioned. Routine husbandry practices can remove important olfactory information, both regarding the self and others in the group. Many species (e.g. non-human primates, felids) scent mark in a bid to demarcate territories, advertise reproductive state, signal resource ownership and convey information on social status (e.g. Drea and Scordato, 2008). Natural scents or urine, faeces, hair and sweat can also provide cues that are both informative, and potentially enriching, in their own right. As a potential solution to the problem of odour removal, Clark and King (2008) recommend cleaning half of an animal's enclosure at a time, thus allowing the retention of some meaningful odour cues (see also Chapter 14 for examples on this issue in laboratory animals).

Further research in this area is very much needed. Many of the studies carried out have utilized small groups of animals (< 5) housed in single institutional settings, and findings are therefore hard to generalize. Relatively few studies have explored the long-term impact of olfactory enrichment and it is unclear whether animals exposed to these stimuli gain anything more than short-term welfare benefits. The results of those studies that have employed 'no odour' as their control must also be carefully scrutinized to ensure that any significant effects reported are not actually spurious results arising from the introduction of any novel odour.

Many institutions housing animals are now paying more attention to the animals' environment and the important relationship between housing and well-being. This chapter demonstrates that olfactory stimulation can be utilized successfully as enrichment for captive animals, although the methodological weaknesses (e.g. low sample size, brief study time) that typically plague research of this nature render firm conclusions difficult to draw. Further research in this area will hopefully ensure that developments continue to be made in our understanding of how ideally to house animals in order to promote both their physical and psychological well-being.

References

Arnould, C. and Signoret, J.P. (1993) Sheep food repellents: Efficacy of various products, habituation, and social facilitation. *Journal of Chemical Ecology* 19, 225–236. DOI: 10.1007/BF00993691

Arnould, C., Orgeur, P., Sempere, A., Signoret, J.P. and Verdier, M. (1993) Food repulsion in three species of ungulates in pasture situation – the effectiveness of the odor of dog feces. *Revue d'Ecologie – La Terre et la Vie* 48, 121–132.

Baker, W., Campbell, R. and Gilbert, J. (1997) Enriching the pride: scents that make sense. *Shape of Enrichment* 6, 1–3.

Beck, A. (2013) Use of pheromones to reduce stress in sheltered cats. *Journal of Feline Medicine and Surgery* 15, 829–830. DOI: 10.1177/1098612X13500882

Boon, M. (2003) Goeldi's monkeys (*Callimico goeldii*): olfactory enrichment to stimulate natural behaviour and greater activity. In: Gilbert, T. (ed.) *Proceedings of the 5th Annual Symposium on Zoo Research*, Marwell Zoo, Owslebury, UK, pp. 212–224.

Bradshaw, R.H., Marchant, J.N., Meredith, M.J. and Broom, D.M. (1998) Effects of lavender straw on stress and travel sickness in pigs. *Journal of Alternative and Complimentary Medicine* 4, 271–275. DOI: 10.1089/acm.1998.4.3-271

Buchanan-Smith, H.M., Anderson, D.A. and Ryan, C.W. (1993) Responses of cotton-top tamarins (*Saguinus oedipus*) to fecal scents of predators and non-predators. *Animal Welfare* 2, 17–32.

Calderisi, D. (1997) Different scents for different responses in predator–prey relationships as a form of enrichment in captive animals. In: Ware, V. and Worley, K. (eds) *Proceedings of the 3rd International Conference on Environmental Enrichment*. Orlando, Florida, pp. 155–161.

Christensen, J.W. and Rundgren, M. (2008) Predator odour *per se* does not frighten domestic horses. *Applied Animal Behaviour Science* 112, 136–145. DOI: 10.1016/j.applanim.2007.08.003

Clark, F. and King, A.J. (2008) A critical review of zoo-based olfactory enrichment. In: Hurst, J.L., Beynon, R.J., Roberts, S.C. and Wyatt, T.D. (eds) *Chemical Signals in Vertebrates 11*. Springer, New York, pp. 391–398. DOI: 10.1007/978-0-387-73945-8_37

Diego, M.A., Jones, N.A., Field, T., Hernandez-Reif, M., Schanberg, S., Kuhn, C., McAdam, V., Galamaga, R. and Galamaga, M. (1998) Aromatherapy positively affects mood, EEG patterns of alertness and math computations. *International Journal of Neuroscience* 96, 217–224.

Doty, R.L. (ed.) (2015) *Handbook of Olfaction and Gustation*. Wiley Blackwell, Hoboken, New Jersey.

Drea, C.M. and Scordato, E.S. (2008) Olfactory communication in the ringtailed lemur (*Lemur catta*): form and function of multimodal signals. In: Hurst, J.L., Beynon, R.L., Roberts, S.C. and Wyatt, T.D. (eds) *Chemical Signals in Vertebrates 11*. Springer, New York, pp. 91–102. DOI: 10.1007/978-0-387-73945-8_8

Driessen, B., Peeters, E. and Geers, R. (2008) Influence of olfactory substances on the heart rate and lying behaviour of pigs during transport simulation. *Animal Welfare* 17, 155–160.

Ellis, S.L.H and Wells, D.L. (2010) The influence of olfactory stimulation on the behaviour of cats housed in a rescue shelter. *Applied Animal Behaviour Science* 123, 56–62. DOI: 10.1016/j.applanim.2009.12.011

Engel, C. (2002) *Wild Health: How Animals Keep Themselves Well and What We Can Learn from Them.* Houghton Mifflin, Boston, Massachusetts.

Estelles, M.G. and Mills, D.S. (2006) Signs of travel-related problems in dogs and their response to treatment with dog-appeasing pheromone. *Veterinary Record* 159, 143.

Falewee, C., Gaultier, E., Lafont, C., Bougrat, L. and Pageat, P. (2006) Effect of a synthetic equine maternal pheromone during a controlled fear-eliciting situation. *Applied Animal Behaviour Science* 101, 144–153. DOI: 10.1016/j.applanim.2006.01.008

Fay, C. and Miller, L.J. (2015) Utilizing scents as environmental enrichment: preference assessment and application with Rothschild giraffe. *Animal Behavior and Cognition* 2, 285–291.

Foss, T.S. (2002) Liquid potpourri and cats-essence of trouble. *Veterinary Technician* 23, 686–689.

Gaultier, E., Falawee, C., Bougrat, L. and Pageat, P. (2005) The introduction of a female tiger (*Panthera tigris*) in a pre-established group of two neutered males: a case study. *Current Issues and Research in Veterinary Behavioral Medicine. Fifth International Veterinary Behavior Meeting*, 14–16 July 2005, Minneapolis, Minnesota, pp. 1–5.

Graham, L., Wells, D.L. and Hepper, P.G. (2005) The influence of olfactory stimulation on the behaviour of dogs housed in a rescue shelter. *Applied Animal Behaviour Science* 91, 143–153. DOI: 10.1016/j.applanim.2004.08.024

Grigg, E.K. and Piehler, M. (2015) Influence of dog appeasing pheromone (DAP) on dogs housed in a long-term kennelling facility. *Veterinary Record Open* 2, e000098.

Kagawa, D., Jokura, H., Ochiai, R., Tokimitsu, I. and Tsubone, H. (2003) The sedative effects and mechanism of action of cedrol inhalation with behavioral pharmacological evaluation. *Planta Medica* 69, 637–641.

Komiya, M., Takeuchi, T. and Harada, E. (2006) Lemon oil vapor causes an anti-stress effect via modulating the 5-HT and DA activities in mice. *Behavioural Brain Research* 172, 240–249. DOI: 10.1016/j.bbr.2006.05.006

Komori, T., Matsumoto, T., Motomura, E. and Shiroyama, T. (2006) The sleep-enhancing effect of valerian inhalation and sleep-shortening effect of lemon inhalation. *Chemical Senses* 31, 731–737. DOI: 10.1093/chemse/bjl015

Kovar, K.A., Gropper, B. and Ammon, H.T.P. (1987) Blood levels of 1,8-cineole and locomotor activity of mice after inhalation and oral administration of rosemary oil. *Planta Medica* 53, 315–319.

Lutz, C.K. and Novak, M.A. (2005) Environmental enrichment for nonhuman primates: theory and application. *Ilar Journal* 46, 178–191.

Martínez-Macipe, M., Lafont-Lecuelle, C., Manteca, X., Pageat, P. and Cozzi, A. (2015) Evaluation of an innovative approach for sensory enrichment in zoos: semiochemical stimulation for captive lions (*Panthera leo*). *Animal Welfare* 24, 455–461.

McGlone, J.J. and Anderson, D.L. (2008) Synthetic maternal pheromone stimulates feeding behavior and weight gain in pigs. *Journal of Animal Science* 80, 3179–3183.

McGlone, J.J., Curtis, S.E. and Banks, E.M. (1987) Evidence for aggression modulating pheromones in prepubertal pigs. *Behavioral and Neural Biology* 47, 27–39.

Mills, D.S. and White, J.C. (2000) Long-term follow up of the effect of a pheromone therapy on feline spraying behaviour. *Veterinary Record* 147, 746–747.

Morrow-Tesch, J. and McGlone, J.J. (1990a) Sensory systems and nipple attachment behaviour in neonatal pigs. *Physiology and Behavior* 47, 1–4. DOI: 10.1016/0031-9384(90)90034-2

Morrow-Tesch, J. and McGlone, J.J. (1990b) Sources of maternal odors and the development of odor preferences in baby pigs. *Journal of Animal Science* 68, 3563–3571.

Moss, M., Cook, J., Wesnes, K. and Duckett, P. (2003) Aromas of rosemary and lavender essential oils differentially affect cognition and mood in healthy adults. *International Journal of Neuroscience* 113, 15–38. DOI: 10.1080/00207450390161903

Newberry, R.C. (1995) Environmental enrichment: Increasing the biological relevance of captive environments. *Applied Animal Behaviour Science* 44, 229–243. DOI: 10.1016/0168-1591(95)00616-Z

Ostrower, S. and Brent, L. (2000) Olfactory enrichment for captive chimpanzees: responses to different odors. *Lab Primate Newsletter* 36, 8–12.

Pageat, P. and Tessier, Y. (1998) Usefulness of a porcine pheromone analogue in the reduction of aggression between weaners on penning: behavioural study. *Proceedings of the 15th International Pig Veterinary Society Congress*, Birmingham, UK.

Pearson, J. (2002) On a roll: novel objects and scent enrichment for Asiatic lions. *Shape of Enrichment* 11, 7–10.

Petherick, J.C. and Blackshaw, J.K. (1987) A review of the factors influencing the aggressive and agonistic behaviour of the domestic pig. *Australian Journal of Experimental Agriculture* 27, 605–611.

Pfister, J.A., Müller-Schwarze, D. and Balph, D.F. (1990) Effects of predator fecal odors on feed selection by sheep and cattle. *Journal of Chemical Ecology* 16, 573–583. DOI: 10.1007/BF01021787

Powell, D.M. (1995) Preliminary evaluation of environmental enrichment techniques for African lions (*Panthera leo*). *Animal Welfare* 4, 361–370.

Resende, L., Pedretti, G., Karla, C. and Andriolo, A. (2011) Influence of cinnamon and catnip on the stereotypical pacing of oncilla cats (*Leopardus tigrinus*) in captivity. *Journal of Applied Animal Welfare Science* 14, 247–254. DOI: 10.1080/10888705.2011.576981

Richardson, J.A. (1999) Potpourri hazards in cats. *Veterinary Medicine* 94, 1010–1012.

Roberts, A. and Williams, J.M.G. (1992) The effect of olfactory stimulation on fluency, vividness of imagery and associated mood – a preliminary study. *British Journal of Medical Psychology* 65, 197–199.

Schuett, E.B. and Frase, B.A. (2001) Making scents: using the olfactory senses for lion enrichment. *Shape of Enrichment* 10, 1–3.

Schwartz, G.E., Whitehorn, D., Hernon, J.C. and Jones, M. (1986) Subjective and respiratory differences of fragrances: interactions with hedonics. *Psychophysiology* 23, 460.

Shaw, D., Annett, J.M., Doherty, B. and Leslie, J.C. (2007) Anxiolytic effects of lavender oil inhalation on open-field behaviour in rats. *Phytomedicine* 14, 613–620. DOI: 10.1016/j.phymed.2007.03.007

Shepherdson, D.J., Mellen, J.D. and Hutchins, M. (1998) *Second Nature: Environmental Enrichment for Captive Animals*. Smithsonian Institution Press.

Sheppard, G. and Mills, D.S. (2003) Evaluation of dog-appeasing pheromone as a potential treatment for dogs fearful of fireworks. *Veterinary Record* 152, 432–436.

Siracusa, C., Manteca, X., Cuenca, R., Alcala, M., Alba, A., Lavin, S. and Pastor, J. (2010) Effect of a synthetic appeasing pheromone on behavioral, neuroendocrine, immune, and acute-phase perioperative stress responses in dogs. *Journal of the American Veterinary Medical Association* 237, 673–681.

Skibiel, A.L., Trevino, H.S. and Naugher, K. (2007) Comparison of several types of enrichment for captive felids. *Zoo Biology* 26, 371–381. DOI: 10.1002/zoo.20147

Spielman, J.S. (2000) Olfactory enrichment for captive tigers (*Panthera tigris*) and lions (*Panthera leo*), using a synthetic analogue of feline facial pheromone. MSc thesis, University of Edinburgh, UK.

Struthers, E.J. and Campbell, J. (1996) Scent-specific behavioral response to olfactory enrichment in captive chimpanzees (*Pan troglodytes*). *XVIth Congress of the International Primatological Society and the XIXth Conference of the American Society of Primatology*, Madison, Wisconsin.

Terlouw, E.M.C., Boissy, A. and Blinet, P. (1998) Behavioural responses of cattle to the odours of blood and urine from conspecifics and to the odour of faeces from carnivores. *Applied Animal Behaviour Science* 57, 9–21. DOI: 10.1016/S0168-1591(97)00122-6

Tod, E., Brander, D. and Waran, N. (2005) Efficacy of dog appeasing pheromone in reducing stress and fear related behaviour in shelter dogs. *Applied Animal Behaviour Science* 93, 295–308. DOI: 10.1016/j.applamin.2005.01.007

Umezu, T., Sakata, A. and Hiroyasu, I. (2001) Ambulation promoting effect of peppermint oil and identification of its active constituents. *Pharmacology Biochemistry and Behavior* 69, 383–390. DOI: 10.1016/S0091-3057(01)00543-3

Warm, J.S. and Dember, W.N. (1990) Effects of fragrances on vigilance performance and stress. *Perfume Flavor* 15, 15–18.

Wells, D.L. (2004) A review of environmental enrichment for kennelled dogs, *Canis familiaris*. *Applied Animal Behaviour Science* 85, 307–317. DOI: 10.1016/j.applanim.2003.11.005

Wells, D.L. (2006) Aromatherapy for travel-induced excitement in domestic dogs. *Journal of the American Veterinary Medical Association* 229, 964–967.

Wells, D.L. (2009) Sensory stimulation as environmental enrichment for captive animals: a review. *Applied Animal Behaviour Science* 118, 1–11. DOI: 10.1016/j.applanim.2009.01.002

Wells, D.L. and Egli, J.M. (2004) The influence of olfactory enrichment on the behaviour of black-footed cats, *Felis nigripes*. *Applied Animal Behaviour Science* 85, 107–119. DOI: 10.1016/j.applanim.2003.08.013

Wells, D.L., Hepper, P.G., Coleman, D. and Challis, M.G. (2007) A note on the effect of olfactory stimulation on the behaviour and welfare of zoo-housed gorillas. *Applied Animal Behaviour Science* 106, 155–160. DOI: 10.1016/j.applanim.2006.07.010

Wyatt, T.D. (2010) Pheromones and signature mixtures: defining species-wide signals and variable cues for identity in both invertebrates and vertebrates. *Journal of Comparative Physiology A* 196, 685–700. DOI: 10.1007/s00359-010-0564-y

Yamada, K., Miura, T., Mimaki, Y. and Sashida, Y. (1996) Effect of inhalation of chamomile oil vapour on plasma ACTH level in ovariectomized rats under restriction stress. *Biological and Pharmaceutical Bulletin* 19, 1244–1246.

Young, R.J. (2003) *Environmental Enrichment for Captive Animals*. Blackwell Science Ltd, Oxford.

Zhang, J.X., Sun, L.X., Bruce, K.E. and Novotony, M.V. (2008) Chronic exposure to cat odor enhances aggression, urinary attractiveness and sex pheromones of mice. *Journal of Ethology* 26, 279–286. DOI: 10.1007/s10164-007-0060-1

12 Olfactory Behaviour in Farm Animals

Caroline Clouard and J. Elizabeth Bolhuis

Adaptation Physiology Group, Wageningen University and Research, The Netherlands

The sense of smell is essential for many animals, as illustrated in the previous chapters. We learned how olfaction plays a role in numerous biological processes, including foraging (Chapter 6), reproduction (Chapter 7), parental care (Chapter 8) and fear (Chapter 10). Olfactory information and communication are also important for farm animals, such as pigs, cattle, sheep and goats – and even chickens and other domestic birds, which seem to have a much more sensitive olfactory system than was imagined in the past (Jones and Roper, 1997; Zidar and Lovlie, 2012). For some biological processes in farm animals, many of which are influenced or even controlled by humans, the importance of odours is well known. For instance, an androgen steroid associated with boar odour is frequently used, as a 'boar spray', to facilitate the detection of oestrus in sows that are to be artificially inseminated (Melrose *et al.*, 1971; Langendijk *et al.*, 2006). Another example of the use of odours in livestock farming is the transfer of the odour of a ewe's own lamb to an alien lamb to facilitate the ewe's acceptance of this alien foster lamb (Price *et al.*, 1984; Martin *et al.*, 1987). Information on olfactory behaviour in farm animals is, however, generally scarce, as is information on the use of odours in farm animal management.

This chapter presents several examples of how olfactory information affects the behaviour of farm animals. Farm animals in intensive production systems are mostly kept at high stocking densities in large groups, and often indoors, which may interfere with the use of their sense of smell. This chapter therefore also describes the influence of farming conditions on how animals perceive and respond to olfactory cues and, finally, it discusses opportunities to improve the welfare and production of farm animals by making use of odours and olfaction.

Olfaction and Social Discrimination in Farm Animals

Most farm animals are social animals for which discrimination between conspecifics is useful for making appropriate behavioural adjustments. Recognition of specific individuals allows animals to further fine-tune their behaviour based on previous encounters, and promotes a stable group structure. Olfaction plays an important role in social discrimination and recognition. Animals generally seem to use a blend of visual, olfactory, auditory and/or tactile cues to discriminate between conspecifics. For example, pigs are able to tell apart two familiar conspecifics, even if these two are siblings, as was shown in a Y-maze test (McLeman *et al.*, 2005). Sheep also use multiple sensory cues to recognize their young during the first days after birth (e.g. Nowak *et al.*, 2011). Even though social discrimination often relies on multisensory perception, olfaction has been found to play a major role in this process. Accordingly, olfactory perception is sufficient for pigs to discriminate between familiar and unfamiliar conspecifics (Kristensen *et al.*, 2001), and even between different familiar group mates from the same sex and litter (McLeman *et al.*, 2008).

In contrast to other sensory cues, odour traces can be used to identify an animal even in its absence. Sources of individual body odours are, for instance, skin secretions, saliva and urine. In an elegant experiment using a habituation–dishabituation procedure, Mendl *et al.* (2002) demonstrated the ability of young pigs to discriminate the odour of urine from different conspecifics. In the habituation–dishabituation procedure, 10-week-old female pigs were exposed to three successive presentations of urine samples for 2 min separated by 15-min intervals. The observations started as soon as the pig nosed the sieve containing the urine, which was held above its head. Pigs in the discrimination group were presented with urine samples from the same individual twice (habituation) and, thereafter, a urine sample from a different individual (dishabituation). Pigs in the discrimination group were tested twice: in one test, they were exposed to sets of urine from familiar pigs (pen mates) and, in the other test, to sets of urine from unfamiliar pigs from other pens. Control pigs, which were presented with urine samples from the same individual three times, showed a stable reduction in the time spent exploring the samples over the three successive presentations (habituation; Fig. 12.1A). In contrast, the discrimination pigs showed a large increase in the time spent investigating the new urine sample during the third presentation (dishabituation; Fig. 12.1A). This clearly indicates that the pigs discriminated between the odour of urine from the first individual and that from the new individual. As shown in Fig. 12.1B, pigs can both discriminate between urine samples from different pen mates and between urine samples from different unfamiliar pigs from other pens.

The ability to discriminate between conspecifics based on olfactory cues present in urine has also been reported in other farm animals. Indeed, not only pigs (Meese *et al.*, 1975), but also calves, goats and sheep (Baldwin, 1977; Baldwin and Meese, 1977) can discriminate between the odours of urine samples from different conspecifics, as shown in an operant conditioning procedure, in which animals have to learn to press a panel switch associated with one odour to obtain a food reward, whereas the other odour stimulus is not rewarded. In a habituation–dishabituation test, horses also distinguished between urine samples from individuals differing in sex and age, but did not show discrimination between samples from horses similar in sex, age and diet (Hothersall *et al.*, 2010). Interestingly, sheep appear to be able to discriminate between individual odours from wool, faeces, saliva and other body fluids too, indicating that, as well as urine, several other body secretions also convey olfactory information that can be used for social recognition (Baldwin and Meese, 1977). Similarly, horses can tell apart the body odour (obtained from rubbing their coats) of different unfamiliar horses (Peron *et al.*, 2014). They also discriminate between the odour of their own faeces and that of

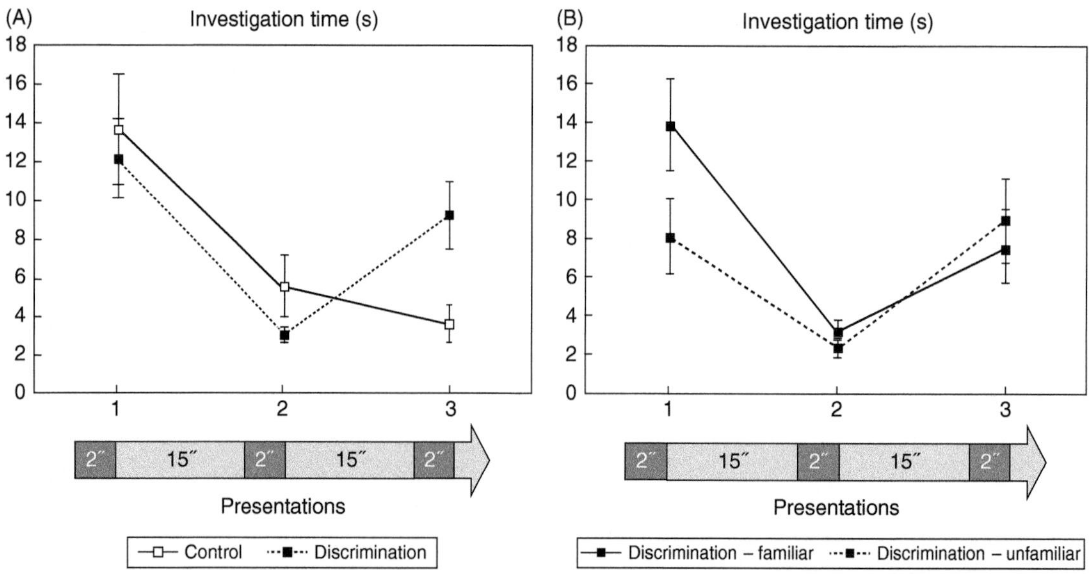

Fig. 12.1. Time spent investigating urine from conspecifics by pigs during three consecutive presentations. All pigs were presented with urine from the same individual during presentations 1 and 2. During presentation 3, (A) Control pigs (n = 11) were presented with urine from the same individual and Discrimination pigs (n = 11) with that from a different individual; (B) Discrimination pigs (n = 11) were presented with urine from an individual that was familiar (from a pen mate) or unfamiliar (from another pen). Urine samples were presented for 2 min, interspersed by 15-min intervals. (The figure is based on data from Mendl *et al.*, 2002.)

conspecifics, and may be able to distinguish between odours of faeces from different individual group mates, as some horses spent more time sniffing the faeces of group mates from which they received a large amount of aggressive behaviours within the herd (Krueger and Flauger, 2011). Chickens also seem to have individual body odours, as mice were clearly able to distinguish the preen oil secretions – produced by the uropygial gland at the base of the chicken's tail – from different red junglefowl birds (Karlsson *et al.*, 2010). Whether chickens use these individual body odours for social discrimination is so far unknown and warrants further investigation, but other bird species have been demonstrated to do so (e.g. blue pretels; Bonadonna and Nevitt, 2004).

Olfaction and Reproduction on Farms

Chemosensory cues do not only carry information about an individual's identity, as explained in the previous section, but also convey temporary information about its reproductive state, such as maturity and sexual receptivity. For instance, ungulates use odours and pheromones to detect oestrous females from a distance. Indeed, rams can discriminate between the urine of sexually receptive and non-receptive ewes (Blissitt *et al.*, 1990). A study on sexual behaviour in chickens suggests that cockerels use olfaction for mate choice, and that secretions from the uropygial gland of the hen may act as a chemical cue in this process (Hirao *et al.*, 2009). Hence, odours and pheromones play a major role in reproductive behaviours (as reviewed in Chapter 7), also in farm animals. Reproduction in intensively kept livestock is mostly controlled by humans as most farm animals are artificially inseminated, and olfactory cues are sometimes used to enhance reproduction efficiency on farms (Rekwot *et al.*, 2001). For example, a nasal spray containing oestrus-specific molecules identified in cow urine can be used to enhance libido and sperm quality in bulls, enabling increased efficiency and safety during semen collection for artificial insemination (Le Danvic *et al.*, 2015). In addition, male pheromones in urine, faeces or glands can stimulate the onset of puberty and reduce the duration of anoestrus after parturition in ewes, goats and sows. For instance, exposing sows to boar odour induces early puberty (Signoret and Lindsay, 1982), while the perception of ram odour in urine, wool and wax (from around the eyes and from the flanks of the ram) stimulate ewes to ovulate early in the breeding season (Izard, 1983). Accurate oestrus detection is crucial for reproductive success on farms applying artificial insemination, and olfactory cues can be used to facilitate the detection of females in heat. As an example, exposure to boar odour induces the characteristic 'standing reflex', indicating acceptance of the male, in oestrous sows (Melrose *et al.*, 1971). In dairy cattle farming, where oestrus detection is challenging, it has been suggested that the assessment of oestrus-specific odours in the cows' urine or other body fluids could help to identify cows in heat. To that end, dogs (Fischer-Tenhagen *et al.*, 2015; Johnen *et al.*, 2015), rodents (Sankar and Archunan, 2005) and bulls (Sankar and Archunan, 2004) have been used in the ongoing search for oestrus-specific molecules in cows.

Another factor vital for reproduction and animal welfare on farms is the establishment of a good mother–offspring relationship in early life, which will impact the quality of maternal care and may have long-term effects on the developing offspring. In ungulates, including goats and sheep, the mother and her offspring establish an exclusive bond in the first hours following parturition, which serves as a basis for maternal behaviour. We learned in Chapter 8 that the early establishment of this mother–offspring bond, which represents a specific part of individual discrimination, relies mainly upon olfactory cues – even though ungulate mothers are likely to use multisensory cues for later discrimination of their offspring. In goats and sheep (Romeyer *et al.*, 1994), but also in pigs (Horrell and Hodgson, 1992), olfaction mediates the initial establishment of this mother–offspring link and the process of offspring recognition. Odour-based strategies could be used to improve maternal behaviour and offspring survival and development on farms. One of the management procedures that could benefit from the use of odours is (cross-)fostering, a common practice in sheep, goat and pig farming to optimize the number of offspring per mother. For instance, ewes can usually provide enough milk for two lambs maximum although the litter sizes can be larger, and fostering an additional lamb to a ewe with a single lamb can increase lamb survival. However, ewes usually reject alien offspring, which can lead to severe injuries or death of the foster lamb. Therefore, strategies based on the manipulation of odours to facilitate fostering and increase the acceptance of alien offspring have been developed for sheep.

A well-established method using odour transfer to facilitate the acceptance of an alien lamb after fostering was developed by Edward Price and his group in the 1980s. This method (see Price *et al.*, 1984) is based on the transfer of the odour of the ewe's own lamb to the alien lamb via the use of a 'stockinette', a knitted jacket covering the neck, body and tail of the lamb, as shown in Fig. 12.2A, and previously worn by the own lamb. In an experiment, lambs from multiparous ewes were fitted with the stockinette within 6 h after parturition. Within 48–72 h after parturition, fostering was performed. First, the stockinettes were removed from the ewe's own lamb and put inside-out on the alien lamb, inside out, or the alien lamb kept its own stockinette (control). Thereafter, the ewe's own lamb, or one of her lambs in case she had twins, was exchanged by the alien lamb wearing the stockinette (substitution procedure). Five 10-min acceptance tests were carried out every 12 h for 48 h after fostering. Acceptance was defined as the ewe allowing the lamb to suckle for at least 20 s with no butts or butt attempts

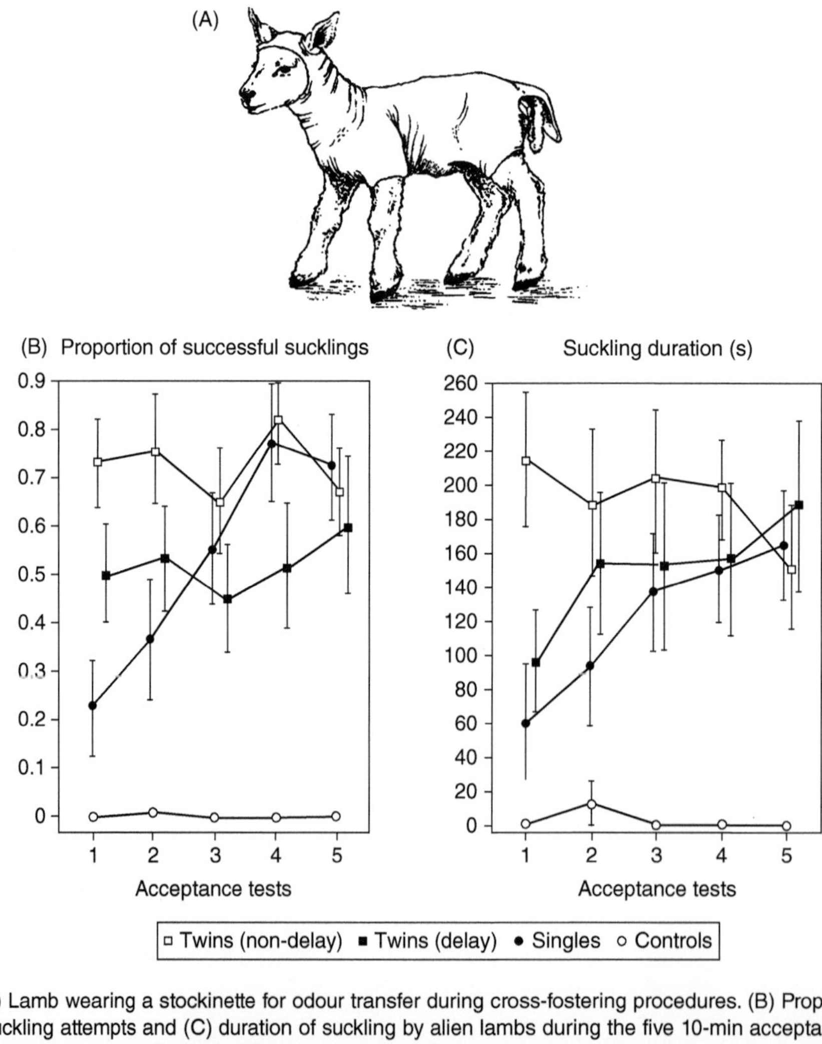

Fig. 12.2. (A) Lamb wearing a stockinette for odour transfer during cross-fostering procedures. (B) Proportion of successful suckling attempts and (C) duration of suckling by alien lambs during the five 10-min acceptance tests within the 48 h after fostering (based on Price *et al.*, 1984). Alien lambs were either wearing a stockinette containing their own odour (Control, n = 6) or the odour of the ewe's own lamb, and were substituted for the own single lamb (Singles, n = 11), for one of the twin lambs (Twins (non-delay), n = 10) or for one of the twin lambs while the other own lamb was removed for the first 12 h after birth (Twins (delay), n = 10).

towards the lamb for two consecutive tests. In total, 84% of the ewes accepted an alien lamb wearing the stockinette worn before by her own lamb. Acceptance was fast, with 38% of the ewes which adopted the alien lamb showing immediate acceptance, and 85% showing acceptance within 48 h of fostering. In contrast, none of the control ewes accepted an alien lamb still wearing its own stockinette, i.e. carrying its own odour, which was accompanied by a lower proportion of successful suckling attempts (Fig. 12.2B) and a shorter duration of suckling by alien lambs (Fig. 12.2C). For ewes with twins, the presence of both the alien lamb and her remaining own lamb from parturition onwards (non-delayed) or the presence of the alien lamb only for the first 12 h, after which the ewe's own lamb was returned (delayed) had no impact on alien lamb acceptance.

Using the same technique to foster an alien lamb while the single own lamb stays with the ewe (add-on fostering), Martin *et al.* (1987) found that 68% of the ewes successfully adopted alien lambs covered with the odour of their own lamb, whereas only 13% of the ewes accepted alien lambs not covered with the odour of their own lamb (control). The lower acceptance rate compared with the substitution fostering described above (Price *et al.*, 1984) was explained by the ewe being able to use visual, auditory and/or olfactory (different from those on the stockinette) cues to compare the alien lamb to her own lamb. As a result, successful add-on fostering appears more difficult to achieve than substitution fostering, where the own lamb is removed from the ewe (e.g. in the event of early postnatal death occurring) – although Price *et al.* (1984) found similar acceptance rates in ewes with single or twin lambs, with even fewer rejection responses in the latter group (Fig. 12.2B and C). Further research is thus needed to determine how the presence of the ewe's own lamb influences the acceptance of an alien lamb during add-on fostering in sheep. Another interesting result found by Martin *et al.* (1987) is that only 20% of the ewes showed acceptance of an alien lamb covered with the odour of their own lamb but with a face differing in colour (black vs white face), thus highlighting the use of multiple sensory cues for offspring recognition in sheep. Price *et al.* (1998) reported that pairing this odour transfer procedure with the use of a common odour for both own and alien lambs improved the efficiency of add-on fostering to rates close to the substitution procedure described by Price *et al.* (1984). Indeed, 80% of the ewes accepted an alien lamb which was covered with the odour of their own lamb and that of an artificial odorant (neatsfoot oil) that was smeared on both own and alien lambs shortly after birth. These findings clearly indicate that odour transfer is an efficient and simple method that can be used to increase acceptance of alien lambs by ewes on farms, for instance when still-born lambs are substituted with orphaned lambs.

Olfaction in Relation to Fear and Stress in Farm Animals

Some odours can be perceived as aversive by farm animals. As demonstrated in Chapter 2, many animals show an (innate) aversive response to predator odours and this appears no different for farm animals. For instance, 7-day-old chicks avoid predatory cat odours (Fluck *et al.*, 1996), and domestic adult chickens have been found to increase their vigilance behaviour and decrease food searching behaviour after exposure to faecal odours from predators (e.g. tigers) compared with faecal odours from herbivores (Zidar and Lovlie, 2012). The potential of predator odours to modify behaviours has also been shown in other farm animal species, like cattle, which show a reduced feed intake when exposed to faecal odours from foxes, coyotes, cougars and bears (Pfister *et al.*, 1990) and increased vigilance when exposed to odours of dog faeces (Terlouw *et al.*, 1998). Similarly, exposure to the odours of wolf and lion urine raised vigilance behaviour in horses, although a physiological fear response (increased heart rate) was found only when the predator odour was paired with a sudden sound (Christensen and Rundgren, 2008).

Other odours than predator odours can be perceived as aversive too, and elicit strong behavioural responses in animals, such as odours from stressed conspecifics. Indeed, as presented in Chapter 10, olfactory cues present in body secretions, like urine, saliva, and/or skin and gland secretions, have been found to convey information about the emotional state of conspecifics (e.g. stress, fear). Although it is unknown whether these olfactory cues, sometimes referred to as 'alarm substances', inadvertently reflect stress or are true signals that have evolved to communicate information (see Chapter 10), it has been demonstrated that they can be perceived by conspecifics and influence their behaviour. Under natural conditions, chemical cues signalling fear and stress may act as a warning to hide or flee, and benefit survival. In intensive farming conditions, however, the social transmission of these negative emotional states may be undesirable and reduce the welfare of the animals exposed to them.

One study reported evidence for the existence of alarm substances in urine from stressed heifers (Boissy *et al.*, 1998). The experimenters collected urine from Aubrac heifers 12 h after tethering (stressed heifers), a procedure that is known to induce a strong transient stress response in cattle, as demonstrated by behavioural and stress hormone measurements; and 30 days after tethering, i.e. when animals were habituated to tethering (unstressed heifers). In a first test, naïve heifers were exposed to a feed dispenser containing concentrate food in a familiar test room (Fig. 12.3A). The food was placed in a bucket with a double bottom in to which 500 ml of warm urine from stressed or unstressed heifers had been poured. Heifers exposed to the urine from stressed conspecifics were more reluctant to start eating, spent less time feeding and tended to have shorter bouts of feeding than animals exposed to urine from unstressed conspecifics (Fig. 12.3B). One week after the first test, the heifers were individually exposed to a traffic cone placed in the centre of the test room. This novel object was sprayed with urine from stressed or unstressed heifers. Urine from the same sample was also sprayed on a plastic panel fixed on the entrance door. Animals exposed to the traffic cone sprayed with urine from stressed heifers approached the novel object later, and tended to take more time to enter the area and to stay immobile longer than animals exposed to the urine from unstressed heifers. They also spent more time in the squares close to the entrance/exit door (squares 1 and 2 in Fig. 12.3C). The authors concluded that heifers can perceive the

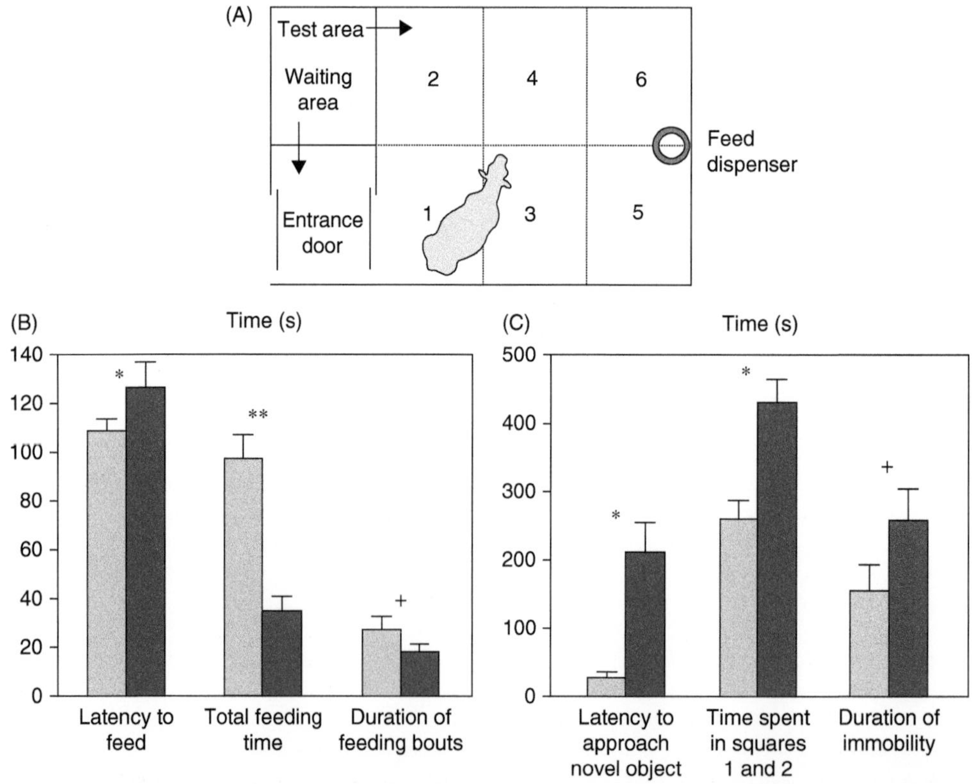

Fig. 12.3. (A) Familiar test room in which the behavioural responses of Aubrac heifers to the odour of urine from non-stressed (light grey; $n = 8$) or stressed (dark grey; $n = 8$) conspecifics were recorded. The urine was (B) presented below the feed in feed dispensers or (C) sprayed on a novel object. ** $P < 0.01$, * $P < 0.05$, + $P < 0.10$. (Based on Boissy *et al.*, 1998.)

negative emotional state of conspecifics through olfactory cues in urine which, in turn, influences their own fear response.

Similarly, Terlouw *et al.* (1998) showed heightened vigilance in heifers exposed to odours of urine from stressed conspecifics, although feeding motivation was unaltered in their study when the feed dispenser was presented with the odours of urine from stressed conspecifics. This perception of olfactory cues that may reflect stress or fear has also been demonstrated in other farm animals. Sows reduce the exploration of an automatic feed dispenser if this feeder is sprayed with urine from stressed pigs, but not when urine from unstressed pigs is used (Vieuille-Thomas and Signoret, 1992). In line with these findings, newly weaned piglets were shown to take longer to approach a feeder sprayed with urine from stressed gilts, and to have shorter feeding bouts (Amory and Pearce, 2000). The piglets also vocalized more when presented with the feeder sprayed with urine from stressed gilts.

As well as urine, blood can also elicit an avoidance response in farm animals. Chicks have been reported to avoid blood from conspecifics (Jones and Black, 1979; Fluck *et al.*, 1996), and to show behavioural indications of fear, such as freezing, lying and low activity. These behavioural responses were absent when the chicks were presented with a red dye, visually close to blood, or when exposed to blood in a sealed Petri dish, suggesting that odours played a crucial role in this fear response (Jones and Black, 1979). Horses also show increased vigilance when exposed to the odour of blood from slaughtered conspecifics (Christensen and Rundgren, 2008). It remains unclear, however, whether this response towards blood is elicited by olfactory cues emitted by the blood donor during blood sampling or just before slaughter, or by other chemical components in blood. It should be noted that the smell of blood has been found to be attractive to pigs and chickens in some studies and may even stimulate cannibalism in these species (Ferguson, 1968; Fraser, 1987).

The social transmission of fear and stress should be taken into consideration in farm animals as it may impair their welfare. For instance, alarm cues emitted by a handful of chickens can lead to panic reactions in a large part of the flock, with injured and, sometimes, dead birds as a consequence (Jones and Roper, 1997). In addition, studies in pigs and chickens indicate that fearful individuals may induce stress or a negative emotional state in their group members (de Haas *et al.*, 2012; Reimert *et al.*, 2013). In modern intensive production systems, farm animals are frequently exposed to stressful events, such as beak trimming (in birds), tail docking, vaccinations, abrupt weaning, rehousing, exposure to unfamiliar conspecifics, transport and (pre-slaughter) handling. Although the exact contribution of olfactory cues to the social transmission of fear in farm animals is unknown, the conditions of intensive production systems, with large numbers of animals packed into one room (or truck) at high stocking densities, could lead to frequent exposure to conspecific odour cues reflecting stress. Farm animals cannot escape from these odours, and often have little opportunity to show an appropriate behavioural response (e.g. fleeing). We do not know how far olfactory cues reflecting stress or fear spread due to their volatility but, unlike visual cues, odours can cross boundaries between pens and thus potentially reach many animals. For instance, it has been suggested that the air flow in abattoirs derived from the animals being handled and slaughtered in the killing area may induce fear in the incoming conspecifics (Jones and Roper, 1997). It is thus crucial to consider the spread of odours when designing animal facilities and ventilation systems.

Use of Familiar Odours to Facilitate Adaptation to Novel Situations on Farms

On farms, animals are exposed to environmental changes, such as alterations in housing conditions, management, group composition or feed, which may cause temporary stress responses, and even affect long-term welfare, health and productivity. The move from rearing to layer farms for laying hens, or from farrowing pens to the nursery for piglets are examples of stressful changes. Experimental studies strongly suggest that the provision of familiar odours may help animals to adapt to new circumstances.

Chickens placed in an unfamiliar environment, for example, have been found to be attracted to an odorant (vanillin), which was present in their home cage during rearing (Jones and Carmichael, 1999). Moreover, Jones and Gentle (1985) reported that the presence of the familiar odour in a novel environment decreases fear responses of chickens, a finding that was later confirmed by Jones *et al.* (2002). Chickens reared with vanillin in their home cage and tested in pairs in an unfamiliar open field moved away sooner and more often when the familiar odorant was present in a Petri dish, reflecting lower fear levels. Elegantly, the authors also demonstrated that the reassuring properties of the vanillin odour was due to familiarity and not to intrinsic anxiolytic effects, as vanillin did not reduce fear responses in naïve chickens reared without this odour.

Animals can be familiarized with odours as early as at the foetal stage. Indeed, odorants may reach the foetus through the foetal blood stream (Hepper, 1988) or appear in the amniotic fluid (e.g. in humans; Mennella *et al.*, 1995), where they stimulate the foetus' olfactory receptors. Mammalian foetuses are able to process and retain chemosensory information during the last part of gestation (Schaal and Orgeur, 1992), and evidence for *in ovo* olfactory learning has been reported in chicks (Sneddon *et al.*, 1998). Odours learnt before birth or hatching can also be used to reassure animals in novel situations. For instance, pigs exposed to anise *in utero* (by adding it to their mother's diet during pregnancy) showed reduced stress and fear responses in a novel environment test in the presence of anise odour (Oostindjer *et al.*, 2009). Offering familiar odours may reduce problems in farming conditions, too. Piglets familiarized with anise odour during early development and re-exposed to this odour at weaning, a stressful event for piglets on farms, had fewer weaning-related problems such as diarrhoea, poor feed intake and a growth dip (Oostindjer *et al.*, 2010), and displayed signs of improved welfare, such as an increase in play behaviour and a decrease in damaging behaviour directed at their group mates (Oostindjer *et al.*, 2011).

A number of studies have tested the effectiveness of maternal pheromones to reduce stress in farm animals. McGlone and Anderson (2002), for instance, investigated the effect of a synthetic putative maternal pheromone on the behaviour and growth of piglets after weaning at 3 or 4 weeks of age. At weaning, piglets were placed in groups of three non-litter mates and exposed to one out of three different treatments, which were applied to the groups of piglets using a camel-hair paint brush: the solution containing the putative pheromone was either put on their snouts (pheromone snout) or on the feeding trough (pheromone feeder), or the solution without the pheromone was applied on the snouts of the piglets (controls). The application of the putative maternal pheromone induced several behavioural changes in the piglets. For instance, piglets spent less time lying (Fig. 12.4A) and aggressing pen mates (Fig. 12.4B) if the pheromone was applied to the feeder compared with controls, with the results being intermediary if the pheromone was applied on the snout of the piglets. The application of the pheromone on the feeder, but not on the snout of the piglets, also increased the time spent active compared with the control condition (Fig. 12.4C). Finally, piglets treated with the pheromone (on the feeder or the snout) spent more time with their heads in the feeder (Fig. 12.4D), compared with control piglets. Moreover, compared with controls, piglets exposed to the pheromone on the feeder or their snouts, respectively, were 19% and 27% heavier and had a 17% and 26% increase in feed efficiency over the 4-week post-weaning period. These findings suggest that the synthetic pheromone stimulated feeding behaviours and reduced agonistic behaviours during the first 48 h after weaning, and improved long-term productivity of growing piglets. Synthetic maternal-like putative pheromones have also been applied to reduce transport stress in pigs (Driessen *et al.*, 2008), and to reduce stress (Madec *et al.*, 2008, 2009) and enhance growth (Madec *et al.*, 2009) in broilers.

Familiar odours can also be added to feed and serve to facilitate the adaptation of animals to a dietary change. Chickens and pigs often show food neophobia when given a new diet (Jones, 1987; Turro *et al.*, 1994; Dong and Pluske, 2007), meaning that they are initially reluctant to eat large amounts of the new feed. Herbivores, such as goats, sheep and cattle, also show a temporary decline in feed intake when they are moved onto a new pasture, or presented with new forage mixtures, which can strongly affect their production (Pain *et al.*, 2005). This phenomenon helps animals to avoid toxic food when foraging under natural conditions, but can lead to undesirable declines in feed intake in animal husbandry. As demonstrated in Chapter 6, animals use odours to recognize and find feed, and have the ability to associate the flavour of a feed with the consequences of its intake. This association leads to feed avoidance or preference and thus shapes the future food selection of animals. Experimental studies have clearly demonstrated the ability of farm animals to form olfactory memories of feed items and to adjust their feeding behaviour accordingly. When 2- to 3-day-old chicks are presented with a feed aromatized with an unfamiliar odour (e.g. orange) followed by the injection of lithium chloride, a substance that

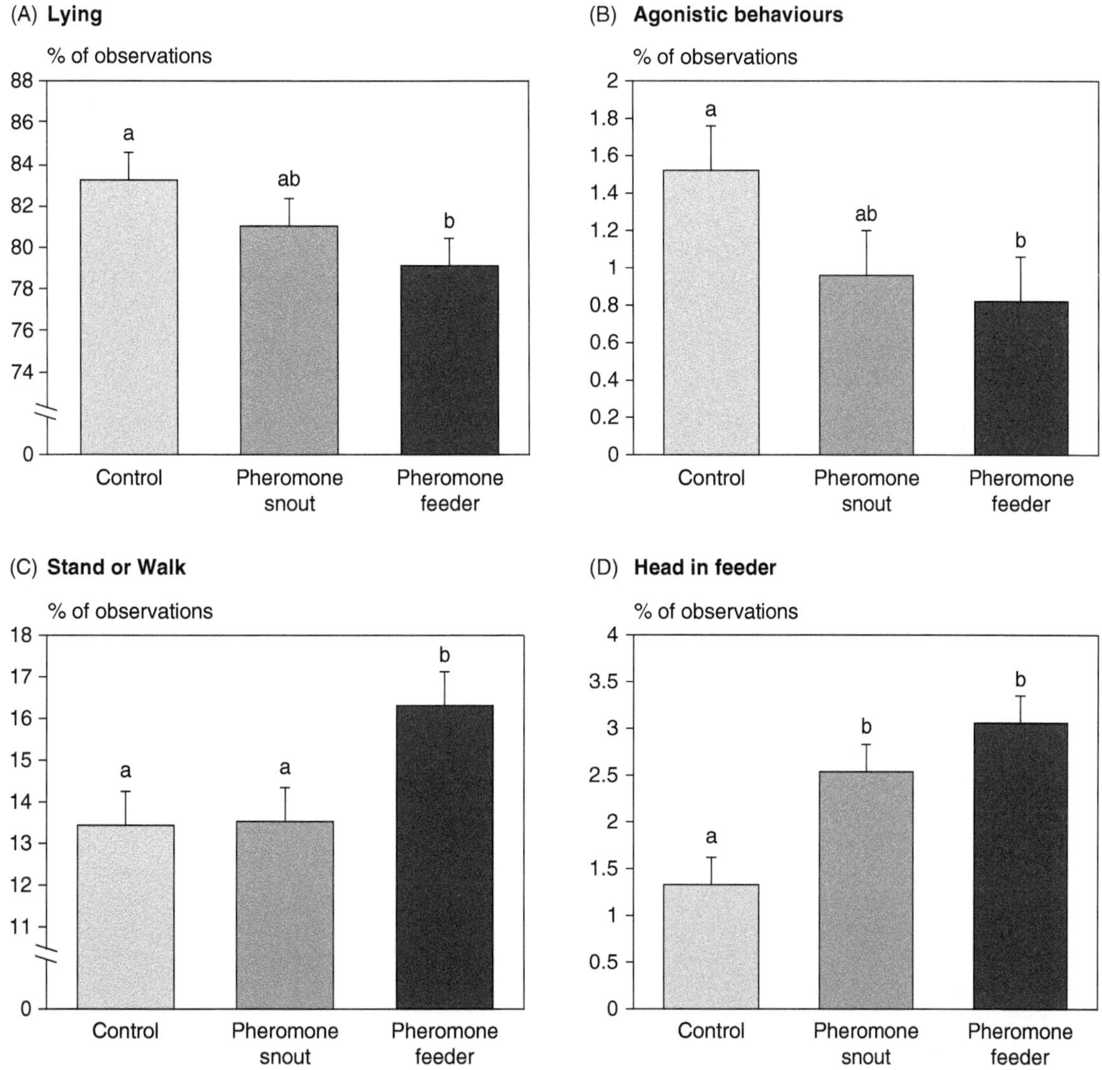

Fig. 12.4. Percentage of observations spent (A) lying, (B) exhibiting agonistic behaviours, (C) standing or walking, and (D) with the head in the feeder during the first 48 h after weaning and regrouping in piglets exposed to a putative maternal pheromone put on the snout of piglets or on the feeder, or to the solvent without the pheromone (control) put on the snout (n = 16 groups per treatment). (The figure is based on data from McGlone and Anderson, 2002.)

induces nausea and a strong intestinal malaise, the chicks subsequently show a strong avoidance of the orange-aromatized feed that was previously associated with these negative post-ingestive consequences (Turro *et al.*, 1994). Similarly, pigs showed long-term aversive responses (i.e. up to 5 weeks after conditioning) to the flavour of a feed (e.g. orange, thyme or cinnamon) previously associated with lithium chloride injections (Gaultier *et al.*, 2011; Clouard *et al.*, 2012). Feed preferences can also be formed by exposing animals to odours before birth or hatching. For instance, newly hatched (4-day-old) chicks exposed to strawberry odour during incubation drank more strawberry-flavoured water and foraged more in strawberry-flavoured litter compared with controls (Sneddon *et al.*, 1998). Others confirmed that olfactory learning during incubation modulates feed preferences of chickens (Bertin *et al.*, 2010, 2012), although the impact of *in ovo* odour exposure on post-hatch preferences appears to depend on the intensity of the odour during learning (Bertin *et al.*, 2010). Adding a familiar odour to new feed types could thus be an efficient strategy to prevent neophobic responses and declines in feed intake in farm animals that experience a dietary switch. The practical implications and opportunities of positive odour learning to stimulate feed intake during abrupt feed changes, e.g. when grazing animals are moved onto a new pasture, have been nicely reviewed by Pain *et al.* (2005).

Interference between Olfaction and Intensive Husbandry Conditions

As mentioned in the previous section, the crucial role of odours for farm animals should be taken into consideration and could be used to alleviate the negative consequences of modern husbandry conditions on animal welfare and productivity. On the other hand, certain husbandry conditions, such as low light intensity, high levels of background noise or poor air quality, can directly impair the ability of animals to use sensory (visual, auditory or olfactory) cues for their normal daily functioning.

Regarding olfaction, current indoor housing systems for farm animals are usually characterized by relatively high levels of aerial noxious gases, such as ammonia or hydrogen sulfide. It is well documented that prolonged exposure to high ammonia concentrations negatively affects health and productivity of, for instance, pigs (Drummond *et al.*, 1980; von Borell *et al.*, 2007) and poultry (Anderson *et al.*, 1964; Reece *et al.*, 1980). Wathes *et al.* (2002) nicely reviewed a series of experiments demonstrating aversive responses of pigs and domestic chickens to atmospheric ammonia. When given a choice between three environments varying in aerial ammonia levels, laying hens spent more time and showed more foraging and resting behaviours in a fresh air environment than in environments with ammonia concentrations of 25 or 45 parts per million (ppm; Kristensen *et al.*, 2000). Jones *et al.* (2005) also reported that broilers avoided atmospheric ammonia concentrations over 10 ppm, and that this choice was independent of their early experience, i.e. broilers reared in fresh air or under high atmospheric ammonia conditions both avoided high levels of ammonia in a multiple choice test. Similarly, growing pigs reared in fresh air preferred the lowest concentration of ammonia when given the choice between different levels (0, 10, 20 or 40 ppm; Jones *et al.*, 1996), and showed aversive responses to a novel environment containing high ammonia levels (43 ppm; Jones *et al.*, 2000b). Animals probably avoid high atmospheric ammonia because of its irritating effect on the respiratory system, but at lower levels, ammonia (and other gases) may well be aversive to farm animals too, simply because of its foul odour (see also Chapter 5). In addition, ammonia could act on the olfactory system *per se* and hamper normal functioning of animals by reacting with other odorants before these reach olfactory receptors, by masking other odour signals or even, in combination with dust, by compromising the sense of smell due to their action on the nasal mucosal and olfactory receptors (Jones *et al.*, 2001). As an example of the latter, female workers on Nebraska pig farms scored worse in an olfactory ability test than controls (Snyder *et al.*, 2003), although others found no evidence for a reduced sense of smell in dairy and pig farm employees in a smaller study in Germany (Gudziol *et al.*, 2007). Unlike humans who usually only enter animal housing during working hours, farm animals kept indoors cannot escape from continuous exposure to relatively high ammonia levels, but the impact of these conditions on their olfactory acuity has been scarcely addressed in research. It has been shown that the ability of pigs to detect *n*-butanol (Jones *et al.*, 2001) or odorized buried food (Jones *et al.*, 2000a) is not compromised if they are simultaneously exposed to atmospheric ammonia at 40 ppm. Unlike acute exposure, chronic exposure (for 24 days) to the same concentration of ammonia induced olfactory impairment, i.e. a fall in olfactory acuity for *n*-butanol, in half of the pigs tested, which was generally not recovered after 24 days of exposure to fresh air (Jones *et al.*, 2001). Chronic exposure to high atmospheric ammonia concentrations may thus affect biological processes. For instance, puberty after exposure to boars was delayed in gilts reared in an environment with ~20 ppm of aerial ammonia (Malayer *et al.*, 1988), possibly reflecting interference with the detection of boar pheromones. In addition, Kristensen *et al.* (2001) showed that pigs reared in an ammoniated environment (at 36 ppm) preferred the odour of a familiar pig to that of an unfamiliar pig, whereas pigs reared in fresh air (3.2 ppm of ammonia) chose to go towards the smell of an unfamiliar pig in a Y-maze. This suggests that although (1-week) chronic exposure to high aerial ammonia levels does not alter the ability of piglets to discriminate between conspecifics, it can influence their behaviour.

Parker *et al.* (2010) and O'Connor *et al.* (2010) provided evidence for a negative influence of atmospheric levels of ammonia that are characteristic of intensive pig production systems (20 ppm) on social behaviour and welfare of growing pigs. They investigated the combined effects of ammonia (20 vs < 5 ppm), background noise (80 vs 53 dB) and light intensity (200 vs 40 lux). These conditions were chosen to reflect the typical commercial conditions and current recommendations, respectively. Behaviour was observed in 16 groups of 14 female pigs each for 14 weeks after weaning and regrouping. Compared with pigs reared in low ammoniated rooms, pigs reared in high ammoniated rooms spent less time feeding in close contact (i.e. contiguous feeding) over the 14-week period, suggesting reduced group synchrony and diminished tolerance to conspecifics. In line with this, these pigs were also more aggressive and showed a lower proportion of benign responses (i.e. no change in behaviours) to aggression in the first week after arrival (Fig. 12.5A and B), although the level of ammonia

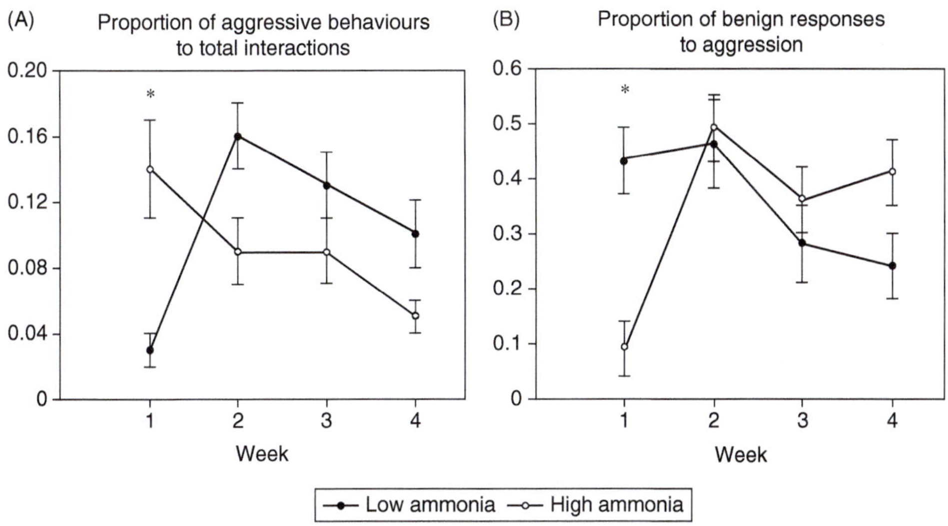

Fig. 12.5. Proportion of (A) aggressive behaviours out of total number of interactions, and of (B) benign responses (i.e. no change in behaviour) to aggression over the 4 weeks following weaning and mixing in piglets reared in an environment with high (20 ppm) or low (< 5 ppm) ammonia levels. (Based on data from Parker *et al.*, 2010.)

did not influence the proportion of submissive responses (out of all responses to aggression). Atmospheric ammonia concentrations also affected positive social interactions after regrouping, with piglets reared in a high ammoniated environment being less playful than those reared in a low ammoniated environment 3 weeks after mixing (O'Connor *et al.*, 2010). Notably, low light intensity *per se* also increased aggression, and the effect of high atmospheric ammonia on aggression was exacerbated under low light conditions. This interaction suggests that interference with cues for social discrimination and communication used by pigs to establish a social order may explain at least part of the effect of high atmospheric ammonia levels (Parker *et al.*, 2010).

Conclusions and Future Directions

The examples presented in this chapter highlight the crucial role of olfaction for many biological functions, such as reproduction, maternal care, social recognition and communication in farm animals, and its potential for practical applications to modify behaviour and improve welfare and productivity in modern husbandry systems. For instance, the ability of farm animals to develop olfactory memories has important implications for the development of odour-based strategies to reduce stress in unfamiliar situations. The use of familiar and reassuring odours to facilitate the transition to new housing conditions, management procedures or diets offers great opportunities to enhance farm animal welfare, but is, to the best of our knowledge, only scantily applied in animal husbandry. An area for research and application that also deserves much more attention is the use of odours as environmental enrichment for livestock. Farm environments and diets are often monotonous, with minimal diversity in sensory stimulation, and this under-stimulation has been shown to impair animal welfare. A plethora of materials have been investigated and applied on farms in an attempt to increase environmental complexity. According to a systematic study, one of the major characteristics determining the attractiveness of enrichment materials for pigs was that the materials should be 'odorous' (Van de Weerd *et al.*, 2003). Yet, the sense of smell is mostly ignored in studies on environmental enrichment for farm animals, as well as in husbandry practice where enrichment strategies, such as radio to provide music, toys and bedding materials, typically target the senses of hearing, sight and/or touch only. The use of odours to create a more dynamic environment (see Chapter 11) for farm animals offers great potential as odours are relatively cheap and take up no space (Nielsen *et al.*, 2015). Finally, a better understanding on how farm animals use olfactory cues to communicate stress or maintain social stability, and to what extent modern husbandry conditions interfere with the sense of smell and thereby hamper important biological processes is also needed. This could be used to formulate new recommendations for the management and housing of farm animals and the design of farm buildings, ventilation systems and slaughter plants.

References

Amory, J.R. and Pearce, G.P. (2000) Alarm pheromones in urine modify the behaviour of weaner pigs. *Animal Welfare* 9, 167–175.

Anderson, D.P., Beard, C.W. and Hanson, R.P. (1964) Adverse effects of ammonia on chickens including resistance to infection with Newcastle Disease virus. *Avian Diseases* 8, 369–379.

Baldwin, B.A. (1977) Ability of goats and calves to distinguish between conspecific urine samples using olfaction. *Applied Animal Ethology* 3, 145–150.

Baldwin, B.A. and Meese, G.B. (1977) Ability of sheep to distinguish between conspecifics by means of olfaction. *Physiology and Behavior* 18, 803–808. DOI: 10.1016/0031-9384(77)90187-1

Bertin, A., Calandreau, L., Arnould, C., Nowak, R., Lévy, F., Noirot, V., Bouvarel, I. and Leterrier, C. (2010) In ovo olfactory experience influences post-hatch feeding behaviour in young chickens. *Ethology* 116, 1027–1037. DOI: 10.1111/j.1439-0310.2010.01820.x

Bertin, A., Calandreau, L., Arnould, C. and Lévy, F. (2012) The developmental stage of chicken embryos modulates the impact of in ovo olfactory stimulation on food preferences. *Chemical Senses* 37, 253–261. DOI: 10.1093/chemse/bjr101

Blissitt, M.J., Bland, K.P. and Cottrell, D.F. (1990) Discrimination between the odors of fresh estrous and non-estrous ewe urine by rams. *Applied Animal Behaviour Science* 25, 51–59. DOI: 10.1016/0168-1591(90)90069-P

Boissy, A., Terlouw, C. and Le Neindre, P. (1998) Presence of cues from stressed conspecifics increases reactivity to aversive events in cattle: evidence for the existence of alarm substances in urine. *Physiology and Behavior* 63, 489–495. DOI: 10.1016/S0031-9384(97)00466-6

Bonadonna, F. and Nevitt, G.A. (2004) Partner-specific odor recognition in an Antarctic seabird. *Science* 306, 835. DOI: 10.1126/science.1103001

Christensen, J.W. and Rundgren, M. (2008) Predator odour per se does not frighten domestic horses.

Applied Animal Behaviour Science 112, 136–145. DOI: 10.1016/j.applanim.2007.08.003

Clouard, C., Jouhanneau, M., Meunier-Salaun, M.C., Malbert, C.H. and Val-Laillet, D. (2012) Exposures to conditioned flavours with different hedonic values induce contrasted behavioural and brain responses in pigs. *PLoS ONE* 7, e37968. DOI: 10.1371/journal.pone.0037968

de Haas, E.N., Kops, M.S., Bolhuis, J.E., Groothuis, T.G.G., Ellen, E.D. and Rodenburg, T.B. (2012) The relation between fearfulness in young and stress-response in adult laying hens, on individual and group level. *Physiology and Behavior* 107, 433–439. DOI: 10.1016/j.physbeh.2012.08.002

Dong, G.Z. and Pluske, J.R. (2007) The low feed intake in newly-weaned pigs: problems and possible solutions. *Asian-Australasian Journal of Animal Sciences* 20, 440–452.

Driessen, B., Peeters, E. and Geers, R. (2008) Influence of olfactory substances on the heart rate and lying behaviour of pigs during transport simulation. *Animal Welfare* 17, 155–160.

Drummond, J.G., Curtis, S.E., Simon, J. and Norton, H.W. (1980) Effects of aerial ammonia on growth and health of young pigs. *Journal of Animal Science* 50, 1085–1091.

Ferguson, W. (1968) Abnormal behavior in domestic birds. In: Fox, M.W. (ed.) *Abnormal Behavior in Animals*. Saunders, Philadelphia, pp. 188–207.

Fischer-Tenhagen, C., Johnen, D., Le Danvic, C., Gatien, J., Salvetti, P., Tenhagen, B.A. and Heuwieser, W. (2015) Validation of bovine oestrous-specific synthetic molecules with trained scent dogs: similarities between natural and synthetic oestrous smell. *Reproduction in Domestic Animals* 50, 7–12. DOI: 10.1111/rda.12440

Fluck, E., Hogg, S., Mabbutt, P.S. and File, S.E. (1996) Behavioural and neurochemical responses of male and female chicks to cat odour. *Pharmacology Biochemistry and Behavior* 54, 85–91. DOI: 10.1016/0091-3057(95)02170-1

Fraser, D. (1987) Attraction to blood as a factor in tail-biting by pigs. *Applied Animal Behaviour Science* 17, 61–68. DOI: 10.1016/0168-1591(87)90008-6

Gaultier, A., Meunier-Salaun, M.C., Malbert, C.H. and Val-Laillet, D. (2011) Flavour exposures after conditioned aversion or preference trigger different brain processes in anaesthetised pigs. *The European Journal of Neuroscience* 34, 1500–1511. DOI: 10.1111/j.1460-9568.2011.07848.x

Gudziol, V., Mackuth, D., Hauswald, B., Knothe, J., Scheuch, K., Zahnert, T. and Hummel, T. (2007) Sense of smell in workers exposed to agricultural odours. *Occupational Medicine – Oxford* 57, 149–151. DOI: 10.1093/occmed/kql156

Hepper, P.G. (1988) Adaptive fetal learning – prenatal exposure to garlic affects postnatal preferences. *Animal Behaviour* 36, 935–936. DOI: 10.1016/S0003-3472(88)80177-5

Hirao, A., Aoyama, M. and Sugita, S. (2009) The role of uropygial gland on sexual behavior in domestic chicken *Gallus gallus domesticus*. *Behavioural Processes* 80, 115–120. DOI: 10.1016/j.beproc.2008.10.006

Horrell, I. and Hodgson, J. (1992) The bases of sow-piglet identification. 1. The identification by sows of their own piglets and the presence of intruders. *Applied Animal Behaviour Science* 33, 319–327. DOI: 10.1016/S0168-1591(05)80069-3

Hothersall, B., Harris, P., Sortoft, L. and Nicol, C.J. (2010) Discrimination between conspecific odour samples in the horse (*Equus caballus*). *Applied Animal Behaviour Science* 126, 37–44. DOI: 10.1016/j.applanim.2010.05.002

Izard, M.K. (1983) Pheromones and reproduction in domestic animals. In: Vandenbergh, J.G. (ed.) *Pheromones and Reproduction in Mammals*. Academic Press, New York, pp. 253–285.

Johnen, D., Heuwieser, W. and Fischer-Tenhagen, C. (2015) How to train a dog to detect cows in heat – training and success. *Applied Animal Behaviour Science* 171, 39–46. DOI: 10.1016/j.applanim.2015.08.019

Jones, E.K.M., Wathes, C.A. and Webster, A.J.F. (2005) Avoidance of atmospheric ammonia by domestic fowl and the effect of early experience. *Applied Animal Behaviour Science* 90, 293–308. DOI: 10.1016/j.applanim.2004.08.009

Jones, J.B., Burgess, L.R., Webster, A.J.F. and Wathes, C.M. (1996) Behavioural responses of pigs to atmospheric ammonia in a chronic choice test. *Animal Science* 63, 437–445.

Jones, J.B., Carmichael, N.L., Wathes, C.M., White, R.P. and Jones, R.B. (2000a) The effects of acute simultaneous exposure to ammonia on the detection of buried odourized food by pigs. *Applied Animal Behaviour Science* 65, 305–319. DOI: 10.1016/S0168-1591(99)00063-5

Jones, J.B., Wathes, C.M., White, R.P. and Jones, R.B. (2000b) Do pigs find a familiar odourant attractive in novel surroundings? *Applied Animal Behaviour Science* 70, 115–126. DOI: 10.1016/S0168-1591(00)00149-0

Jones, J.B., Wathes, C.M., Persaud, K.C., White, R.P. and Jones, R.B. (2001) Acute and chronic exposure to ammonia and olfactory acuity for *n*-butanol in the pig. *Applied Animal Behaviour Science* 71, 13–28. DOI: 10.1016/S0168-1591(00)00168-4

Jones, R.B. (1987) Food neophobia and olfaction in domestic chicks. *Bird Behaviour* 7, 78–81.

Jones, R.B. and Black, A.J. (1979) Behavioral responses of the domestic chick to blood. *Behavioral and Neural Biology* 27, 319–329.

Jones, R.B. and Carmichael, N.L. (1999) Domestic chicks are attracted to a familiar odorant in a novel test situation: a brief report. *Applied Animal Behaviour Science* 61, 351–356. DOI: 10.1016/S0168-1591(98)00204-4

Jones, R.B. and Gentle, M.J. (1985) Olfaction and behavioral modification in domestic chicks (*Gallus domesticus*). *Physiology and Behavior* 34, 917–924. DOI: 10.1016/0031-9384(85)90014-9

Jones, R.B. and Roper, T.J. (1997) Olfaction in the domestic fowl: a critical review. *Physiology and Behavior* 62, 1009–1018. DOI: 10.1016/S0031-9384(97)00207-2

Jones, R.B., Facchin, L. and McCorquodale, C. (2002) Social dispersal by domestic chicks in a novel environment: reassuring properties of a familiar odourant. *Animal Behaviour* 63, 659–666. DOI: 10.1006/anbe.2001.1943

Karlsson, A.C., Jensen, P., Elgland, M., Laur, K., Fyrner, T., Konradsson, P. and Laska, M. (2010) Red junglefowl have individual body odors. *Journal of Experimental Biology* 213, 1619–1624. DOI: 10.1242/jeb.040279

Kristensen, H.H., Burgess, L.R., Demmers, T.G.H. and Wathes, C.M. (2000) The preferences of laying hens for different concentrations of atmospheric ammonia. *Applied Animal Behaviour Science* 68, 307–318. DOI: 10.1016/S0168-1591(00)00110-6

Kristensen, H.H., Jones, R.B., Schofield, C., White, R.P. and Wathes, C.M. (2001) The use of olfactory and other cues for social recognition by juvenile pigs. *Applied Animal Behaviour Science* 72, 321–333. DOI: 10.1016/S0168-1591(00)00209-4

Krueger, K. and Flauger, B. (2011) Olfactory recognition of individual competitors by means of faeces in horse (*Equus caballus*). *Animal Cognition* 14, 245–257. DOI: 10.1007/s10071-010-0358-1

Langendijk, P., Soede, N.M. and Kemp, B. (2006) Effects of boar stimuli on the follicular phase and on oestrous behaviour in sows. *Society of Reproduction and Fertility supplement* 62, 219–230.

Le Danvic, C., Gerard, O., Sellem, E., Ponsart, C., Chemineau, P., Humblot, P. and Nagnan-Le Meillour, P. (2015) Enhancing bull sexual behavior using estrus-specific molecules identified in cow urine. *Theriogenology* 83, 1381–1388. DOI: 10.1016/j.theriogenology.2015.02.004

Madec, I., Gabarrou, J.F., Saffray, D. and Pageat, P. (2008) Broilers (*Gallus gallus*) are less stressed if they can smell a mother odorant. *South African Journal of Animal Science* 38, 201–206.

Madec, I., Gabarrou, J.F., Moulin, D., Bougrat, L., Lecuelle-Lafont, C., Gigante, J. and Pageat, P. (2009) Effects of a semiochemical on feed conversion index and related indicators on fast-growing broilers housed during forty-two days. *Poultry Science* 88, 251–255. DOI: 10.3382/ps.2007-00405

Malayer, J.R., Brandt, K.E., Green, M.L., Kelly, D.T., Sutton, A.L. and Diekman, M.A. (1988) Influence of manure gases on the onset of puberty of replacement gilts. *Animal Production* 46, 277–282.

Martin, N.L., Price, E.O., Wallach, S.J.R. and Dally, M.R. (1987) Fostering lambs by odor transfer – the add-on experiment. *Journal of Animal Science* 64, 1378–1383.

McGlone, J.J. and Anderson, D.L. (2002) Synthetic maternal pheromone stimulates feeding behavior and weight gain in weaned pigs. *Journal of Animal Science* 80, 3179–3183.

McLeman, M.A., Mendl, M., Jones, R.B., White, R. and Wathes, C.M. (2005) Discrimination of conspecifics by juvenile domestic pigs, *Sus scrofa*. *Animal Behaviour* 70, 451–461. DOI: 10.1016/j.anbehav.2004.11.013

McLeman, M.A., Mendl, M.T., Jones, R.B. and Wathes, C.M. (2008) Social discrimination of familiar conspecifics by juvenile pigs, *Sus scrofa*: development of a non-invasive method to study the transmission of unimodal and bimodal cues between live stimuli. *Applied Animal Behaviour Science* 115, 123–137. DOI: 10.1016/j.applanim.2008.06.010

Meese, G.B., Conner, D.J. and Baldwin, B.A. (1975) Ability of the pig to distinguish between conspecific urine samples using olfaction. *Physiology & Behavior* 15, 121–125. DOI: 10.1016/0031-9384(75)90289-9

Melrose, D.R., Reed, H.C.B. and Patterson, R.L. (1971) Androgen steroids associated with boar odour as an aid to detection of oestrus in pig artificial insemination. *British Veterinary Journal* 127, 497–502.

Mendl, M., Randle, K. and Pope, S. (2002) Young female pigs can discriminate individual differences in odours from conspecific urine. *Animal Behaviour* 64, 97–101. DOI: 10.1006/anbe.2002.3040

Mennella, J.A., Johnson, A. and Beauchamp, G.K. (1995) Garlic ingestion by pregnant women alters the odor of amniotic fluid. *Chemical Senses* 20, 207–209. DOI: 10.1093/chemse/20.2.207

Nielsen, B.L., Jezierski, T., Bolhuis, J.E., Amo, L., Rosell, F., Oostindjer, M., Christensen, J.W., McKeegan, D., Wells, D.L. and Hepper, P. (2015) Olfaction: an overlooked sensory modality in applied ethology and animal welfare. *Frontiers in Veterinary Science* 2, 69. DOI: 10.3389/fvets.2015.00069

Nowak, R., Keller, M. and Lévy, F. (2011) Mother–young relationships in sheep: a model for a multidisciplinary approach of the study of attachment in mammals. *Journal of Neuroendocrinology* 23, 1042–1053. DOI: 10.1111/j.1365-2826.2011.02205.x

O'Connor, E.A., Parker, M.O., McLeman, M.A., Demmers, T.G., Lowe, J.C., Cui, L., Davey, E.L., Owen, R.C., Wathes, C.M. and Abeyesinghe, S.M. (2010) The impact of chronic environmental stressors on growing pigs, *Sus scrofa* (Part 1): stress physiology, production and play behaviour. *Animal* 4, 1899–1909. DOI: 10.1017/S1751731110001072

Oostindjer, M., Bolhuis, J.E., van den Brand, H. and Kemp, B. (2009) Prenatal flavor exposure affects flavor recognition and stress-related behavior of piglets. *Chemical Senses* 34, 775–787. DOI: 10.1093/chemse/bjp063

Oostindjer, M., Bolhuis, J.E., van den Brand, H., Roura, E. and Kemp, B. (2010) Prenatal flavor exposure affects

growth, health and behavior of newly weaned piglets. *Physiology and Behavior* 99, 579–586. DOI: 10.1016/j.physbeh.2010.01.031
Oostindjer, M., Bolhuis, J.E., Simon, K., van den Brand, H. and Kemp, B. (2011) Perinatal flavour learning and adaptation to being weaned: all the pig needs is smell. *PLoS ONE* 6, e25318. DOI: 10.1371/journal.pone.0025318
Pain, S., Franklin-McEvoy, J., Cox, A. and Revell, D.K. (2005) The odour profile of feedstuffs as modulators of feed preference in herbivores. *Recent Advances in Animal Nutrition in Australia* 15, 229–236.
Parker, M.O., O'Connor, E.A., McLeman, M.A., Demmers, T.G.M., Lowe, J.C., Owen, R.C., Davey, E.L., Wathes, C.M. and Abeyesinghe, S.M. (2010) The impact of chronic environmental stressors on growing pigs, *Sus scrofa* (Part 2): social behaviour. *Animal* 4, 1910–1921. DOI: 10.1017/S1751731110001084
Peron, F., Ward, R. and Burman, O. (2014) Horses (*Equus caballus*) discriminate body odour cues from conspecifics. *Animal Cognition* 17, 1007–1011. DOI: 10.1007/s10071-013-0717-9
Pfister, J.A., Mullerschwarze, D. and Balph, D.F. (1990) Effects of predator fecal odors on feed selection by sheep and cattle. *Journal of Chemical Ecology* 16, 573–583. DOI: 10.1007/BF01021787
Price, E.O., Dunn, G.C., Talbot, J.A. and Dally, M.R. (1984) Fostering lambs by odor transfer – the substitution experiment. *Journal of Animal Science* 59, 301–307.
Price, E., Dally, M., Erhard, H., Gerzevske, M., Kelly, M., Moore, N., Schultze, A. and Topper, C. (1998) Manipulating odor cues facilitates add-on fostering in sheep. *Journal of Animal Science* 76, 961–964.
Reece, F.N., Lott, B.D. and Deaton, J.W. (1980) Ammonia in the atmosphere during brooding affects performance of broiler-chickens. *Poultry Science* 59, 486–488.
Reimert, I., Bolhuis, J.E., Kemp, B. and Rodenburg, T.B. (2013) Indicators of positive and negative emotions and emotional contagion in pigs. *Physiology and Behavior* 109, 42–50. DOI: 10.1016/j.physbeh.2012.11.002
Rekwot, P.I., Ogwu, D., Oyedipe, E.O. and Sekoni, V.O. (2001) The role of pheromones and biostimulation in animal reproduction. *Animal Reproduction Science* 65, 157–170. DOI: 10.1016/S0378-4320(00)00223-2
Romeyer, A., Poindron, P. and Orgeur, P. (1994) Olfaction mediates the establishment of selective bonding in goats. *Physiology and Behavior* 56, 693–700. DOI: 10.1016/0031-9384(94)90229-1
Sankar, R. and Archunan, G. (2004) Flehmen response in bull: role of vaginal mucus and other body fluids of bovine with special reference to estrus. *Behavioural Processes* 67, 81–86. DOI: 10.1016/j.beproc.2004.02.007
Sankar, R. and Archunan, G. (2005) Discrimination of bovine estrus-related odors by mice. *Journal of Ethology* 23, 147–151. DOI: 10.1007/s10164-004-0140-4
Schaal, B. and Orgeur, P. (1992) Olfaction in utero: can the rodent model be generalized? *The Quarterly Journal of Experimental Psychology Section B – Comparative and Physiological Psychology* 44, 245–278.
Signoret, J.P. and Lindsay, D.R. (1982) The male effect in domestic mammals. Effect on LH secretion and ovulation. Importance of olfactory cues. In: Breiphol, W. (ed.) *Olfaction and Endocrine Regulation*. IRL Press, London, pp. 63–72.
Sneddon, H., Hadden, R. and Hepper, P.G. (1998) Chemosensory learning in the chicken embryo. *Physiology and Behavior* 64, 133–139. DOI: 10.1016/S0031-9384(98)00037-7
Snyder, M.C., Leopold, D.A., Chiu, B.C., Von Essen, S.G. and Liebentritt, N. (2003) The relationship between agricultural environments and olfactory dysfunction. *Journal of Agricultural Safety and Health* 9, 211–219.
Terlouw, E.M.C., Boissy, A. and Blinet, P. (1998) Behavioural responses of cattle to the odours of blood and urine from conspecifics and to the odour of faeces from carnivores. *Applied Animal Behaviour Science* 57, 9–21. DOI: 10.1016/S0168-1591(97)00122-6
Turro, I., Porter, R.H. and Picard, M. (1994) Olfactory cues mediate food selection by young chicks. *Physiology and Behavior* 55, 761–767. DOI: 10.1016/0031-9384(94)90057-4
Van de Weerd, H.A., Docking, C.M., Day, J.E.L., Avery, P.J. and Edwards, S.A. (2003) A systematic approach towards developing environmental enrichment for pigs. *Applied Animal Behaviour Science* 84, 101–118. DOI: 10.1016/S0168-1591(03)00150-3
Vieuille-Thomas, C. and Signoret, J.P. (1992) Pheromonal transmission of an aversive experience in domestic pig. *Journal of Chemical Ecology* 18, 1551–1557. DOI: 10.1007/bf00993228
von Borell, E., Ozpinar, A., Eslinger, K.M., Schnitz, A.L., Zhao, Y. and Mitloehner, E.M. (2007) Acute and prolonged effects of ammonia on hematological variables, stress responses, performance, and behavior of nursery pigs. *Journal of Swine Health and Production* 15, 137–145.
Wathes, C.M., Jones, J.B., Kristensen, H.H., Jones, E.K.M. and Webster, A.J.F. (2002) Aversion of pigs and domestic fowl to atmospheric ammonia. *Transactions of the ASAE* 45, 1605–1610.
Zidar, J. and Lovlie, H. (2012) Scent of the enemy: behavioural responses to predator faecal odour in the fowl. *Animal Behaviour* 84, 547–554. DOI: 10.1016/j.anbehav.2012.06.006

13 Olfactory Behaviour in Zoo Animals

Róisín Campbell-Palmer and Frank Rosell

Department of Natural Sciences and Environmental Health, University of Southeast Norway, Bø i Telemark, Norway

Many different animal species are held in zoological collections, wildlife parks and aquaria for public entertainment, education, animal rehabilitation and/or conservation purposes. In more recent times concern over zoo animal behaviours and their welfare has risen greatly, though this can vary depending on socioeconomic, cultural and political circumstances of the countries in which they are based (Seidensticker and Forthman, 1998). Melfi (2009) considers the study of zoo animal welfare research to be less advanced than that of livestock, laboratory and companion animal sectors. Zoo behavioural studies are increasing but can be criticized for small sample size, experimental design with numerous confounding variables, and differences in individual animal life histories (Swaisgood and Shepherdson, 2005; Plowman, 2008). The energy and resources invested in addressing stereotypic behaviours in zoo animals have largely been directed towards environmental enrichment as a means to improve welfare (Swaisgood and Shepherdson, 2005). Research opportunities may be compromised by competing interests such as public perception, availability of keeper time, and the willingness of captive collections to participate and publish findings, especially if they are critical of welfare standards. It may also be due to anthropomorphic perception of which stimuli are more important to a species and where resource investment should be prioritized. These constraints should not, however, discount the value of zoo studies, especially when investigating uncommonly kept/understudied or endangered species in which sample size may be a limiting factor.

Behaviour can be used to assess an animal's welfare, with zoo animal behaviours often described as 'normal' or 'abnormal'. It is important to note that this may not be known for all species held in zoo environments, and that behavioural cues need to be validated as animals may mask signs of poor welfare (Hill and Broom, 2009). The welfare experienced by zoo animals can be affected by a range of sometimes competing factors. These include visitor effects, animal husbandry procedures (e.g. cleaning routines, diets and feeding regimes, animal movements for breeding programmes), enclosure provisions and social groupings.

Olfaction has typically been a neglected area in animal behaviour and welfare research, including zoo-based studies, despite the importance of scent to many species (Nielsen *et al.*, 2015). For those undertaken, there is a clear bias in zoo olfactory enrichment studies, favouring large cat species (46%) and primates (16%), while reptiles represent only 3% of studies (Clark and King, 2008). Such research bias means the welfare needs of many species may not be fully understood or proactively addressed. This taxa bias does not reflect the diversity of species held in zoological collections (Melfi, 2009). Birds for example, possess a functional and in some cases highly developed olfactory system (Caro *et al.*, 2015), and have been shown in a laboratory setting to adjust their feeding behaviour in response to mammal faecal odour cues (Roth *et al.*, 2008). The investigation of familiar odours being utilized to improve welfare and productivity of poultry (Jones *et al.*, 2002) demonstrates promising application to a zoo environment, such as reducing neophobia and fear responses when introducing birds to a new enclosure or unfamiliar feed. Studies on zoo birds have revealed their olfactory abilities, which have previously been underestimated, such as the behavioural responses of Kea, *Nestor notabilis*, and Kaka, *Nestor meridionalis*, to scent station experiments (Gsell *et al.*, 2012). Like many zoo studies, sample size was small and individual responses were variable but both parrot species demonstrated a capacity to detect novel scent and discriminate between scent and controls. Studies in wild birds have indicated that olfaction plays an important role in, for example, food location (Roper, 1999), partner recognition (Bonadonna and Nevitt, 2004) and burrow

  Olfaction in Animal Behaviour and Welfare (ed. B.L. Nielsen)

location (Bonadonna and Bretagnolle, 2002). Despite this, odour enrichment is not routinely implemented in zoo settings as a means to improve bird welfare.

Environmental enrichment is one of the main husbandry practices employed in the zoo community to address captivity-induced stereotypic behaviours, to encourage behavioural diversity and species typical behavioural patterns, as well as to increase positive use of the captive environment and promote psychological and physiological well-being (Swaisgood and Shepherdson, 2005; Wells, 2009). Swaisgood and Shepherdson (2005) define the addition of scents or scented material to an animal's enclosure as olfactory enrichment. Any olfactory enrichment should aim to increase stimulation and behavioural complexity, in a species-appropriate manner, without presenting any risk or 'overtaxing' an individual's capacity to cope (Clark and King, 2008; Hill and Broom, 2009).

Novelty in the Captive Environment

Many zoo enclosures can be monotonous environments with predictable routines, and therefore lacking in sensory simulation (Marriner and Drickamer, 1994). Fay and Miller (2015) investigated exploration and activity levels, along with enclosure use by Rothschild giraffes (*Giraffa camelopardalis rothschildi*) housed at San Diego Zoo Safari Park, USA. They carried out preference testing of six different scents followed by an assessment of how scent enrichment impacted on behaviour once applied to the main enclosure. The experiment involved a pairwise presentation of essential oils (rose, ginger, vanilla, orange, mint and banana extract), with distilled water acting as a control. Recordings of giraffe behaviour and duration of time spent interacting with either the right or left scent-infused container were made and compared with baseline data prior to scent presentation (collected for 4 weeks before the introduction of scent). There was no directional bias, with the giraffes spending most time and the greatest number of bouts investigating the rose scent, although differences were observed in individual scent preference, with one individual preferring banana scent while another preferred orange (Fig. 13.1).

Though most behavioural states were not affected by the addition of scents, giraffes spent significantly less time standing still and resting during both the scent and control conditions. Giraffes spent more time in the area of enclosure closest to where the scent condition was presented when compared with the baseline condition (pre-scent exposure). The authors concluded that such olfactory enrichment may decrease predictability and introduce novelty into a zoo environment. The practical application of olfactory enrichment could be used to encourage

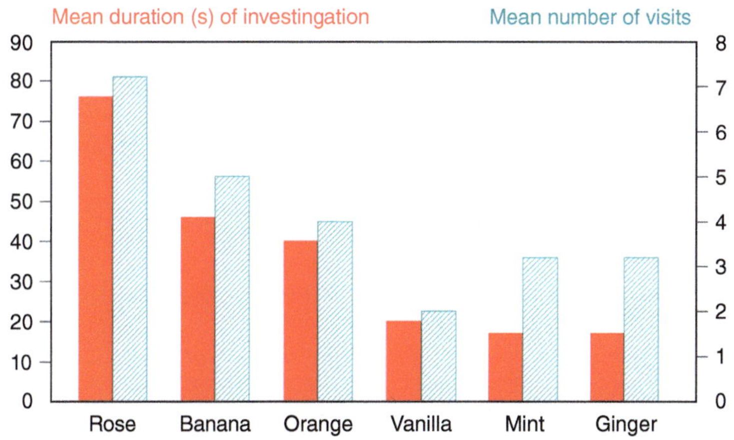

Fig. 13.1. Behavioural response of giraffes, in terms of mean duration investigating (solid bars) and number of visits spent with (hatched bars) each of the different scents presented. This study found that rose scent generated not only the highest number of visits but also the longest investigation time compared with all other scents presented. (Adapted from Fay and Miller, 2015.)

animals to utilize a more visible part of an enclosure and hence increase the visitor experience and educational opportunities by witnessing animals engaging in species-specific behaviours (Miller, 2012).

Stereotypies are repetitive behavioural patterns, with no obvious function (Mason, 1991); their performance is often linked with poor welfare, though their expression may be as a result of previous suboptimal conditions (Mason and Latham, 2004). An important aspect to note, demonstrated by this study and others, is that individual differences in responses exist within the same species and this should be taken into consideration when evaluating results (Clark and King, 2008). This is especially true when assessing welfare in the zoo environment. The introduction, especially of a novel scent, may result in having adverse effects such as generating anxiety or even inducing stereotypies (Clark and King, 2008). Nevertheless, olfactory enrichment has been found to have positive effects in a number of zoo animals, including promoting activity levels, reducing predictability, increasing enclosure exploration and decreasing inactivity (Wells and Egli, 2004; Clark and King, 2008; Wells, 2009).

Stressful Scents and Their Absence

Zoos incorporate a large number of species living in close proximity that would not occur in nature. Individuals are exposed to a range of conspecific scents they may not normally be in contact with, including predator and prey species odours, along with novel scents they may never typically encounter. Additionally, there will be a vast range of artificial odours associated with the captive environment. Odours have the potential to be chronic sources of stress in captivity (Morgan and Tromborg, 2007). Exposure to predator odour has been demonstrated to induce stress responses in captive cotton-top tamarins (*Saguinus oedipus*, Buchanan-Smith *et al.*, 1993), and also lead to reduced food gathering and avoidance behaviours in captive water voles (*Arvicola terrestris* [*Arvicola amphibius*], Barreto and Macdonald, 1999). In addition, the detection of odour from distressed conspecifics has also been linked to stress responses (Vieuille-Thomas and Signoret, 1992). It is worth noting, particularly in a captive environment, that not all exposures to acute stress are negative and can in fact be beneficial by increasing behavioural repertoires (Breazile, 1987; Moodie and Chamove, 1990). Brief and controlled exposure to stressful situations could be important in the retention of behaviours key to reintroduction success. The tendency of animals to avoid unfamiliar stimuli, neophobia, may be undesirable in the captive environment but in the wild retention of this behaviour may protect animals from numerous potential threats (Clark and King, 2008).

Wombats use scent marking and defecation in social communication in the wild (Gaughwin, 1979; Taylor, 1993). Descovich *et al.* (2012) exposed 12 captive southern hairy-nosed wombats (*Lasiorhinus latifrons*), housed in four groups of unrelated individuals, at Rockhampton Botanical Gardens and Zoo, Australia, to conspecific and predator scents (male and female wombat, and dingo faecal samples, along with a control of a plastic imitation canine faeces). Their behaviour was analysed (via burrow and external enclosure cameras), including enclosure use and inter-individual distances (5-min sampling). Descovich *et al.* (2012) found that two behaviours (approach to treatment area and retreating from conspecifics) were significantly affected by scent treatment (Fig. 13.2). Wombats approached the dingo faeces more than the control or female wombat faeces. They also avoided other wombats, within their enclosure, more so when male wombat faeces were present compared with the control or female wombat odours.

This study demonstrated that the southern hairy-nosed wombat was capable of distinguishing between the scents, with avoidance behaviour exhibited when predator (dingo) and male conspecific faeces were presented. Though wombats initially approached the dingo scent, the authors suggest this was due to the novelty of the scent and the need for them to confirm the scent identity. The residual retreat response exhibited by individuals that had previously approached the dingo scent was presumed by the authors to demonstrate that the wombats had recognized this as a predator scent. Confrontations, especially with dominant male wombats, can be risky and potentially lethal, especially for sub-adults trying to establish new territories and mating opportunities. As wombats actively avoided male faeces, this should be carefully considered in animal management as exposure to male scent may be perceived as stressful. Another study indicates that such male recognition initiates more intensive investigation: olfactory studies in zoo-housed sub-adult brown bears

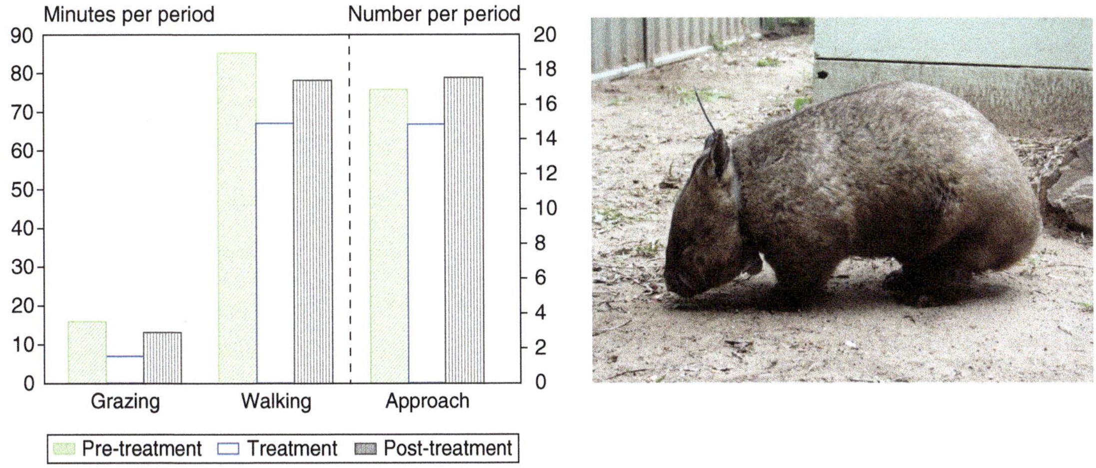

Fig. 13.2. During the treatment period, presence of dingo faeces significantly increased the approach by male wombats to the treatment area compared with when control (plastic, imitation canine faeces) or female wombat faeces were present, and wombats retreated twice as often from male than from female wombats. Period effects were also evident in three behaviours (grazing, walking and approaching another wombat) regardless of treatment type. These behaviours occurred less during the treatment period than the pre-treatment or post-treatment periods. These data demonstrated that wombats could differentiate faeces from different species (wombat versus dingo) and between the sex of conspecifics. (Data from Descovich *et al.*, 2012, and photo by Kristin A. Descovich.)

(*Ursus arctos*), presented with anal gland secretions from male bears, demonstrated greater investigative and cautious behaviours, indicating a recognition of sex and potentially reproductive status. Adult males present more of a risk and hence this may be an evaluation of potential threat (Jojola *et al.*, 2012).

Descovich *et al.* (2012) conclude that the use of faeces as captive wombat enrichment may not significantly improve welfare, as it did not encourage exploratory behaviour or reduce stereotypic pacing. This study demonstrated, however, that exposure to faeces (of all types) caused wombats to adjust their behavioural patterns, including increased hiding behaviours and use of burrows, which are considered as protective behaviour and risk avoidance (Triggs, 2009). The reaction and interest towards scents used in behavioural enrichment will be influenced by the animal's motivation to approach the odour and its familiarity with it (Clark and King, 2008). Therefore, the olfactory influences of the location of an enclosure within a zoo, or olfactory stimuli from neighbouring enclosures, may influence an animal's well-being and behavioural pattern. Such effects should also be considered when relocating individuals to different enclosures within a zoo. Neophobia towards scents in their environment can be exhibited by captive animals, with potentially undesirable consequences, such as reduced behavioural repertoire, feeding, and increased aggression and stereotypies that may challenge well-being (Clark and King, 2008). Similarly, the regular removal of scent marks and an individual's odour within their 'territory', through regularly cleaning of enclosures, has proven stressful in laboratory animals (Castelhano-Carlos and Baumans, 2009; see also Chapter 14). Along with scent removal, the addition of odours from cleaning products may influence animal behaviour, positively or negatively, and therefore impact on physiology and health (López-Salesansky *et al.*, 2015). Daily handling associated with cage cleaning in young rats (*Rattus norvegicus*) has been demonstrated to reduce anxiety; therefore, cleaning regime can have a significant effect on stress profiles (Burn *et al.*, 2008). The sex of human handlers can also induce differing stress responses, as demonstrated in captive mice (*Mus musculus*) and rats (Sorge *et al.*, 2014), as male-related stimuli (e.g. T-shirts worn by men) induced a robust physiological stress response for both species. Such findings require further investigation in zoo environments, as reactions to various scents have also been demonstrated to wane over repeated exposure (Wells and Egli, 2004).

Use of Biologically Relevant and Irrelevant Scents to Expand Behavioural Repertoire

Clark and King (2008) determined that faeces and urine are the most commonly presented odour sources used in olfactory enrichment studies, with essential oils utilized in ~25% of studies. Natural prey scent has been demonstrated to increase activity and behavioural diversity in a number of species (Chamove and Anderson, 1989; e.g. zoo-housed lions, *Panthera leo*, Schuett and Frase, 2001), though primate faecal scent increased captive jaguar (*Panthera onca*; stereotypic pacing in, Clark, 2004). Response variations exist both across species examined and dependent on type of scent used (whether biologically relevant or not). The use of essential oils, though mostly biologically meaningless to many species, potentially may improve captive animal welfare (Wells, 2009), though the long-term impact has not been established. Clark and King (2008) highlight that justifications for using particular scents are not often given in the literature, and the health risks of presenting biological scents should be carefully considered.

Nilsson *et al.* (2014) tested four carnivore species at Kolmården Wildlife Park, Sweden (South American bush dogs, *Speothos venaticus*; Siberian tigers, *Panthera tigris altaica*; Asian wild dogs, *Cuon alpinus*; and African wild dogs, *Lycaon pictus*), and their reaction to wooden logs impregnated with either mammalian blood or the blood odour component, and two control odours (Fig. 13.3). Each of the four odour stimuli was presented five times in total to each of the four species in a pseudo-randomized order, with continuous sampling used to record the occurrence of each interaction with a wooden log.

This study found that the two biologically relevant prey-associated odours instigated significantly more behavioural interactions compared with the two control odours. The South American bush dogs and the Siberian tigers spent significantly longer per interaction with the odorized logs when presented with the real blood odour compared with the other three stimuli. Interestingly, a single blood odour component was as efficient in eliciting behavioural responses as when whole blood was presented. One of the species tested (the bush dog) did exhibit a significant decrease in the number of interactions with the scented logs over time, although none stopped interacting with the logs completely even after 20 days of exposure. Wells and Egli (2004) determined that the behaviour of captive black-footed cats (*Felis nigripes*) was also significantly influenced by olfactory enrichment; however, this effect decreased with repeated exposure of tested scents. In this study, more episodes of exploratory behaviour were observed, and importantly there was no increase in negative behavioural types such as stereotypies. However, as highlighted by Wells (2009), it should be noted that an increase in activity levels and the display of species-specific behaviours does not automatically imply that greater well-being has been achieved. Olfactory enrichment is complicated, especially in a zoo environment, so more robust testing is required.

Encouraging Reluctant Breeders

A major justification for the existence of zoos is the role they play in the captive breeding of endangered species, not only acting to preserve genetic biodiversity but also reducing the need for zoos to collect animals from the wild to maintain captive populations. Applied behavioural research has greatly aided captive breeding success of endangered animals (Swaisgood, 2007). However, behavioural problems have been highlighted as the main cause of failure in captive breeding programmes (Synder *et al.*, 1996). Captive breeding via natural matings is notoriously unsuccessful for certain species, e.g. clouded leopards (*Neofelis nebulosa*, Yamada and Durrant, 1989) and cheetahs (*Acinonyx jubatas*, Marker-Kraus and Grisham, 1993).

Swaisgood *et al.* (2000) demonstrated the importance of social odours in their studies on chemical communication in captive giant pandas (*Ailuropoda melanoleuca*) at Wolong Breeding Centre, China. Giant pandas rely heavily on chemical communication in the wild. This is not only to advertise reproductive status, but is also thought to regulate agonistic interactions and inter-individual spacing (Schaller *et al.*, 1985). Olfactory modulation of aggression and influence on sexual motivation can play a significant role in captive husbandry and

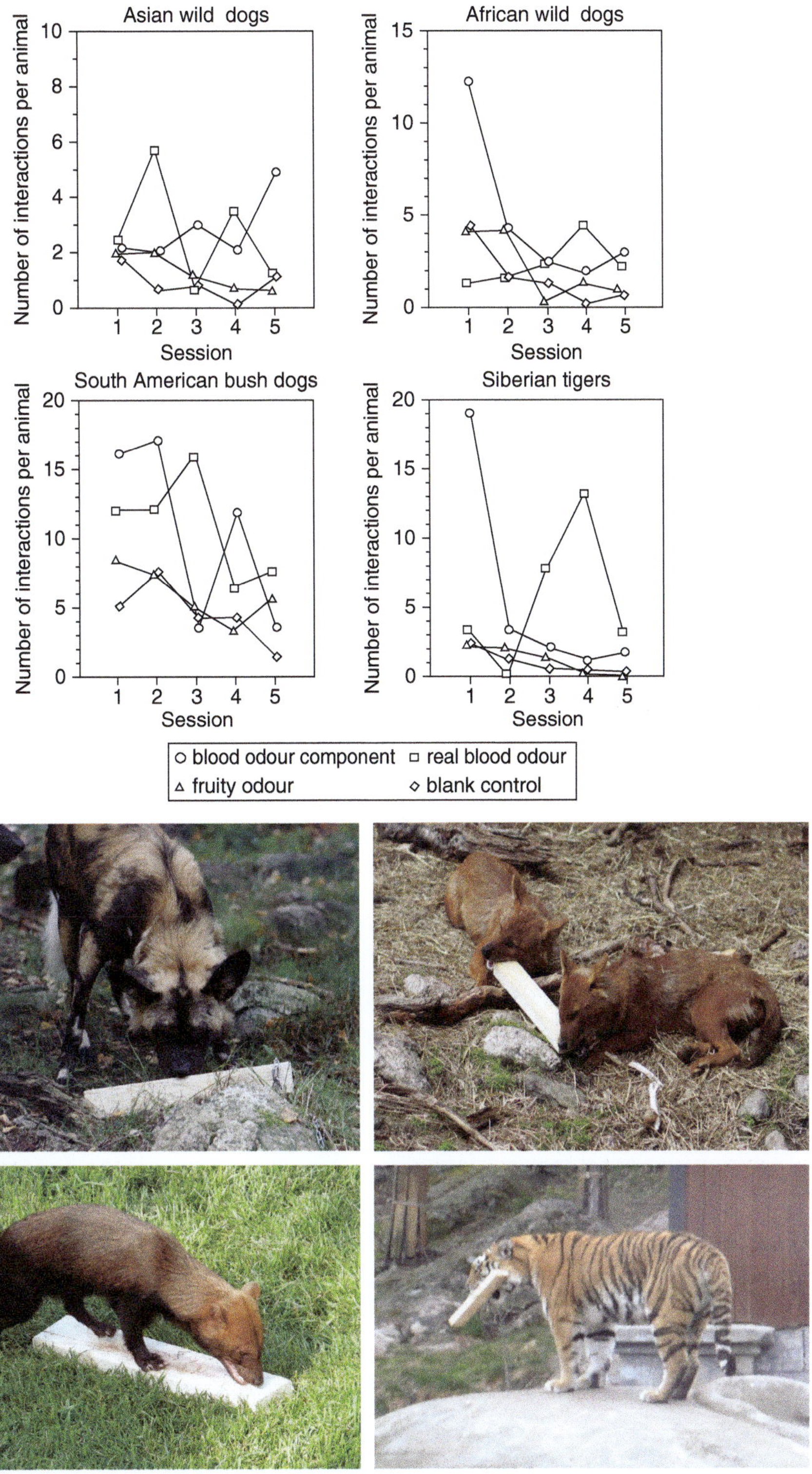

Fig. 13.3. Number of interactions with odorized logs in different carnivore species (Asian wild dogs, African wild dogs, South American bush dogs and Siberian tigers), across five sessions per odour stimulus and species (*circles*:

(*Caption continued*)

breeding programmes, as in giant pandas where excessive aggression and lack of sexual motivation result in the majority of mating attempts failing (He *et al.*, 1994; Lindburg *et al.*, 1997).

Swaisgood *et al.* (2000) investigated the relationship between chemical communication and reproductive behaviour in pandas, through systematic exposure to naturally deposited conspecific scents, and evaluated the contextual influences on chemosensory responsiveness (such as sex and reproductive state). The experimental design involved a 'visitor' panda being introduced to a 'resident' enclosure (in which the resident panda had been living for a minimum of 10 days and then removed), and comparing the behaviour observed for 30 min to that seen in control conditions (visitor panda in its own enclosure with no human visitors 48 h before the experiment). Particular attention was paid to olfactory behaviours in pandas, including chemosensory investigation, flehmen (deep inhalations accompanied by lip curling), licking, scent marking, body rubbing, foot scraping and vocalizations. This study demonstrates that pandas significantly increase these behaviours when exposed to scent from conspecifics, and this was particularly influenced by oestrus status of the observed panda. Oestrous females scent marked, urinated and vocalized significantly more than non-oestrous females across all experimental conditions. Sex of the signallers was recognized by the pandas, with male panda vocalization patterns indicating that they can discriminate female reproductive status on the basis of scent exposure. Given the degree of behavioural response to conspecific scents in this study, chemical communication is clearly demonstrated to be vitally important in the captive breeding husbandry of this species (Fig. 13.4). Male aggression and lack of sexual motivation have been determined as the primary cause of reproductive failure in captive breeding programmes (He *et al.*, 1994; Lindburg *et al.*, 1997). This study demonstrates that the exposure of potential breeding partners to each other's scent prior to introduction may increase reproductive success, through reduction of male aggression, increase in their libido, along with increased receptivity in females (Goodwin *et al.*, 1979; Johnston, 1979, 1990).

This scent exposure study could serve as a template, particularly in carnivore captive breeding programmes. Providing potential breeders with olfactory access before actual introduction has also been shown to increase captive breeding of cheetahs at San Diego Zoo (Lindburg, 1999). Chemosignals are regularly employed to stimulate breeding in agricultural mammals (Rekwot *et al.*, 2001), and the potential implications for zoo husbandry are only beginning to be realized (Swaisgood and Schulte, 2010).

Manipulating Mate Choice, Overcoming Incompatibility and Distance

Few studies have rigorously tested or incorporated the role of olfaction and mate selection in captive breeding programmes where animal pairings tend to be selected through matchings via studbooks designed to maximize genetic diversity. The exclusion of mate choice may result in incompatible pairs and reduce breeding success, as females that can select their mates tend to produce more viable offspring (Drickamer *et al.*, 2000) than females with no or more limited choice.

The practicality of enabling zoo animals to exhibit mate choice may be difficult, especially in highly endangered species in which a limited number of individuals exist in captive collections spread across the globe. Inter-zoo transfers may be logistically complex, expensive and challenge welfare as individuals are transported solely for breeding purposes. Also, to maximize outbreeding and increase genetic diversity, matings between specific individuals may be crucial for population management and viability. Therefore, active manipulation of mate choice and compatibility may be essential to improve breeding success. Fisher *et al.* (2003) successfully used odour cues to experimentally modify mate choice in pygmy lorises (*Nycticebus pygmaeus*), a small nocturnal primate and a vulnerable species that can be difficult

Fig. 13.3. Continued.
blood odour component, *trans*-4,5-epoxy-(*E*)-2-decenal; *squares*: real blood odour, horse blood; *triangles*: fruity odour, iso-pentyl acetate; *diamonds*: blank control, diethyl phthalate). Please note that the *y*-axes are on different scales. There was considerable variability between sessions. The results showed a significant decrease in the number of interactions per animal across sessions in the South American bush dogs, but not in the Asian wild dogs, African wild dogs and Siberian tigers. (Based on data from Nilsson *et al.*, 2014, and photos by Matthias Laska.)

Fig. 13.4. Chemical communication is obviously important in giant panda reproduction. Natural matings between captive giant pandas are rare. Exposure to conspecific scent as part of their routine captive husbandry is therefore becoming increasingly incorporated in captive panda breeding programmes. (Photos by Ron R. Swaisgood.)

to breed in captivity. Scent marks appear to be an important form of communication in this primate and the authors found that females use competitive counter-marking to assess male quality. Through manipulation of exposure of females to male odours, the authors succeeded in creating a preference towards certain males regardless of their actual qualities.

Previous studies have shown that females of solitary mammalian species prefer to mate with males whose scent marks they have encountered more often (Daly, 1977; Tang-Martinez *et al.*, 1993) and/or whose scents indicate quality (Clark *et al.*, 1992; Rich and Hurst, 1999). This can facilitate manipulation in a captive environment. Roberts and Gosling (2004) demonstrated mate choice manipulation in harvest mice (*Micromys minutus*) at Chester Zoological Gardens, UK, by successfully increasing the attractiveness of previously unpreferred males through the use of preferred male odours in female choice tests (Fig. 13.5). They used artificial 'branches' to transfer scent marks between harvest mice enclosures. Their study was composed of two main tests: (i) female mice preferences among a series of males were determined via a series of two-choice tests; (ii) the degree of familiarity between females and odour of potential mates was varied. Females could see and smell, though not directly interact with males. Males were also ranked according to rates of scent marking, with territory holding or dominant males signalling more often (Gosling *et al.*, 2000; Gosling and Roberts, 2001). In the second part of the study, the degree of familiarity was varied, so that females spent more time with unpreferred males in the second trial than they had in the first trial. Results showed that females paired with familiar males demonstrated reduced aggression (counted as number of fights, bites or aggressive lunges) compared to females paired with unfamiliar males.

As discussed by Roberts and Gosling (2004), the use of scent transfer is a cheap and non-invasive tool for manipulating the environment of captive animals, and may serve to promote mating through familiarity of potential mates before physical exposure. Through artificially introducing the scent of potentially competitive males, one may stimulate initially unpreferred males to make a greater investment in scent marking (i.e. increase frequency), which constitutes a greater investment in energy, and potentially exposes an individual to increased predation. It was found that the degree of familiarity of females with the odour of potential mates influences the amount of time that a female will spend with that male and how she acts towards him. Importantly, this degree of familiarity was thought to affect only female behaviour during the initial phase of the choice tests. Visual characteristics may play an important role: when females

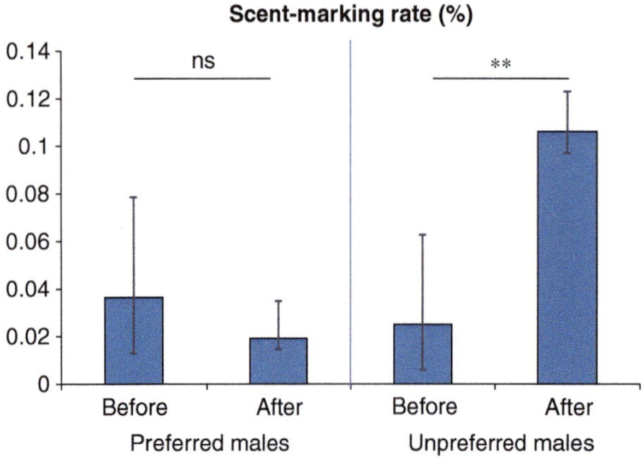

Fig. 13.5. Scent-marking rates of harvest mouse males before and after manipulation of the males' social context, by introduction of a competitor's odour into the home enclosure. Although it was expected that males preferred by females would scent mark at higher intensities than unpreferred males, the two groups did not differ significantly before the manipulation of the male's social context. After manipulation, however, the scent marking rates of initially unpreferred males increased and were now higher than those of initially preferred males (**, $P < 0.01$). However, initial scent marking rates did not predict female preference in the first choice test. Thus, the authors cannot yet attribute changes in preferences of female harvest mice to changes in marking rates. (Data from Roberts and Gosling, 2004, and photo by Jeroen van der Kooij.)

could both see and smell males there was a lack of expressed preference. This suggests that familiarity had a comparably weak influence and needs further investigation. However, teasing apart visual and olfactory cues may be difficult and it is important to note that as humans we may underestimate the reliance on olfactory stimulation.

As shown above, scent manipulation may be utilized to encourage mating between human-selected pairs, counteracting possible female perceptions of less desirable males. In captive breeding programmes, particularly for endangered species, mate selection is typically determined by conservation managers with a view to maximize outbreeding and genetic diversity, as opposed to allowing animals to select their own mates. However, the ability to express mate choice in captive animals has been demonstrated to increase significantly long-term population viability (Grahn *et al.*, 1998; Wedekind, 2002).

Training for Release into the Wild

Another reason for the existence of zoos is the role they can play in the future reintroduction of species. Many reintroductions still fail due to behavioural issues, and it is therefore vital to establish before individuals can be released that they are behaviourally competent in order to optimize their chance of survival. One example is the failure to recognize predators and respond appropriately when they are encountered. Prey species exposed to predator scents react with species-typical behavioural patterns, such as reduced activity and avoidance behaviours (Apfelbach *et al.*, 2005). The captive environment is relatively unchallenging and therefore life skills can be lost through long periods of captivity and/or multiple generations being raised in very different environments from natural conditions. Training for release is the term commonly used for the management practices employed prior to release to ensure, as far as possible, that the release candidate is prepared for survival. These may include exposing animals to stressful stimuli, which may challenge concepts of good welfare in captivity (Reading *et al.*, 2013). However, different sources of predator scent, such as fur versus faeces, have been shown to have greater and long-lasting effects, including suppression of reproduction (Apfelbach *et al.*, 2005), so application must be carefully considered.

Crane and Mathis (2011) investigated the use of olfactory cues in a 'head-starting' programme for hellbenders (*Cryptobranchus alleganiensis*) at Saint Louis Zoo, USA. Head-starting involves captive rearing of individuals past their most vulnerable early life stages before their release (IUCN/SSC, 2013). A major challenge to head-starting programmes for

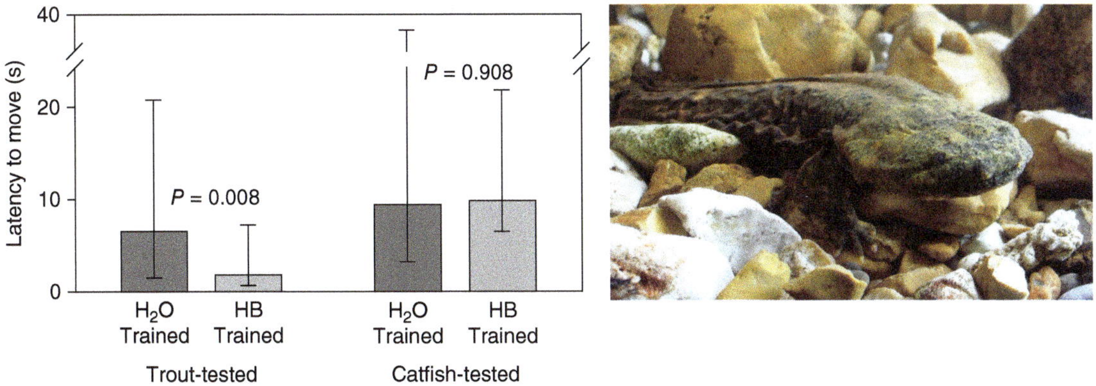

Fig. 13.6. Latency to move for hellbender larvae previously trained with trout odour paired with either adult hellbender (HB) distress secretion (light grey bars) or blank river water (H_2O; dark grey bars) and then tested with either trout (n = 24 per group) or suckermouth catfish (n = 12 per group) odour. Vertical bars and error bars represent medians and quartiles, respectively. (Data from Crane and Mathis, 2011, and photo by Adam L. Crane.)

hellbenders is the high rates of post-release predation. Larval hellbenders produce distress secretions when in contact with native predators but demonstrate only weak responses towards introduced predators such as non-native rainbow (*Oncorhynchus mykiss*) and brown (*Salmo trutta*) trout, which may significantly impact on hellbender reintroduction success (Gall and Mathis, 2009). Crane and Mathis (2011) utilized classical conditioning to train hellbenders to respond to chemical cues from these non-native predators coupled with hellbender distress secretion.

Trained larvae were tested ~24 h after training, in a different testing arena. The larvae that had been trained with trout–hellbender odour were significantly faster to move relative to the larvae previously trained with the trout–H_2O odour (control). To test this, the behavioural responses of a subset of trained larvae were observed when exposed to the unfamiliar tropical suckermouth catfish (*Hypostomus plecostomus*). When exposed to the catfish stimuli, the behavioural responses did not differ from the responses demonstrated when larvae were exposed to the control (trout secretions with water; Fig. 13.6). The results demonstrate that larvae can be trained to associate chemical cues from trout as dangerous and instigate an alarm response, based on adult distress secretions. Such predator-recognition learning may be utilized by zoos to increase post-release survival in species restoration programmes.

From the zoo-based behavioural studies presented here we can determine that olfactory stimuli are clearly an important modality in the captive environment, and that odours are often overlooked in relation to their influence on welfare. Olfactory cues are often ignored or poorly understood in the zoo environment. Lessons learnt from other animal husbandry sectors, in particular livestock management (Chapter 12) and laboratory animal research (Chapter 14), can help to highlight the importance of olfactory stimuli for animal welfare, and in particular present a real application opportunity when it comes to zoo-housed animals. Sources of odour (such as scent marks, urine, faeces or pieces of fur) facilitate manipulation (they are easy to collect, store and transport), and could be important management tools to increase the complexity of the captive environment, influence captive breeding success, and train zoo animals for release back to the wild. Zoos hold a wide range of species, particularly non-mammalian, therefore, it is vital to determine that any scent manipulation or olfactory enrichment should be carefully assessed for impact on species-specific well-being.

References

Apfelbach, R., Blanchard, C.D., Blanchard, R.J., Hayes, R.A. and McGregor, I.S. (2005) The effects of predator odors in mammalian prey species: a review of field and laboratory studies. *Neuroscience Behavioural Review* 29, 1121–1144. DOI: 10.1016/j.neubiorev.2005.05.005

Barreto, C. and Macdonald, D.W. (1999) The response of water voles, *Arvicola terrestris*, to the odours of predators. *Animal Behaviour* 57, 1107–1112. DOI: 10.1006/anbe.1998.1042

Bonadonna, F. and Bretagnolle, V. (2002) Smelling home: a good solution for burrow-finding in nocturnal petrels? *Journal of Experimental Biology* 205, 2519–2523.

Bonadonna, F. and Nevitt, G.A. (2004) Partner-specific odour recognition in an Antarctic seabird. *Science* 306, 835. DOI: 10.1126/science.1103001

Breazile, J.E. (1987) Physiological basis and consequences of distress in animals. *Journal of American Veterinary Medicine Association* 191, 1212–1215.

Buchanan-Smith, H.M., Anderson, D.A. and Ryan, C.W. (1993) Responses of cotton-top tamarins (*Saguinus oedipus*) to faecal scents of predators and non-predators. *Animal Welfare* 2, 17–32.

Burn, C.C., Deacon, R.M. and Mason, G.J. (2008) Marked for life? Effects of early cage-cleaning frequency, delivery batch, and identification tail-marking on rat anxiety profiles. *Developmental Psychobiology* 50, 266–277. DOI: 10.1002/dev.20279

Caro, S.P., Balthazart, J. and Bonadonna, F. (2015) The perfume of reproduction in birds: chemosignaling in avian social life. *Hormones and Behavior* 68, 25–42. DOI: 10.1016/j.yhbeh.2014.06.001

Castelhano-Carlos, M.J. and Baumans, V. (2009) The impact of light, noise, cage cleaning and in-house transport on welfare and stress of laboratory rats. *Laboratory Animals* 43, 311–327. DOI: 10.1258/la.2009.0080098.

Chamove, A.S. and Anderson, J.R. (1989) Examining environmental enrichment. In: Segal, E.F. (ed.) *Housing, Care and Psychological Wellbeing of Captive and Laboratory Primates*. Noyes Publications, Park Ridge, New Jersey, pp. 183–202.

Clark, F. (2004) Olfactory enrichment for the captive jaguar (*Panthera onca*). BSc thesis, University of Southampton, UK.

Clark, F. and King, A.J. (2008) A critical review of zoo-based olfactory enrichment. In: Hurst, J.L., Beynon, R.J., Roberts, C., and Wyatt, T.D. (eds) *Chemical Signals in Vertebrates*, 11th edn. Springer, New York, pp. 391–398.

Clark, M.M., Tucker, L., and Galef, B.G. (1992) Stud males and dud males: intra-uterine position effects on the reproductive success of male gerbils. *Animal Behaviour* 43, 215–221

Crane, A.L. and Mathis, A. (2011) Predator-recognition training: a conservation strategy to increase postrelease survival of hellbenders in head-starting programs. *Zoo Biology* 30, 611–622. DOI: 10.1002/zoo.20358

Daly, M. (1977) Some experimental tests of the functional significance of scent-marking by gerbils. *Journal of Comparative Psychology* 91, 1082.

Descovich, K.A., Lisle, A.T., Johnston, S., Nicolson, V. and Phillips, C.J.C. (2012) Differential responses of captive southern hairy-nosed wombats (*Lasiorhinus latifrons*) to the presence of faeces from different species and male and female conspecifics. *Applied Animal Behaviour Science* 138, 110–117. DOI: 10.1016/j.applanim.2012.01.017

Drickamer, L.C., Gowaty, P.A. and Holmes, C.M. (2000) Free female mate choice in house mice affects reproductive success and offspring viability and performance. *Animal Behaviour* 59, 371–378. DOI: 10.1006/anbe.1999.1316

Fay, C. and Miller, L.J. (2015) Utilizing scents as environmental enrichment: preference assessment and application with Rothschild giraffe. *Animal Behaviour and Cognition* 2, 285–291. DOI: 10.12966/abc.08.07.2015.

Fisher, H.S., Swaisgood, R.R. and Fitch-Snyder, H. (2003) Counter marking by male pygmy lorises (*Nycticebus pygmaeus*): do females use odour cues to select mates with high competitive ability? *Behavioural Ecology and Sociobiology* 53, 123–130. DOI: 10.1007/s00265-002-0552-5

Gall, B.G. and Mathis, A. (2009) Innate predator recognition in larval hellbenders (*Cryptobranchus alleganiensis*) and the problem of introduced trout. *Ethology* 116, 47–58. DOI: 10.1111/j.1439-0310.2009.01718.x

Gaughwin, M.D. (1979) Occurrence of flehmen in a marsupial – hairy nosed wombat (*Lasiorhinus latifrons*). *Animal Behaviour* 27, 1063–1065. DOI: 10.1016/0003-3472(79)90054-X

Goodwin, M., Gooding, K.M. and Regnier, F. (1979) Sex pheromone in the dog. *Science* 203, 559–561. DOI: 10.1126/science.569903

Gosling, L.M. and Roberts, S.C. (2001) Scent-marking by male mammals: cheat-proof signals to competitors and mates. *Advances in the Study of Behaviour* 30, 169–217. DOI: 10.1016/S0065-3454(01)80007-3

Gosling, L.M., Roberts, S.C., Thornton, E.A. and Andrew, M.J. (2000) Life history costs of olfactory status signalling in mice. *Behavioural Ecology and Sociobiology* 48, 328–332. DOI: 10.1007/s002650000242

Grahn, M.A., Langefors, A. and von Schantz, T. (1998) The importance of mate choice in improving viability in captive populations. In: Caro, T. (ed.) *Behavioural Ecology and Conservation Biology*. University of Oxford Press, Oxford, pp. 341–363.

Gsell, A.C., Hagelin, J.C. and Brunton, D.H. (2012) Olfactory sensitivity in kea and kaka. *Emu* 112, 60–66. DOI: 10.1071/mu11052

He, T., Zhang, K., Zhang, H., Wei, R., Tang, C., Zhang, G. and Cheng, M. (1994) Training male giant pandas for natural mating. In: Zhang, A. and He, G. (eds) *Proceedings of the International Symposium on the Protection of the Giant Panda*. Chengdu Foundation of Giant Panda Breeding, Chengdu, China, pp. 188–192.

Hill, S.P. and Broom, D.M. (2009) Measuring zoo animal welfare: theory and practice. *Zoo Biology* 28, 531–544. DOI: 10.1002/zoo.20276.

IUCN/SSC (2013) Guidelines for Reintroductions and Other Conservation Translocations. Version 1.0. IUCN Species Survival Commission, Gland, Switzerland, 57 pp.

Johnston, R.E. (1979) Olfactory preferences, scent marking and 'proceptivity' in female hamsters. *Hormones and Behaviour* 13, 21–29. DOI: 10.1016/0018-506X(79)90032-1
Johnston, R.E. (1990) Olfactory communication in golden hamsters: from behaviour to molecules to neural mechanisms. In: Dewsbury, D.E. (ed.) *Contemporary Trends in Comparative Psychology*. Sinauer, Sutherland, Massachusetts, pp. 381–409.
Jojola, S.M., Rosell, F., Warrington, I., Swenson, J.E. and Zedrosser, A. (2012) Subadult brown bears (*Ursus arctos*) discriminate between unfamiliar adult male and female anal gland secretion. *Mammalian Biology* 77, 363–368. DOI: 10.1016/j.mambio.2012.05.003
Jones, R.B., Facchin, L. and McCorquodale, C. (2002) Social dispersal by domestic chicks in a novel environment: reassuring properties of a familiar odourant. *Animal Behaviour* 63, 659–666. DOI: 10.1006/anbe.2001.1943
Lindburg, D.G. (1999) Zoos as arks: issues in ex situ propagation of endangered wildlife. In: Strum, S.C., Lindburg, D.G. and Hamburg, D. (eds) *The New Physical Anthropology: Science, Humanism, and Critical Reflection*. Prentice Hall, Upper Saddle River, New Jersey, pp. 201–213.
Lindburg, D.G., Huang, X.M. and Huang, S.Q. (1997) Reproductive performance of male giant panda in Chinese zoos. In: Zhang, A. and He, G. (eds) *Proceedings of the International Symposium on the Protection of the Giant Panda*. Chengdu Foundation of Giant Panda Breeding, Chengdu, China, pp. 67–71.
López-Salesansky, N., Mazian, N.H., Whitfield, L.E., Wells, D.J. and Burn, C.C. (2015) Olfactory variation in mouse husbandry and its implications for refinement and standardization: UK survey of non-animal scents. *Laboratory Animal* 50, 286–295. DOI: 10.1177/0023677215614296.
Marker-Kraus, L. and Grisham, J. (1993) Captive breeding of cheetahs in North American zoos: 1987–1991. *Zoo Biology* 12, 5–18. DOI: 10.1002/zoo.1430120104
Marriner, L.M. and Drickamer, L.C. (1994) Factors influencing stereotyped behaviour of primates in a zoo. *Zoo Biology* 13, 267–275. DOI: 10.1002/zoo.1430130308
Mason, G.J. (1991) Stereotypies: a critical review. *Animal Behaviour* 41, 1015–1037.
Mason, G.J. and Latham, N. (2004) Can't stop, won't stop: is stereotypy a reliable animal welfare indicator? *Animal Welfare* 13, 57–69.
Melfi, V.A. (2009) There are big gaps in our knowledge, and thus approach, to zoo animal welfare: a case for evidence-based zoo animal management. *Zoo Biology* 28, 574–588. DOI: 10.1002/zoo.20288
Miller, L.J. (2012) Visitor reaction to pacing behaviour: influence on the perception of animal care and interest in supporting zoological institutions. *Zoo Biology* 31, 242–248. DOI: 10.1002/ZOO.20411
Moodie, E.M. and Chamove, A.S. (1990) Brief threatening events beneficial for captive tamarins. *Zoo Biology* 9, 275–286.
Morgan, K.N. and Tromborg, C.T. (2007) Sources of stress in captivity. *Applied Animal Behaviour Science* 102, 262–302. DOI: 10.1016/j.applanim.2006.05.032
Nielsen, B.L., Jezierski, T., Bolhuis, J.E., Amo, L., Rosell, F., Oostindjer, M., Christensen, J.W., McKeegan, D., Wells, D.L. and Hepper, P. (2015) Olfaction: an overlooked sensory modality in applied ethology and animal welfare. *Frontiers in Veterinary Science* 2, 69. DOI: 10.3389/fvets.2015.00069.
Nilsson, S., Sjöberg, J., Amundin, M., Hartmann, C., Buettner, A. and Laska, M. (2014) Behavioural responses to mammalian blood odour and a blood odour component in four species of large carnivores. *PLoS ONE* 9(110), e112694. DOI:10.137/journal.pone.0112694.
Plowman, A.B. (2008) BIAZA statistics guidelines: toward a common application of statistical tests for zoo research. *Zoo Biology* 27, 226–233. DOI: 10.1002/zoo.20184.
Reading, R.P, Miller, B. and Shepherdson, D. (2013) The value of enrichment to reintroduction success. *Zoo Biology* 32, 332–341. DOI: 10.1002/zoo.21054
Rekwot, P.I., Ogwu, D., Oyedipe, E.O. and Sekoni, V.O. (2001) The role of pheromones and biostimulation in animal reproduction. *Animal Reproduction Science* 65, 157–170. DOI: 10.1016/S0378-4320(00)00223-2
Rich, T.J., and Hurst, J.L. (1999) The competing countermarks hypothesis: reliable assessment of competitive ability by potential mates. *Animal Behaviour* 58, 1027–1037. DOI: 10.1006/anbe.1999.1217
Roberts, S.C. and Gosling, L.M. (2004) Manipulation of olfactory signalling and mate choice for conservation breeding: a case study of harvest mice. *Conservation Biology* 18, 548–556. DOI: 10.1111/j.1523-1739.2004.00514.x
Roper, T.J. (1999) Olfaction in birds. *Advances in the Study of Behaviour* 28, 247–332. DOI: 10.1016/S0065-3454(08)60219-3
Roth, T.C., II, Cox, J.G. and Lima, S.L. (2008) Can foraging birds assess predation risk by scent? *Animal Behaviour* 76, 2021–2027. DOI:10.1016/j.anbehav.2008.08.022
Schaller, G.B., Hu, J., Pan, W. and Zhu, J. (1985) *The Giant Panda of Wolong*. The University of Chicago Press, Chicago, Illinois.
Schuett, E.B. and Frase, B.A. (2001) Making scents: using the olfactory senses for lion enrichment. *The Shape of Enrichment* 10, 1–3.
Seidensticker, J. and Forthman, D.L. (1998) Evolution, ecology and enrichment: basic considerations for wild animals in zoos. In: Shepherdson, D.J., Mellen, J.D., Hutchins, M. and Maple, T. (eds) *Second Nature: Environmental Enrichment for Captive Animals*. Smithsonian Institution, Washington, DC, pp. 15–29.

Sorge, R.E., Martin. L.J., Isbester, K.A., Sotocinal, S.G., Rosen, S. *et al.* (2014) Olfactory exposure to males, including men, causes stress and related analgesia in rodents. *Nature Methods* 11, 629–632. DOI: 10.1038/nmeth.2935

Swaisgood, R.R. (2007) Current status and future directions of applied behavioral research for animal welfare and conservation. *Applied Animal Behavioural Science* 102, 139–162. DOI: 10.1016/j.applanim.2006.05.027

Swaisgood, R.R. and Schulte, B.A. (2010) Applying knowledge of mammalian social organization, mating systems and communication to management. In: Kleiman, D.G., Thompson, K.V. and Baer, C.K. (eds) *Wild mammals in Captivity: Principles and Techniques for Zoo Management*. University of Chicago Press, Chicago, Illinois, pp. 329–343.

Swaisgood, R.R. and Shepherdson, D.J. (2005) Scientific approaches to enrichment and stereotypies in zoo animals: what's been done and where should we go next? *Zoo Biology* 24, 499–518. DOI: 10.1002/zoo.20066

Swaisgood, R.R., Lindburg, D.G., Zhou, X. and Owen, M.A. (2000) The effects of sex, reproductive condition and context on discrimination of conspecific odours by giant pandas. *Animal Behaviour* 60, 227–237. DOI: 10.1006/anbe.2000.1464

Synder, N.F.R., Derrickson, S.R., Beissinger, S.R., Wiley, J.W., Smith, T.B., Toone, W.D. and Miller, B. (1996) Limitations of captive breeding in endangered species recovery. *Conservation Biology* 10, 338–348.

Tang-Martinez, Z., Mueller, L.L. and Taylor, G.T. (1993) Individual odours and mating success in the golden hamster *Mesocricetus auratus*. *Animal Behaviour* 45, 1141–1151. DOI: 10.1006/anbe.1993.1138

Taylor, R.J. (1993) Observations on the behaviour and ecology of the common wombat, *Vombatus ursinus*, in northeast Tasmania. *Australian Mammal* 16, 1–7.

Triggs, B. (2009) *Wombats*, 2nd edn. CSIRO publishing, Collingwood, Victoria.

Vieuille-Thomas, C. and Signoret, J.P. (1992) Pheromonal transmission of an aversive experience in domestic pig. *Journal of Chemical Ecology* 18, 1551–1557. DOI: 10.1007/BF00993228

Wedekind, C. (2002) Sexual selection and life-history decisions: implications for supportive breeding and the management of captive populations. *Conservation Biology* 16, 1204–1211. DOI: 10.1046/j.1523-1739.2002.01217.x

Wells, D.L. (2009) Sensory stimulation as environmental enrichment for captive animals: a review. *Applied Animal Behaviour Science* 118, 1–11. DOI: 10.1016/j.applanim.2009.01.002

Wells, D.L. and Egli, J.M. (2004) The influence of olfactory enrichment on the behaviour of captive black-footed cats, *Felis nigripes*. *Applied Animal Behavioural Science* 85, 107–119. DOI: 10.1016/j.applanim.2003.08.013

Yamada, J.K. and Durrant, B.S. (1989) Reproductive parameters of clouded leopards (*Neofelis nebulosa*). *Zoo Biology* 8, 223–231.

14 Olfactory Behaviour in Laboratory Animals

Olivia Le Moëne and Anders Ågmo

University of Tromsø, Norway

Introduction

Considering the use of research animals within the European Union (Fig. 14.1), in this chapter we concentrate on rats, mice and guinea pigs, the typical laboratory rodents, and the most commonly used lagomorph, the rabbit. These four species cover close to 80% of the research animals used. Even though domestic animals like horses, cows and sheep (see Chapter 12) as well as animals often found in zoos (Chapter 13), like monkeys, hyenas and snakes, are used in the laboratory now and then, they are usually not considered to be laboratory animals. This is also the case for pets like parrots, canaries, dogs and cats (Chapter 15). Even though all these species have been employed in research, their use has much declined and is now marginal. Some species of birds, notably the Japanese quail, the European starling, zebra finches, and pigeons and doves, are indeed used in the laboratory, although in very small numbers compared with rodents. It was long considered that birds relied on visual and auditory information far more than on olfaction. However, evidence for an important role of olfactory stimuli in bird behaviour has started to accumulate (Caro *et al.*, 2015), although very little is known about the consequences of odorant exposure for bird well-being.

We will first describe experimental studies of the possible roles of odours for the behaviour of rodents and lagomorphs, and then we will briefly mention some examples of how odours have been used experimentally in these species. Another section of the chapter will be dedicated to a discussion of how the standard laboratory conditions affect olfactory stimulation and the potential consequences thereof. However, we will begin with a short note on fish. Although the use of fish in behavioural studies is still very limited, it is increasing. The zebrafish is becoming popular, and some other fish species have attracted the attention of ethologists and behavioural ecologists, particularly the species used for intensive farming.

A note on the role of olfaction in fish

Both teleosts (which comprise 96% of all fish) and elasmobranchs (which include sharks) have well developed olfactory systems. Interestingly, they seem to possess odour receptor genes specialized for the detection of water-soluble molecules. These genes have been lost in mammals. Many studies have shown that fish respond to olfactory stimuli in different ways, and such stimuli are important for reproduction. It has been shown that zebrafish can also learn to respond to previously neutral odours through classical conditioning. The role of odours in the well-being of fish is almost entirely unknown. There is some indication, though, that olfactory stimuli may be involved in 'anxiety-like' responses. Anosmic zebrafish respond far less to an environmental stressor than smelling controls (Abreu *et al.*, 2016). It appears, then, that a potential role of olfaction in the well-being of fishes should not be ignored.

Mammalian Behaviour and Olfaction

The importance of olfactory stimuli for sexual attraction, food choice, parental behaviour and disease avoidance has been discussed in earlier chapters and we will not offer more information on these issues here. The principles outlined in the earlier chapters are generally applicable also to the laboratory animals that are the subjects of this chapter. In addition to the role of odours in the phenomena mentioned above, olfaction is involved

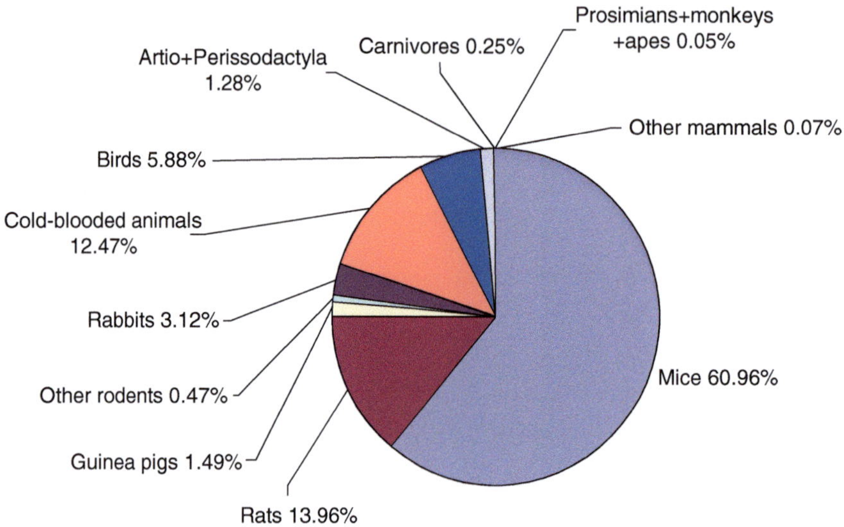

Fig. 14.1. Percentage of animals used in scientific research by classes in the member states of the European Union, data from 2011. (This figure belongs to the public domain.)

in several other behavioural or cognitive functions. In the wild, olfactory cues may be important for age appraisal, gregariousness, group membership appraisal, pain indication and social status appraisal, and may serve to communicate some emotional states (for a review, see the classic paper by Doty, 1986). The many behaviours and behavioural patterns that depend on a functioning olfactory system are important for survival, and anosmic animals are likely to show reduced fitness.

Laboratory strains of rodents and domesticated lagomorphs have preserved most, if not all, of the olfactory capacities of their wild ancestors. However, a fundamental question is whether any of these capacities are important for the animals' daily life in laboratory conditions. It is not evident that a rat or a mouse housed alone or with a few other conspecifics in a rather small cage has any need of olfaction for identifying group members, not even for identifying a potential mate. The group (normally consisting of two to five animals of the same strain, gender and age) is usually established by the experimenter by random assignment of animals to each cage. The animal cannot escape from the thus established groups, and encounters with members of other groups occur only for specific experimental purposes under carefully controlled circumstances. Whenever mating is desired, the experimenter or breeder will provide a mate considered suitable, except in the rather infrequent studies of mate choice. Olfaction cannot be used for the choice of food or drink since there is no choice available, and so on. It would appear that in standard laboratory conditions, olfaction is of no use to the animals. However, this is not necessarily the case.

The wild relatives of laboratory animals often choose habitats with a rich odoriferous environment. Some of the most dense rat populations are found in garbage dumps, for example. There, the environment can be supposed to provide an extremely varied mixture of intense odours and this is also the case in another habitat much appreciated by rodents: sewage systems. These examples may be extreme, but it can be maintained that wild rodents and lagomorphs are exposed to a variety and intensity of odours far superior to what is found in the laboratory. Even if odours have a very limited role in determining the behaviour displayed in a laboratory cage, their absence influences the animals' olfactory capacity and may affect the animals' emotional state.

How to Evaluate an Animal's Emotional State?

If we are seriously interested in evaluating the role of olfaction in animal well-being, we will probably need a reliable and objective measure of emotions. Psychologists interested in human emotions have devised several instruments with which to obtain quantitative estimates of an individual's emotional state. They are all based on the human capacity of symbolic language. In its simplest form, we can just verbally ask any human how he or she feels, and then record the response, for example, on a scale ranging from 0 (feel terrible) to 5 (feel euphoric). Non-verbal measures of human emotions (or other manifestations of well-being) are rarely used, and when they are used, they are a source of disagreement. It needs also to be pointed out that there is no general agreement about a definition of human well-being and happiness, and much less so as to how it can be measured (for a discussion of this fascinating issue, see Kapteyn *et al.*, 2015).

Unfortunately, verbal reports of well-being are difficult to obtain from non-human animals. We need, then, to look for other means to know and quantify an animal's emotional state or well-being. It goes without saying that the confusion concerning human well-being and happiness is nothing compared to the quandary in which the corresponding discussion of non-human animals finds itself. Most scientists in the area simply beat around the bush rather than trying to come to grips with this fundamental issue. We will follow this noble tradition, even though we will mention a few of the efforts that have been made to operationalize and quantify well- or ill-being.

One measure that has long been considered an indicator of ill-being, elevated serum concentrations of corticosteroids, is more likely an indicator of enhanced arousal, regardless of the emotion that is associated with the arousal. Copulating rats, for example, have extremely high serum concentrations of corticosterone (Koolhaas *et al.*, 2011). Yet rats engage enthusiastically in sexual behaviour whenever given the opportunity, and there is considerable evidence showing that this behaviour is pleasurable for male and female rats (for a review, see Paredes, 2009). Also, corticosterone itself has rewarding properties, as shown many years ago (Piazza *et al.*, 1993), and for these reasons cannot be used as an indicator of ill-being. Many other measures purported to assess emotional state are just as unreliable as corticosterone. Changes in body weight, rate of growth, food or drink consumption, or general activity can be caused by a variety of factors with different emotional consequences. For example, morphine reduces locomotor activity in rats, whereas amphetamine enhances it. Both drugs, however, cause a positive effect. Access to a running wheel reduces body weight in mice (Venezia *et al.*, 2016), which could be interpreted as a manifestation of a poor emotional state. However, wheel running is well known to induce a positive effect (Belke and Wagner, 2005). These simple examples should illustrate the difficulties in determining a non-human animal's emotional state. However, even though there is no perfect measure available, we have accumulated a substantial amount of knowledge about specific aspects of emotions and their control.

Purported Consequences of Olfactory Deprivation and Stimulation

Deprivation

Anosmia caused by removing the olfactory bulbs is used as a model of depression (e.g. Harkin *et al.*, 2003). The main behavioural consequences of bulbectomy are increased locomotor activity in a novel environment, deficits in avoidance behaviour and hyperreactivity to stressors. These behavioural alterations are not necessarily manifestations of ill-being. Moreover, peripheral anosmia, caused by destruction of the epithelium, does not mimic the effects of bulbectomy (Song and Leonard, 2005). It must be remembered that the olfactory bulbs are part of the brain, and that they have many functions in addition to being a relay station for olfactory information from the nasal and vomeronasal epithelia. Peripheral anosmia, on the other hand, leaves the brain intact, and is, consequently, a more accurate test of the importance of olfactory stimulation. The consequences of anosmia for the well-being of rats are, therefore, not entirely clear.

Data from mice, however, suggest that olfactory deprivation does indeed have consequences for well-being. Inactivation of the main olfactory epithelium through deletion of a gene (*Cnga2*) leads to increased anxiety-like behaviour in several procedures. Behaviour in tests supposed to detect depression-like behaviour was unaltered (Glinka *et al.*, 2012). Another study reported that destruction of the nasal epithelium with $ZnSO_4$ made the mice display depression-like behaviour (Ahn *et al.*, 2016). At the same time, anosmia had anxiolytic-like effects in other tests. These contradictory observations are difficult to reconcile. We conclude that available experimental data from anosmic mice are insufficient for a firm conclusion as to the role of olfactory stimulation in well-being.

In rabbits, peripheral anosmia has a peculiar effect: it accelerates the appearance of maternal behaviour in virgin rabbits exposed to pups (Chirino *et al.*, 2007). This effect was explained by proposing that the odour of new-borns is aversive to females that have not been exposed to the hormonal changes typical of pregnancy and parturition. Anosmia did not affect the females' food intake, ambulatory activity or scent marking. These observations could suggest that olfactory stimulation is not essential for the well-being of rabbits, but more data are needed before such a conclusion can be considered substantiated.

Instead of producing anosmia, olfactory stimulation can be reduced in other ways. One such way is to remove all germs from the environment and from the experimental subject itself. Because most odorants are produced by bacterial transformation of compounds inside and outside of the organism, the absence of bacteria leads to a drastic reduction of olfactory stimulation. Rats raised without microbiota are more fearful (show more anxiety-like behaviours) than conventionally raised rats (Crumeyrolle-Arias *et al.*, 2014), whereas the opposite has been reported in mice (Heijtza *et al.*, 2011). Since the absence of intestinal bacteria has several central and peripheral effects, it is difficult to attribute potential behavioural changes to reduced olfactory stimulation. It is known, though, that the olfactory epithelium is altered in germ-free mice (François *et al.*, 2016).

The rather limited number of experimental studies of the effects of reduced olfactory stimulation in rodents and lagomorphs makes any final conclusion concerning consequences for well-being impossible. It should be noted, however, that none of the studies mentioned here was explicitly designed for evaluating effects on well-being. Consequently, important aspects of the behavioural effects of anosmia may have been ignored.

Intentional Stimulation with Odorants

Laboratory animals respond to experimental olfactory stimulation in a variety of ways. Predator odours, either natural in the form of faeces or urine, or synthetic, mainly in the form of 2,5-dihydro-2,4,5-trimethylthiazoline (TMT), are frequently used to activate reactions of fear or avoidance. Likewise, putatively agreeable odours like lavender oil are frequently used to modulate fear responses. There is evidence showing that TMT indeed acts through stimulation of receptors in the olfactory epithelium (Hacquemand *et al.*, 2010), whereas lavender oil does not. Anosmic mice respond as much as control mice to inhaled lavender odour (Chioca *et al.*, 2013; Takahashi *et al.*, 2014), and lavender oil preserves all its action when administered intraperitoneally (Kumar, 2013; Schuwald *et al.*, 2013). In the latter case, olfactory stimulation should be minimal or entirely absent. There are many other compounds that have been reported to affect emotional reactions after inhalation. However, these compounds have not been tested in anosmic animals and it would thus be premature to attribute their actions to olfactory stimulation. They could just as well be absorbed by the nasal epithelium and then enter the systemic circulation or the brain (as is the case after intranasal administration of many drugs), or they could be transported into the lungs and there be absorbed through the alveolar epithelium. The finding that an airborne compound is effective after administration through the nose does not by itself mean that the compound is acting on the olfactory system. This is not always taken into account, which may inflate the importance of olfactory processes.

Non-intentional Stimulation with Odorants

It is well known that some odours cause avoidance and some cause approach. Rats and mice avoid peppermint odour, whereas they readily approach the odour of 2,6,10-dodecatrien-1-ol-3,7,11-trimethyl(Z), a compound extracted from female rat genitals. The emotional consequences of exposure to attractive and aversive odours remain largely unknown, but according to general behavioural principles it can be assumed that odours making an animal approach the source have positive incentive properties, meaning that they are experienced as agreeable. On the contrary, odours causing withdrawal may be experienced as unpleasant (see also Chapter 4). In the wild, rodents and lagomorphs are probably exposed to both kinds of odours, although the exposure might be extremely variable from one situation to another. Nevertheless, it is reasonable to assume that animals will not remain in an environment full of aversive odours, and vice versa. Thus, in the wild, animals will select a 'comfortable' olfactory environment. In the laboratory, animals are usually not able to select their environment because they are confined to a small cage. As a result of this, animals may be exposed to aversive odours if these happen to be present in the animal facilities. Since animals themselves produce odours that are potentially aversive to some conspecifics, this might well occur. It has been reported that aversive odours indeed activate a negatively charged emotional response involving the autonomic nervous system, while agreeable odours produce the opposite response (Horii *et al.*, 2013). The fact that animals may be exposed to agreeable odours generated by conspecifics (like sexual attractants) does not seem to be of much concern, because it cannot be associated with reduced well-being. Moreover, such exposure does not seem to affect subsequent responses to these odours. Male and female rodents may well be housed in the same room, consequently being persistently exposed to odours from the opposite sex, without losing their responsiveness to such odours in experimental tests. Moreover, odours that may be associated with a positive effect are generally and reasonably supposed to enhance well-being, which is a most desirable effect.

It is difficult to determine the potential impact of non-intentional exposure to aversive odours. Nevertheless, it is quite possible that such exposure has a more profound impact on well-being than the reduced contact with odorants found in most facilities housing laboratory animals.

In this section we have briefly mentioned some of the available evidence concerning the importance of olfactory stimulation for the well-being of animals in the laboratory. We will now turn to an analysis of the living conditions in a typical animal facility and their possible relationship both to well-being and to the external validity of results obtained in such facilities.

Olfactory Environment in the Laboratory

For obvious ethical and hygienic reasons, animal facilities are strictly regulated and controlled. Aspects such as food quality, daily procedures and maintenance are submitted to international and national conventions. These conventions provide a high level of standardization of procedures worldwide, therefore allowing for an easier comparison between studies and a better investigation of differences in experimental results from one facility to another. However, by this high degree of standardization, the lab environment can also fail to provide adequate sensory cues, especially olfactory. It is therefore crucial to update our knowledge of laboratory conditions to ensure the welfare of the animals used, as well as the scientific validity and reliability of the results obtained. In this section, we will discuss three issues: (i) what the specific conditions found in the lab are; (ii) how they interact with the expression of animals' behaviour; and (iii) whether the standard lab conditions are suitable for research.

By comparison with other sensory modality such as audition, olfaction has been overlooked in behavioural studies (Nielsen *et al.*, 2015). Indeed, when studying the olfactory modality, the first problem arising is finding ways to quantify an odour. Because of this inherent difficulty, it is not easy to provide standardized protocols on the use of olfactory stimulations. Among the large numbers of studies using olfactory stimuli, the vast majority give very few details about the system used to distribute the odour, the quantity of substance used, the airflow intensity or the substance's evaporation rate. Understanding olfactory behaviours in the lab comes through mastering the range of olfactory cues present, and there is an urgent need for standardization and rigorous specification of olfactory procedures.

Odours spread dynamically in space, making the control of air flow/ventilation a key feature in the regulation of odours. In closed environments, ventilation is often expressed in terms of air changes per hour (ACH). The European Commission recommends a rate of 15–20 ACH in animal facilities (EU Directive 2010/63/EU) to ensure that sufficient fresh air of an appropriate quality is provided. In many facilities, the reported ACH by cage is actually higher, as commercially available systems run at air changes ranging from 25–120 times per hour (Krohn *et al.*, 2003). For humans, housing is generally considered adequate and habitable if it contains at least one room of a volume superior or equal to 20 m^3, and the latest international guidelines recommend 0.6 ACH (International Passive House Association, 2013). The animals used in the lab are submitted to high rates of ventilation preventing odours from spreading. In a natural environment, the olfactory cues would consist of a coherent flow of information represented by dynamic odours logically fading as a function of time and distance. This flow is interrupted in artificial environments and consequently prevents the animals from experiencing coherent chemosensory information, as odours do not necessarily match the identity or the timing of the presence of other individuals.

Along with high ventilation, daily routines in the lab include regular cleaning and removal of soiled bedding, causing a brutal change in the animals' olfactory environment. Finally, the lab provides a small range of food types that in no way resemble the wide range of olfactory stimuli that rodents would be exposed to in the wild. Taken together, the standards of the lab, including ventilation, cleaning and feeding procedures, provide a very poor olfactory environment. Laboratory animals live in this depleted environment from birth. This lack of fluctuation in environmental odours is likely to play an important role in the development of olfactory senses and abilities, at the neural, physiological and behavioural levels. This can be critical in the later response to experimental designs. Indeed, when submitted to experimental odours, animals may, for example, display signs of neophobia, potentially biasing the behavioural effects recorded. Contrary to odours unintentionally present in the lab environment, experimental odours are often delivered through an olfactometer, allowing for a more precise control and distribution of the odours in terms of intensity and duration. However, because of the lack of olfactory tracking in the laboratory, experimental results cannot be rigorously compared to standard conditions. Taken together, these considerations highlight the need for a diverse olfactory environment from birth in animals raised for experimental use, and for a specification in olfactory procedures all along laboratory animals' lifespan.

Consequences for Behaviour

According to Toth and collaborators (2011), an inadequate environment can lead to inappropriate behaviours. Animal facilities providing a small range of olfactory stimulations may, consequently, fail to provide a naturalistic habitat relevant for the needs of lab animals.

A main characteristic of the lab is the frequent removal of olfactory cues by regular cleaning. In doing so, the spatial and sensory markers are disturbed and this can lead to social disruption (Baumans, 2005a). For example, in male mice housed in pairs or in groups, the replacement of soiled cage material by completely clean material provoked a peak in aggression between the males (Van Loo *et al.*, 2000). This result may seem counterintuitive at first, but can be explained by the removal of the odour traces associated with the established hierarchy among the males, which then have to renew their marks of dominance. Oppositely, soiled bedding contains potentially aggression-eliciting pheromones enhancing this type of behaviour (Van Loo *et al.*, 2000). The same authors also found that providing clean bedding supplemented with used nesting material decreased aggression among the male mice compared with other types of bedding, possibly because, as demonstrated before (Jones and Nowell, 1975), home-cage odours are aggression-inhibiting. The notable peak in aggression following cage cleaning can be regarded as inappropriate behaviour since it is triggered by an external factor disturbing the established hierarchy and familiar home-cage odour. Similarly, increased aggression has been reported in highly ventilated cages, with male mice displaying more and longer episodes of fighting than mice housed in low ventilation cages (Oliva *et al.*, 2010), which is attributable to the constant removal of odours related to maintenance of the social dominance hierarchy. The frequency of cage cleaning is also associated with cannibalism, with a higher percentage of pups cannibalized in more frequently cleaned cages compared with lower frequencies. In a study by Burn and Mason (2008), the percentage of rat pups cannibalized was higher in twice-weekly cleaned cages than in cages cleaned every 7 or 14 days (Fig. 14.2). In the same way, a higher percentage of rat pairs cannibalized their pups in twice-weekly cleaned cages. Cannibalism is a

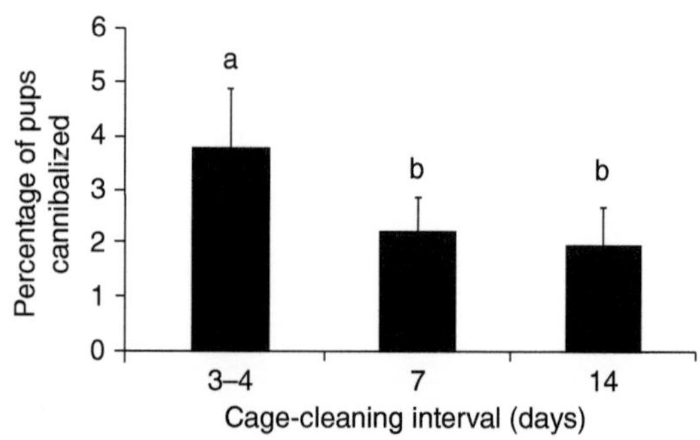

Fig. 14.2. The effect of the cage cleaning frequency on the aggression of rat pairs towards their pups. Frequent removal of olfactory clues increases parents' aggression towards the pups. A higher proportion (mean + SE) of rat pups are cannibalized by the parents when their cages are cleaned every 3–4 days compared with lower cleaning frequencies, every 7 or 14 days. Bars annotated with different letters are significantly different ($P < 0.05$). (Adapted from Burn and Mason, 2008.)

behaviour observed in the wild, particularly when resources are missing or in the case of high stress on the parent or parental pair. Any laboratory facility should aim to provide a familiar, non-threatening environment, allowing a longer lifespan and less stress to the animals, not only for the sake of animal welfare, but also to ensure valid experimental results. However, the high cannibalism rates reported in this study indicate that the laboratory olfactory environment can trigger the same aggressive behaviour that can be seen in the wild. Cannibalism seems irrelevant in a familiar environment providing primary resources *ad libitum*. Therefore, the presence of cannibalism in animal facilities suggests that animal housing does not reach the familiarity or comfort criteria at which it aims.

The strength of ventilation and therefore the potential prevention of odour accumulation have also been found to influence primary behaviours such as feeding and reproductive behaviours. Rats housed in traditional open-top cages, therefore with low ventilation, eat and drink more compared with animals housed in individually ventilated cages (Kostomitsopoulos *et al.*, 2011). Dependent on the context, increases in food and water intake can generally be interpreted as improved well-being (Baumans, 2005b). Also in individually ventilated cages (high ventilation), fewer mouse pups were born compared with classic open rack cages or Scantainer (low ventilation cabinet; Tsai *et al.*, 2003). The mechanisms underlying these changes in food and water intake and in reproductive parameters are still unclear but could suggest that olfactory factors influence the behaviour of the animals, due to an odour-poor environment or perhaps rapid odour turnover. However, there are many differences other than intensity of ventilation that also could explain the effects seen.

Laboratory Environment and External Validity

Understanding the influence of the lab environment on experimental results is crucial for the validity of data. As we have seen already, the lab provides a poor olfactory environment that may influence the animals' behaviour. Very few studies have attempted to provide an ethological description of olfactory behaviours of typical lab animals in the wild. A few projects studied lab animals under naturalistic-like conditions and have so far found very promising results in social and reproductive behaviours, suggesting that animals in naturalistic environments expressed different behavioural patterns, and different cortical sensory maps (Polley *et al.*, 2004; Chu and Ågmo, 2014). Clearly, the use of complex naturalistic habitats makes it difficult to control what key feature(s) of these kinds of environment is relevant for the observed results compared with standard cages. Their implementation must therefore be carefully evaluated before being applied to scientific experimentation. However, under standard conditions, the lab provides an incomplete range of olfactory cues, and even with certain types of (olfactory) enrichment, these odours are only supplemented to a certain extent.

As shown above, scientific studies conducted in the lab often take place in a simplified olfactory milieu. This is particularly important since a recent study showed that a simplified environment can lead to experimental differences non-existent in more complex environments. For example, mice reared in a structurally enriched environment were not sensitive to the deleterious effect of diesel exposure on gene expression in the olfactory bulb (Yokota *et al.*, 2013). Most studies prior to this one had been carried out in a standard laboratory environment and this study proposed to investigate the effects of diesel exhaust particles (DEP) under environmental enrichment conditions. They used 31 litters from 16 different rearing cages. Ten pregnant dams were allocated to a standard cage environment (CE), while ten others were allocated to an enriched environment (EE). The EE was composed of a larger cage containing several toys, regularly moved to different locations and exchanged with new toys. One dam and her pups were assigned to one of the environmental conditions throughout the perinatal period until weaning. After weaning, at post-natal day 27, the pups were placed either in a control chamber (C) or in a diesel exhaust chamber (D). The diesel exhaust chamber exposed the mice to 8 h/day of DEP (90 $\mu g/m^3$). This protocol allowed for four experimental groups: CE-C, CE-D, EE-C and EE-D. The gene expression pattern in the olfactory bulb was determined using a microarray technique, and 112 genes were found to be differentially expressed, according to both the diesel exposure and the hosting environment, confirming the toxicity of DEP exposure (Fig. 14.3). Hierarchical clustering analysis classified these 112 genes into three major clusters according to their expression patterns. After analysis of gene expression, exposure to DEP at environmentally relevant concentration resulted in a significant toxicity to gene

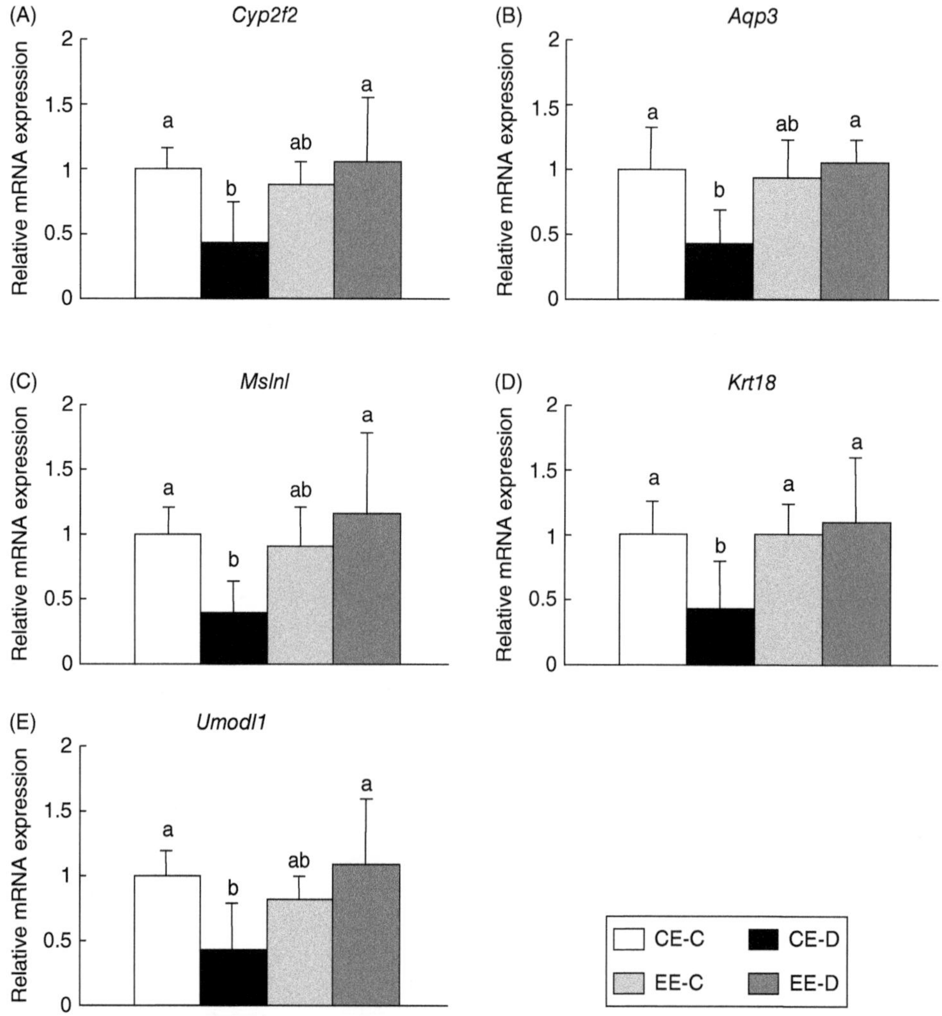

Fig. 14.3. The data show mRNA expressions for (A) *Cyp2f2*, (B) *Aqp3*, (C) *Mslnl*, (D) *Krt18* and (E) *Umodl1* in the olfactory bulb of mice in four different experimental groups: mice bred in an enriched environment (EE) or control environment (CE), subsequently exposed to diesel particles (D) or to a control (C). The expression levels of *Cyp2f2*, *Aqp3*, *Mslnl*, *Krt18* and *Umodl1* of mice reared in a standard cage environment were significantly downregulated by exposure to diesel exhaust. There was no significant effect of exposure to diesel exhaust on gene expression of mice reared in environmental enrichment. The data are expressed as relative target gene expression compared with *Gapdh* expression (mean ± SD). Bars annotated with different letters are significantly different ($P < 0.05$). (Adapted from Yokota *et al.*, 2013.)

expression; however, the dysregulation of gene expression levels was only present when individuals had been housed in the standard cage environment (Fig. 14.3). This finding highlights two interesting facts: (i) not only does olfactory enrichment affect later olfaction; but (ii) environmental conditions can alter the sensitivity to olfactory disturbances. Consequently, when facing environmental threats, individuals differ depending on their previous olfactory experience. This finding suggests that olfactory differences observed in the lab may not be extrapolated to real-life situations, nor are they applicable to individuals born and raised in a rich, varied olfactory environment. This makes housing environment a key feature in analysing and comparing results obtained in different facilities and also in the potential generalization of such results to other species, including humans.

By minimizing the range of stimulations available, animal facilities offer easily reproducible environments. Depleted environments have become a standard because of the higher reliability and external validity of data they initially allow to be collected. However, every animal facility potentially presents its own specifications regarding the key features that represent ventilation rate, housing design and environmental odours. In such strictly controlled and regulated environments, slight changes can make a significant difference as every original feature becomes salient in a deprived milieu. Nevertheless, with the latest findings on the consequences of environmental depletion, added to the general trend for physical EE, scientific research may be facing a paradigm switch, according to Hanno Würbel (2000), who emphasized the 'standardization fallacy'. The initiation of more and more enrichment procedures could partly reverse the standardization trend and considerably influence the experimental protocols and results. Even if EE is regularly used in experimental protocols, the breeding protocols do not change, when environmental variety should be initiated in animals' lives from birth. Moreover, EE is mainly physical, involving larger spaces, and supplementation with toys or different bedding materials. It is notable that olfactory enrichment has, to our knowledge, never been documented long term. For now, protocols allowing for a comparison between standard and enriched housing greatly contribute to the scientific debate.

The lab environment could act as a sensitizer on the animals by providing them with abnormally low levels of sensory stimulation. First, rats have been found to have a higher frequency and duration of olfactory behaviours, particularly sniffing the air inside their cage when living in standard cages compared with physically enriched ones (Abou-Ismail *et al.*, 2010). Notably, the enrichment used in this study was only physical, not social or olfactory. Thus, the more the housing environment is deprived, the more exploratory olfaction is observed, perhaps an indication that exploratory behaviour is used to mitigate the boredom experienced in standard cages. This finding is coherent with many others suggesting that when lab animals are provided with enough resources they lose interest in slight changes in their environment.

Animals raised in barren animal facilities can develop compensatory processes to counteract the lack of changes in their olfactory environment. This was found by Angely and Coppola (2010), who looked at the olfactory capacities of adult mice submitted to perinatal unilateral naris occlusion (blocking of one nostril) followed by contralateral olfactory bulbectomy. Considering that very few studies have focused on retained olfactory capacities after surgical impairment of olfactory organs and tissues, Angely and Coppola decided to investigate this. They studied olfactory capacity in mice subjected to one of the most common practices of odour deprivation in rodents, the uni- or bilateral naris occlusion, which has the advantages of being both simple and reversible. Especially, the extent of behavioural deficit caused by perinatal unilateral naris occlusion (UNO) in adults is poorly known. Half of the mice pups underwent UNO performed by cauterization of one nostril on the day after birth, and followed by a contralateral bulbectomy, while the other half only underwent unilateral bulbectomy. The authors followed a standard habituation–dishabituation procedure and an operant conditioning paradigm to compare odour discrimination and generalization in adult mice (see Chapter 4 for details of these tests). Two types of odorants were used in the procedures, after prior verification of the absence of odour preference or avoidance in the mice: iso-amyl acetate (IA), which has a banana-like odour, and ethyl butyrate (EB), which smells of pineapple. For the habituation–dishabituation procedure, six habituation trials consisted of exposing the mice to a teabag containing 40 µl of IA on a filter paper, for 50 s each, separated by 3-min inter-trial intervals. On the seventh trial, the IA was replaced by 40 µl of EB. The test was repeated daily using decreasing dilutions of EB in IA as follows: 10%, 2%, 0.4%, 0.08%, 0.016% and 0.0032%. This test phase ended when the mice failed to show dishabituation to the seventh mixture, indicating that they could no longer detect a difference from pure IA. As a result, both UNO mice (n = 10) and controls (n = 9) showed significant dishabituation on the seventh trial to pure EB and 10% EB. However, in the next dilution level, 2% EB, both groups failed to dishabituate, but the UNO group approached significance ($P < 0.06$), surprisingly suggesting potentially stronger olfactory discrimination abilities in the UNO group. Therefore, a second cohort of 4-month-older mice (UNO, n = 8; control, n = 7) was tested with the same protocol and the UNO mice this time significantly dishabituated to 2% EB while the control group did not.

For the odour discrimination procedure in the study by Angely and Coppola (2010), the conditioning was set in three phases. In the first phase, the mice had to distinguish between two dishes in a test-cage, one dish containing mineral oil and the other containing 5 µl EB (the rewarded stimuli, S+) to access a 45 mg sucrose reward placed in front of the EB dish directly on the sand covering the floor of the test-cage. This phase continued until the mice consistently approached the EB dish and consumed the reward. In the second phase, the sucrose reward was buried into the sand so that the mice had to dig to retrieve it. In the third and last phase, the filter paper soaked with mineral oil was replaced by a filter paper with IA (unrewarded odour, S–). Once the mice were conditioned to dig next to the EB dish, the sucrose was removed to avoid potential odorant bias and the test phase could start. After a first block of trials was completed, during which the mice consistently investigated the S+ dish more than the S–, the experiment continued with the S+ dish being replaced by increasing dilutions of EB in IA. If a mouse failed twice to investigate the S+ dish more, it was considered to have reached its threshold of discrimination. In this task again, UNO mice (n = 4) tended to outperform the control mice (n = 4) since the highest thresholds in UNO mice matched the lowest threshold of control ones. Moreover, and despite the small number of animals tested, the UNO group had a significantly lower discrimination threshold than the control mice (0.22% ± 0.10 vs 1.6% ± 0.40). In this study, and contrary to what was expected, UNO mice performed better at the discrimination task after the dishabituation paradigm than control mice. Actually, in accordance with other studies, these results suggest that compensatory processes are implemented to preserve the olfactory function in the case of stimulus deprivation. This means that individuals placed in sensory-deprived environments may be able to develop acute sense in the modality concerned, conversely to what could be expected.

The simplified environments available in the standard lab do not allow individual animals to adapt properly as they are constantly submitted to slight variations in their housing conditions without any control over them. This situation rarely occurs in naturalistic conditions, which may call into question the scientific validity of some of the data obtained in the lab, especially with regard to olfactory behaviours. We can assume that the standard experimental lab fails to provide an adequate olfactory environment adapted to the rodents. As a result, animals may overdevelop their olfactory abilities, making them oversensitive to environmental disturbances. At the same time, frequent disruptions in their olfactory environment increase the aggression rate towards cage-mates and new-borns, and potentially increase their level of stress, illustrated by lower food and water intake.

Olfaction, despite being an important sense for mammals, is understimulated in the lab. In the past decades, the use of EE has allowed a more complex picture of olfactory behaviours in the lab, but these studies only provide a comparison of a poor olfactory environment with a slightly enhanced one. The current lab conditions do not allow for a complete understanding of the olfactory modality role in behaviour under naturalistic conditions. Apart from the ethical issues involved, the validity of some results obtained under lab conditions could be questioned when it comes to their application to real-life conditions. Regarding the benefits from physical EE for the understanding of behaviours and the external validity of results obtained in the lab, the same reasoning could be extrapolated to sensory enrichment. Breeding and testing lab animals in diversified environments could become the new standard in the coming years. However, before jumping to future debates, there is a current need for the inclusion of odorant environments in regulations applying to lab animals, involving a review of exposures to positive and negative stimuli.

Finally, one should keep in mind that laboratory animals are genetically different from wild ones. This could potentially mean that lab animals do not require as much environmental diversity as their wild conspecifics. Thus, the genetic adaptation of captive animals to a closed, standardized milieu would hopefully lower the risk of lab conditions being poorly suited for behavioural research. The development of an accurate olfactory neuroethology of laboratory rodents would greatly improve our knowledge and assessment of the relevance of lab conditions and experiments for actual biological functions, as well as experimental treatments and animal care.

References

Abou-Ismail, U.A., Burman, O.H., Nicol, C.J. and Mendl, M. (2010) The effects of enhancing cage complexity on the behaviour and welfare of laboratory rats.

Behavioural Processes 85, 172–180. DOI: 10.1016/j.beproc.2010.07.002

Abreu, M.S., Giacomini, A.C.V.V., Kalueff, A.V. and Barcellos, L.J.G. (2016) The smell of 'anxiety': behavioral modulation by experimental anosmia in zebrafish. *Physiology and Behavior* 157, 67–71. DOI: 10.1016/j.physbeh.2016.01.030

Ahn, S., Shin, H.-W., Mahmood, U., Khalmuratova, R., Jeon, S.-Y., Jin, H.R., Choi, J.-S., Kim, H.-S. and Kim, D.W. (2016) Chronic anosmia induces depressive behavior and reduced anxiety via dysregulation of glucocorticoid receptor and corticotropin-releasing hormone in a mouse model. *Rhinology* 54, 80–87. DOI: 10.4193/Rhino15.209

Angely, C.J. and Coppola, D.M. (2010) How does long-term odor deprivation affect the olfactory capacity of adult mice? *Behavioral and Brain Functions* 6, 26. DOI: 10.1186/1744-9081-6-26

Baumans, V. (2005a) Environmental enrichment for laboratory rodents and rabbits: requirements of rodents, rabbits, and research. *ILAR Journal* 46, 162–170.

Baumans, V. (2005b) Science-based assessment of animal welfare: laboratory animals. *Scientific and Technical Review of the Office International des Epizooties* 24, 503–514.

Belke, T.W. and Wagner, J.P. (2005) The reinforcing property and the rewarding aftereffect of wheel running in rats: a combination of two paradigms. *Behavioural Processes* 68, 165–172. DOI: 10.1016/j.beproc.2004.12.006

Burn, C.C. and Mason, G.J. (2008) Effects of cage-cleaning frequency on laboratory rat reproduction, cannibalism, and welfare. *Applied Animal Behaviour Science* 114, 235–247. DOI: 10.1016/j.applanim.2008.02.005

Caro, S.P., Balthazart, J. and Bonadonna, F. (2015) The perfume of reproduction in birds: chemosignaling in avian social life. *Hormones and Behavior 68,* 25–42.

Chioca, L.R., Antunes, V.D., Ferro, M.M., Losso, E.M. and Andreatini, R. (2013) Anosmia does not impair the anxiolytic-like effect of lavender essential oil inhalation in mice. *Life Sciences* 92, 971–975. DOI: 10.1016/j.lfs.2013.03.012

Chirino, R., Beyer, C. and González-Mariscal, G. (2007) Lesion of the main olfactory epithelium facilitates maternal behavior in virgin rabbits. *Behavioral Brain Research* 180, 127–132. DOI: 10.1016/j.bbr.2007.02.039

Chu, X. and Ågmo, A. (2014) Sociosexual behaviours in cycling, intact female rats (*Rattus norvegicus*) housed in a seminatural environment. *Behaviour* 151, 1143–1184. DOI: 10.1163/1568539X-00003177

Crumeyrolle-Arias, M., Jaglin, M., Bruneau, A., Vancassel, S., Cardona, A., Dauge, V., Naudon, L. and Rabot, S. (2014) Absence of the gut microbiota enhances anxiety-like behavior and neuroendocrine response to acute stress in rats. *Psychoneuroendocrinology* 42, 207–217. DOI: 10.1016/j.psyneuen.2014.01.014

Doty, R.L. (1986) Odor-guided behavior in mammals. *Experientia* 42, 257–271.

EU Directive (2010 63/EU) of the European Parliament and of the council of 22 September 2010 on the protection of animals used for scientific purposes. *Official Journal of the European Union* 276, 33–74.

François, A., Grebert, D., Rhimi, M., Mariadassou, M., Naudon, L., Rabot, S. and Meunier, N. (2016) Olfactory epithelium changes in germfree mice. *Scientific Reports* 6, 24687. DOI: 10.1038/srep24687

Glinka, M.E., Samuels, B.A., Diodato, A., Teillon, J., Mei, D.F., Shykind, B.M., Hen, R. and Fleischmann, A. (2012) Olfactory deficits cause anxiety-like behaviors in mice. *Journal of Neuroscience* 32 6718–6725. DOI: 10.1523/jneurosci.4287-11.2012

Hacquemand, R., Jacquot, L. and Brand, G. (2010) Comparative fear-related behaviors to predator odors (TMT and natural fox feces) before and after intranasal $ZnSO_4$ treatment in mice. *Frontiers in Behavioral Neuroscience* 4, 188. DOI: 10.3389/fnbeh.2010.00188

Harkin, A., Kelly, J.P. and Leonard, B.E. (2003) A review of the relevance and validity of olfactory bulbectomy as a model of depression. *Clinical Neuroscience Research* 3, 253–262. DOI: 10.1016/S1566-2772(03)00087-2

Heijtza, R.D., Wang, S.G., Anuar, F., Qian, Y., Björkholm, B., Samuelsson, A., Hibberd, M.L., Forssberg, H. and Pettersson, S. (2011) Normal gut microbiota modulates brain development and behavior. *Proceedings of the National Academy of Sciences USA* 108, 3047–3052. DOI: 10.1073/pnas.1010529108

Horii, Y., Nagai, K. and Nakashima, T. (2013) Order of exposure to pleasant and unpleasant odors affects autonomic nervous system response. *Behavioural Brain Research* 243, 109–117. DOI: 10.1016/j.bbr.2012.12.042

International Passive House Association (2013) Passive house certification criteria. Available at: https://www.passivehouse-international.org (accessed 14 November 2016).

Jones, R. and Nowell, N. (1975) Effects of clean and soiled sawdust substrates and of different urine types upon aggressive behavior in male mice. *Aggressive Behavior* 1, 111–121.

Kapteyn, A., Lee, J., Tassot, C., Vonkova, H. and Zamarro, G. (2015) Dimensions of subjective well-being. *Social Indicators Research* 123, 625–660. DOI: 10.1007/s11205-014-0753-0

Koolhaas, J., Bartolomucci, A., Buwalda, B.D., De Boer, S., Flügge, G., Korte, S.M., Meerlo, P., Murison, R., Olivier, B. and Palanza, P. (2011) Stress revisited: a critical evaluation of the stress concept. *Neuroscience and Biobehavioral Reviews* 35, 1291–1301. DOI: 10.1016/j.neubiorev.2011.02.003

Kostomitsopoulos, N., Dontas, I.A., Alexakos, P., Lelovas, P., Galanos, A., Paronis, E., Balafas, E., Paschidis, K. and Kostakis, A. (2011) Growing male rats in individually ventilated and open-top cages. *Journal of the American Association for Laboratory Animal Science* 50, 879–883.

Krohn, T.C., Hansen, A.K. and Dragsted, N. (2003) The impact of cage ventilation on rats housed in IVC systems. *Laboratory Animals* 37, 85–93. DOI: 10.1258/00236770360563714

Kumar, V. (2013) Characterization of anxiolytic and neuropharmacological activities of Silexan. *Wiener Medizinische Wochenschrift* 163, 89–94.

Nielsen, B.L., Jezierski, T., Bolhuis, J.E., Amo, L., Rosell, F., Oostindjer, M., Christensen, J.W., McKeegan, D., Wells, D.L. and Hepper, P. (2015) Olfaction: an overlooked sensory modality in applied ethology and animal welfare. *Frontiers in Veterinary Science* 2, 69. DOI: 10.3389/fvets.2015.00069

Oliva, A.M., Salcedo, E., Hellier, J.L., Ly, X., Koka, K., Tollin, D.J. and Restrepo, D. (2010) Toward a mouse neuroethology in the laboratory environment. *PLoS ONE* 5, e11359. DOI: 10.1371/journal.pone.0011359

Paredes, R.G. (2009) Evaluating the neurobiology of sexual reward. *ILAR Journal* 50, 15–27.

Piazza, P.V., Deroche, V., Deminiere, J.-M., Maccari, S., Le Moal, M. and Simon, H. (1993) Corticosterone in the range of stress-induced levels possesses reinforcing properties: implications for sensation-seeking behaviors. *Proceedings of the National Academy of Sciences USA* 90, 11738–11742. DOI: 10.1073/pnas.90.24.11738

Polley, D.B., Kvasnák, E. and Frostig, R.D. (2004) Naturalistic experience transforms sensory maps in the adult cortex of caged animals. *Nature* 429, 67–71. DOI: 10.1038/nature02469

Schuwald, A.M., Nöldner, M., Wilmes, T., Klugbauer, N., Leuner, K. and Müller, W.E. (2013) Lavender oil-potent anxiolytic properties via modulating voltage dependent calcium channels. *PLoS ONE* 8, e59998. DOI: 10.1371/journal.pone.0059998

Song, C. and Leonard, B.E. (2005) The olfactory bulbectomised rat as a model of depression. *Neuroscience and Biobehavioral Reviews* 29, 627–647. DOI: 10.1016/j.neubiorev.2005.03.010

Takahashi, M., Yamanaka, A., Asanuma, C., Asano, H., Satou, T. and Koike, K. (2014) Anxiolytic-like effect of inhalation of essential oil from *Lavandula officinalis*: investigation of changes in 5-HT turnover and involvement of olfactory stimulation. *Natural Product Communications* 9, 1023–1026.

Toth, L.A., Kregel, K., Leon, L. and Musch, T.I. (2011) Environmental enrichment of laboratory rodents: the answer depends on the question. *Comparative Medicine* 61, 314–321.

Tsai, P.P., Oppermann, D., Stelzer, H.D., Mähler, M. and Hackbarth, H. (2003) The effects of different rack systems on the breeding performance of DBA/2 mice. *Laboratory Animals* 37, 44–53. DOI: 10.1258/002367703762226683

Van Loo, P.L.P., Kruitwagen, C.L.J.J., Van Zutphen, L.F.M., Koolhaas, J.M. and Baumans, V. (2000) Modulation of aggression in male mice: influence of cage cleaning regime and scent marks. *Animal Welfare* 9, 281–295.

Venezia, A.C., Guth, L.M., Sapp, R.M., Spangenburg, E.E. and Roth, S.M. (2016) Sex-dependent and independent effects of long-term voluntary wheel running on Bdnf mRNA and protein expression. *Physiology and Behavior* 156, 8–15. DOI: 10.1016/j.physbeh.2015.12.026

Würbel, H. (2000) Behaviour and the standardization fallacy. *Nature Genetics* 26, 263. DOI: 10.1038/81541

Yokota, S., Hori, H., Umezawa, M., Kubota, N., Niki, R., Yanagita, S. and Takeda, K. (2013) Gene expression changes in the olfactory bulb of mice induced by exposure to diesel exhaust are dependent on animal rearing environment. *PLoS ONE* 8, e70145. DOI: 10.1371/journal.pone.0070145

15 Olfactory Behaviour in Companion Animals

Peter G. Hepper and Deborah L. Wells

Animal Behaviour Centre, School of Psychology, Queen's University Belfast, Northern Ireland, UK

Introduction

Companion animals form a major part of today's society, performing a wide variety of roles and comprising a vast array of species. The Pet Food Manufacturers Association's annual survey of pet ownership in the UK indicates that in 2016 there were 57 million pets in the UK (Pet Food Manufacturers Association, 2016). The UK's population of people is estimated at just over 65 million as of June 2015 (Office of National Statistics, 2016). Some 40% of households (11 million) have a pet. Fish are the most numerous pets owned across the country, whilst dogs are present in more households (see Table 15.1), and a similar pattern is found in the US (PODS, 2012).

Although the existence of pets is widespread and owners may put much effort into looking after and understanding their pet, the olfactory behaviour of pets is somewhat overlooked compared with other sensory modalities. Many view certain pets, for example fish and birds, as having a very poor or non-existent sense of smell. For other pets, their use of olfaction is viewed in its negative aspect and considered in terms of the presence of (bad) smells. Owners report their dogs or cats as smelling after their animal has rolled in some pungent substance and coated itself with the smell. Other times when odours may be noticed include when a female is in oestrus or when the owner has to visit the vet with their pet to treat blocked anal glands. Despite many toys being available for pets very few aim to utilize the animals' sense of smell to encourage play. One exception is the use of catnip (see Chapter 11) in a few toys for cats.

This chapter will provide an overview of the olfactory behaviour of companion animals. The focus will be on cats and dogs (representing mammals and being the most studied), birds (representing a group that for many years was considered incapable of using odours), reptiles (concentrating on the tortoise and representing years of personal experience) and fish (representing the most numerous of all pets). The aim will be to highlight how olfaction influences and guides the behaviour of these animals. The chapter will very briefly review olfactory structures and the role of olfactory receptor genes (but see Chapter 1 for more details), as these set the limits of the individual's olfaction and the organs that may produce odours. The main focus will be on the behaviour of the animals and how its olfactory world is essential for normal behaviour and survival.

Considerations Regarding Olfaction in Companion Animals

Olfaction, perhaps because of our biased use of visual and auditory sensory modalities, has been comparatively neglected in many companion animals. Indeed some whole classes of animals were initially considered to be unable to sense olfactory stimuli and hence little effort was undertaken to examine their olfactory world. For example, J.J. Audubon in the 1820s 'proved' that turkey vultures did not use smell to locate carrion (this is now known to be incorrect) and so began the view that birds had no olfactory sense. This view started to be challenged in the 1960s (e.g. Wenzel, 1968). Thus, the evidence base for examining olfactory behaviour in companion animals is somewhat patchy across different animal classes.

Similarly, within animal classes, evidence of olfactory behaviour is available from studies undertaken on only a few groups and species (e.g. when

Table 15.1. The number of pets and number of households containing a pet in the UK. (Pet Food Manufacturers Association, 2016.)

	Pet population (million)	% households with …
Any pet	57	40
Indoor fish	16	8
Outdoor fish	20	5
Dogs	8.5	24
Cats	7.5	17
Rabbits	0.8	2
Domestic fowl	0.5	<1
Guinea pigs	0.7	1
Indoor birds	0.6	2
Hamsters	0.4	1
Lizards	0.3	1
Snakes	0.3	1
Tortoises/turtles	0.3	1

comparing studies undertaken on dogs to those undertaken on cats, there has been much more research of the olfactory behaviour of dogs) and extreme caution must be exercised in generalizing from one species to another. Companion animals present a significant further problem that may not be present in non-companion animals. Some companion animals, such as dogs and cats, have been selectively bred to produce a multitude of breeds with specific characteristics. At the time of writing this chapter, the International Progressive Cat Breeders Alliance recognized 73 different cat breeds. This is dwarfed by the Fédération Cynologique Internationale, which recognizes (definitive and provisional) over 350 different dog breeds. It would be a false assumption to assume that the olfactory system of each different breed is identical. Exactly how breed differences influence olfactory behaviour and function has yet to be fully explored.

As this chapter is concerned with companion animals, there is an important final consideration that must be taken into account, which is entirely obvious to the owners of pets, but appears overlooked in the writings of many researchers. That is, the question of individual differences. Any owner of a pet will immediately be able to identify the individual characteristics of their pet(s) that make it a unique individual and different from any other animal of the same species and breed. Just as parents readily identify differences between their children, pet owners similarly identify unique characteristics of their pet, no matter what species – fish through to dogs. Indeed, six out of ten pet owners in the US consider their pets to be family members (PODS, 2012). The literature to date has given scant regard to the presence of individual differences in olfactory abilities and behaviour, rather focusing on broad general statements, e.g. dogs have an excellent sense of smell. Dogs undoubtedly do have a good sense of smell, but whether they all have the same good sense of smell is largely unknown (examples of this are given in Chapter 9). These factors, together with the relative scarcity of research on companion animal olfaction, should be kept in mind when reading the following sections.

Olfactory Structures

It is impossible to provide a comprehensive review of olfactory structures across companion animals within the space of this chapter. Readers are referred to Stoddart (1980) for an excellent review. We simply note here that at the general level there is a remarkable similarity between the overall structure of the olfactory system of all companion animals, punctuated by various adaptations that maximize the opportunity to receive olfactory stimuli within the environment that the animal inhabits. For example, individuals have a set of paired openings through which olfactory stimuli are drawn in towards the olfactory apparatus. For air-breathing animals, the exit for this stimulus is via openings into the respiratory system, whereas for aquatic species the exits are to the external world.

The olfactory receptors are contained within the olfactory epithelium, which is sized and situated to be stimulated maximally by the odours within the stimulus flow. However, there are large between-species differences in the size of the olfactory epithelium, e.g. German shepherds 150 cm^2, cats 21 cm^2, humans 3–5 cm^2. This impacts on the number of receptors available to detect olfactory stimuli (see also Fig. 1.1 in Chapter 1).

One major difference between the olfactory systems of fish and birds is that these classes do not possess an accessory olfactory bulb or vomeronasal organ (or if present it is vestigial), whereas these organs are present in dogs and cats. It is unclear whether the vomeronasal organ of the dog, despite containing sensory receptors, serves a function, since the 'flehmen' response (see later) is not observed in this species.

In the tortoise, a small part of the ventral portion of the olfactory sac serves as a vomeronasal organ

(Kruzhalov and Boiko, 1987). The olfactory system of birds is less complex and organized compared with dogs and cats, suggesting that these animals may have a poorer overall olfactory experience (but see examples in Chapters 2 and 4). Fish, however, have a complex olfactory bulb structure, which suggests that these animals are capable of recognizing odour patterns.

Living in an aquatic environment does not present a barrier, as once thought, to the perception of olfactory stimuli. Whereas air-breathing animals sniff air into their nose, fish use their accessory sacs to pump water containing the stimulus to the olfactory epithelium. They can also, by entering and re-entering water streams or pools containing odour, drive the stimulus to the receptors as a result of their motion.

Importantly, all companion animals have evolved efficient mechanisms suited to their environment to sample any odours present and deliver these in a highly effective manner to their olfactory receptors (Lawson *et al.*, 2012).

Olfactory Receptor Genes

An overview of how olfaction works is given in Chapter 1 of this book, and this section is just a brief reminder of some of the aspects with special emphasis on companion animals. Following the discovery of olfactory receptor genes (ORs) by Buck and Axel (1991), researchers have attempted to determine the number of ORs (both intact and pseudo-genes) present in different species. It is thought that the ability to detect and discriminate odours appears, in part, to be determined by the OR repertoire of the individual. In general, the greater the number of ORs, the more odours can be detected, although the exact link between ORs and odour detection has yet to be fully elucidated.

The number of (intact) ORs in some of the common pet species have been reported as more than 800 for dogs, 677 for cats, 133 for canaries and 155 for zebrafish (Nei *et al.*, 2008; Niimura, 2014). There are wide differences across species within a class, for example, the kakapo has 550 ORs, much higher than in the canary (Steiger *et al.*, 2008).

There are other classes of genes associated with olfaction, namely vomeronasal receptor genes (VRs), and these similarly demonstrate differences between species. For example, dogs have eight intact VRs, whereas the cat has almost three times as many (Montague *et al.*, 2014). This may, in part, contribute to or be the reason why cats exhibit flehmen responses (where odours are transported to the vomeronasal organ), whereas dogs do not exhibit this behaviour (see later). Finally, trace amine-associated receptors (TAARs) may be particularly important for the detection of odours in water, which may account for the increased number of intact receptors present in fish (109) compared with other species, e.g. the dog (15; Gloriam *et al.*, 2005; Zucchi *et al.*, 2006).

It is likely that the proportion of different types of ORs found in companion animals is related to the different abilities of individual species to detect odours in their environment. The presence of TAARs, for example, means that fish will experience a different olfactory landscape compared with dogs. An individual's repertoire of ORs will predispose that individual to detect certain types of odours present in their environment. Most likely evolution has driven the development of ORs to detect particularly relevant odours for the individual's survival.

Olfactory Detection and Discrimination

Examination of the performance of the olfactory system has focused on two features: how sensitive is the system, i.e. what is the lowest concentration of odour that can be detected, and how discriminating is the system, i.e. what different odours can it sense? Tests of these features are dealt with in more detail in Chapter 4.

The sensitivity of the olfactory system has typically been investigated via learning paradigms. Individuals are trained to exhibit a particular response to an odour and then its concentration is reduced until the point at which the animal no longer responds. At this stage it is assumed that the animal can no longer detect the odour. Most of these studies have been undertaken with dogs, perhaps because of the role canine olfactory detection has in forensic and health situations (see Chapter 9). Taken together, the results indicate that dogs respond to odours at concentrations 1000–1,000,000 lower than humans (e.g. Krestel *et al.*, 1984; Walker *et al.*, 2006). However, the results of such studies are significantly influenced by the paradigm used and the training provided. Hence even within dogs as a species, different paradigms produce very different results for detection thresholds of the same odour, such as detection thresholds for amyl acetate of 1–2 parts/trillion vs 52–36,000 parts/trillion (Walker *et al.*, 2006; Krestel *et al.*, 1984, respectively). Such task-related differences make it very difficult to compare studies undertaken on different species. Davis (1973) compared studies using similar odours and techniques and observed that pigeons had much higher thresholds for odour detection than did dogs.

The assessment of the limits of the olfactory system to discriminate between odours has usually been undertaken by more biologically relevant studies, which have examined whether the individual responds differentially to odours from different sources, e.g. from males and females. Such studies make the assumption that the different odour sources produce distinct and detectable odours, and have demonstrated just how finely tuned the olfactory system is, being able to detect incredibly minute variations between odours.

One of the many uses of the dog's olfactory system is its ability to determine the direction of an odour trail from footprints left by an individual. Such a capability may be useful in the location of prey, or possibly mates. Early anecdotal reports suggested that dogs could follow odour trails (e.g. Romanes, 1887), and more recent experimental studies have confirmed this (e.g. Wells and Hepper, 2003). Whilst this ability is well established, there has been little progress in understanding how this is achieved. Hepper and Wells (2005) attempted to determine how much information was needed for a dog to determine the direction of a trail. A row of 21 carpet squares was laid out and an individual walked from one end to the other, leaving one footprint per square, thereby providing a directional trail of odour. Six trained dogs were brought in at right angles to the trail at its centre 60 min after the trail had been deposited and left to determine the direction of the trail. Dogs were successful at doing this. When the order of the squares was randomized, thereby losing the correct directionality of the trail, the dogs were unable to follow direction. Further, when individuals walked the trail in plastic overshoes, thus maintaining the cue of ground disturbance, but removing any cue based on individual odour, the dogs were unable to determine directionality. The final study reduced the number of footsteps available to dogs to make the determination of direction. When provided with five consecutive footprints, dogs were able to determine directionality; however, when offered only three footprints, the animals were unable to do so (see Fig. 15.1). The most likely mechanism that dogs used to achieve this was a comparison of the odour presented by each individual footstep and from this determined direction. The dog may use a difference in the intensity of the odour, following the strongest or a difference in decay, following the least decayed. When timing the deposition of the footsteps there was approximately 1.9 s between the deposition of the first footstep and the fifth footstep, approximately 0.49 s between each footstep. This study demonstrates the remarkable olfactory capacity of the dog; it can detect a change in an odour brought about by exposure to air for less than 2 s. As we review studies examining the influence of odours on

Fig. 15.1. The dog's ability to determine the direction of an odour trail can be used to assess the ability of the dog's nose to discriminate between different odours. Hepper and Wells (2005) found that a difference of less than 2 s between footsteps in terms of their freshness is sufficient for a dog to discriminate and hence determine the direction of the trail.

animals' behaviour, more examples of the ability of the individual's olfactory system to discriminate between different odours will be presented.

Odour Production

Having demonstrated the presence of anatomical and neurological systems to detect odours, consideration should now turn to whether animals produce odours that may have a role in guiding their behaviour. Animals may produce odours from their waste products, urine and faeces, and also from more specialized structures, glands, which may be found all over the animal's body. Different animal classes will have evolved specific means of producing and presenting odours to suit their environment and to maximize the delivery and exposure of the odour to others. However, just because an individual produces an odour from a particular source does not necessarily mean it carries information. It is important to evaluate whether an odour actually has some communication function(s), and if the odour or information is specific to a certain point in the individual's life.

Fish

Goldfish appear to recognize members of their own species and Levesque *et al.* (2011) examined the response of goldfish (*Carassius auratus*) to a possible species-identifying pheromone at different life stages to determine if the odour has functional significance and when.

In this study, the odour cue was obtained from whole body washings of goldfish at three different stages of their life: juvenile, immature and mature. The response of goldfish to these odours was tested at the same three developmental stages. The study examined the position preference of goldfish living in a circular maze in which odour stimuli could be added to opposite sides of the maze. For any trial, fish were exposed to two different olfactory stimuli and the position of the fish in relation to each stimulus was noted. The first set of studies (comparing odour vs a control odour – just water) demonstrated that no matter what life stage the odour was drawn from, or at what stage the fish was tested, all subjects exhibited a preference for the odour of other fish (see Fig. 15.2A for the response of juvenile fish). Further, when the trial compared the fish's positional preference when given a choice of two different fish odours (e.g. juvenile vs immature), fish, at any life stage, did not exhibit a preference for one particular odour over another. This demonstrates that fish from juvenile to mature produce an odour that is attractive to other fish and, although the odour may still change in quality as the fish matures, it is equally attractive across the various life stages. An additional problem faced by studies of olfactory behaviour in aquatic animals is that the stimulus has both olfactory (sensed by the olfactory apparatus) and taste (sensed by the taste apparatus) components. To demonstrate that

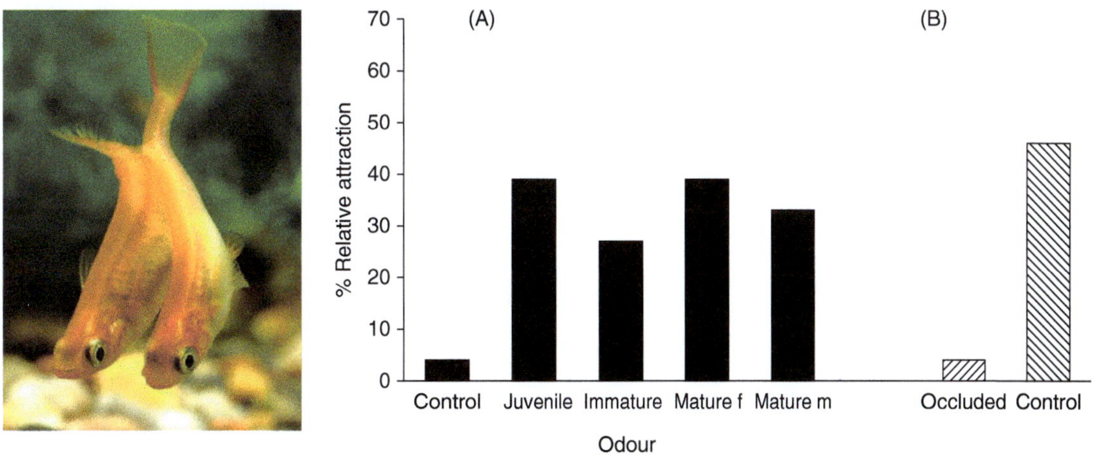

Fig. 15.2. The relative attraction exhibited by juvenile goldfish to (A) washings of odours from goldfish at different life stages and (B) the influence of nostril occlusion on the relative attraction of immature goldfish to washings from immature goldfish. (f = female, m = male; redrawn from Levesque *et al.*, 2011.)

the stimulus being responded to is an odour and not a taste, it is necessary to test for this. In this study, fish were tested with their nostrils occluded and tested again with the occlusion removed. The preference disappeared when the fish's nostrils were occluded, but returned when occlusion was removed (see Fig. 15.2B). The final aspect of the study started to explore which aspects of the body washing contained the active ingredients that led to the preference. It was found that polar and non-polar fractions of the body washings were attractive but that bile acids (often thought a source of odour) were not.

The above study used whole body washings and hence the odours that triggered the fish's preferential swimming response may have been produced from any part of the body or components from excretion or elimination. Male fish have been observed inspecting the gill and urogenital regions of female fish, suggesting these are sites of importance in odour production (Kobayashi *et al.*, 2002). Indeed, specific pheromones released by females that attract males have been identified as being produced from these locations, e.g. 17α,20β-dihydroxy-4-pregen-3-one is released by the gills and 17α,20β-dihydroxy-4-pregen-3-one-20 sulfate is released in urine. Fish also produce odours to indicate stress (Brown *et al.*, 2012) and also when they are physically damaged (Wisenden and Sargent, 1997).

Tortoise

For the tortoise, as with all other species, faeces serve as a potential source of odour information. Due to the arid conditions in which many tortoises live they stay hydrated by reducing urination to a minimum by using and re-using urine. As such, the urination (excretion of urates) of tortoises is unlikely to serve as a source of odour cues. Desert tortoises have specialized glands, chin glands (mental glands), which may play an important role in mating (Auffenberg, 1977; Alberts *et al.*, 1994). Glands of males are generally bigger than those of females and increase in size and productivity during the mating season, most likely linked to increases in plasma testosterone. These glands are similar to the sebaceous gland observed in mammals.

Birds

The skin of birds is much thinner than that of mammals and this is most probably a result of the bird needing to be as light as possible to facilitate flight. However, despite this, it should be considered as a sebaceous gland and hence has the potential to produce odours (Rajchard, 2010). The feathers may also contribute to any odour. Birds possess a large exocrine gland at the base of the tail, the uropygial gland (preen gland), which, in addition to producing an odour, produces the oil used in preening. Although not a companion animal, the crested auklet, produces an odour described as 'citrus/tangerine', which can be detected by humans over 1 km away from the bird (Hagelin, 2007).

Dogs and cats

Both cats and dogs possess a wide range of glands all over their body that produce odours. Both species possess eccrine (sweat) glands on their feet, which may deposit an odour when walking or when undertaking specific marking activities, e.g. scratching. Dogs have a variety of glands around their anal region: anal sac (containing both sebaceous and apocrine glands), circumanal glands and supracaudal glands. This should not surprise anyone who has observed two dogs meeting, given the attention paid by the dogs to each other's anal region suggesting this is an important source of information. Cats have a similar array of glands, although the circumanal glands are smaller than those of dogs.

Cats have a variety of glands located on their head, submandibular glands under the chin, waxy secretion from the ears, temporal glands on the side of their head and glands (perioral) at the side of their mouths. The cat has a wide range of marking behaviours (see later) undertaken by rubbing its head against objects and, once deposited, these scent marks are investigated by other cats (Pageat and Gaultier, 2003).

Thus, all companion animals possess a wide array of glands that may produce odours and these may interact with other odours from urine, faeces and skin to produce sources of information that are used to influence the behaviour of conspecifics and other animals.

Olfaction – a Life-long Influence

The olfactory system is one of the earliest senses to become functional, detecting odours before birth or hatching. Studies have revealed that in all the major vertebrate groups, individuals detect and learn about odours prior to birth, hatching or emergence. This is the case for a variety of species, including rainbow trout (Sneddon, 2002), frogs (Hepper and Waldman, 1992), crocodiles (Sneddon *et al.*, 2001), chickens (Sneddon *et al.*, 1998), dogs (Wells and Hepper, 2006) and humans (Hepper, 1995). Interest in the possible prenatal emergence of olfaction is not restricted to simple exploration of the development of the olfactory system, but also to the influence of prenatal exposure on olfactory guided preferences in later life.

For example, Hepper *et al.* (2012) explored the immediate, medium and long-term effects of prenatal olfactory exposure in the cat. Unborn kittens were exposed to one of two novel chemicals (vanillin or 4-ethylguaiacol) via the mother's diet, mimicking a possible natural route of exposure. Kittens tested as new-borns in a two-choice odour preference test exhibited a preference for the familiar odour (i.e. the odour they had experienced via the mother's diet). At 9–10 weeks of age, kittens were given a choice of a food treat (chicken) flavoured with the familiar or unfamiliar odour. Kittens preferred the food treat that was flavoured with the familiar odour. To examine whether prenatal exposure persisted into adulthood, cats were tested at 6 months of age and given two bowls of food, one that was flavoured with the odour that the cat had been exposed to *in utero* and an unfamiliar flavour. Cats ate more of the food flavoured with the familiar odour (Fig. 15.3).

This study and others (see above) have indicated that prenatal exposure may exert a life-long influence on the olfactory preferences of virtually all vertebrates that have been tested. We speculate that this reflects an evolutionary adaptation to ensure the individual chooses a safe diet. If the mother eats the food then it is safe and thus the mother is transmitting to her, as yet unborn offspring, the odour (flavour) of safe food. It may also benefit the initiation of suckling (see Chapters 2 and 8).

Olfactory Guided Behaviour

The preceding sections demonstrate that companion animals possess functioning olfactory systems and have a variety of sources for producing odours and this suggests that scent plays a significant role in the behaviour of these animals. In the following sections we consider the role of olfaction in social recognition, sexual behaviour and scent marking, concluding with examples of specific situations in which odours play a significant role for particular species.

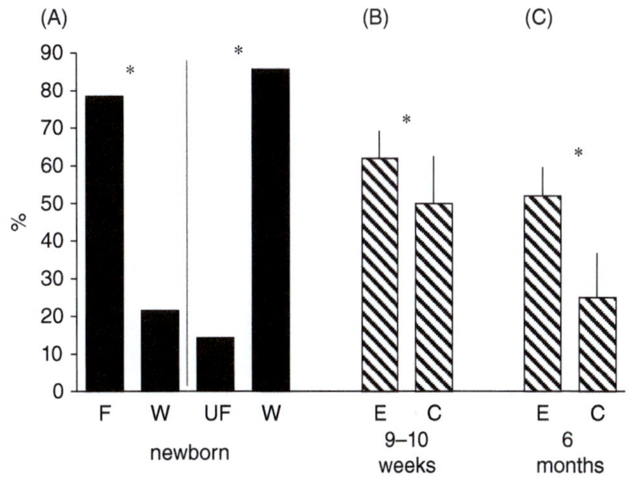

Fig. 15.3. The percentage number of: (A) new-born kittens that preferred the prenatally exposed odour (F) when paired with water (no odour, W), and the unfamiliar odour (UF) when paired with the water (W); (B) times 9- to 10-week-old kittens prenatally exposed to the odour (E) choose the familiar flavoured treat compared with kittens that had not prenatally experienced the odour (C); and (C) times 6-month-old kittens prenatally exposed to the odour (E) choose the familiar flavoured food compared with kittens that had not prenatally experienced the odour (C). Asterisk indicates significant differences ($P < 0.05$) between column pairs. (Redrawn from Hepper *et al.*, 2012.)

Social recognition

The ability to recognize other members of the same species is a fundamental cornerstone of sociality. For solitary species, this may simply mean recognition of mature conspecifics of the opposite sex to enable reproduction. As sociality increases, the range of other individuals capable of being recognized expands and may include parents, offspring, littermates, kin, members of the group and individuals. Such abilities present significant advantages in terms of the individual's fitness (Hepper, 1986a).

One key ability is that of parents recognizing their offspring, especially in cases where parental care is required to enable the young to mature, e.g. through feeding and protection. Caring for offspring incurs significant costs to parents and hence it is important that their efforts are directed towards their own offspring. Research now suggests that odours may enable individuals to recognize their own offspring and hence respond to them preferentially. This is demonstrated quite clearly by Golüke *et al.* (2016), who examined whether the female adult zebra finch was able to recognize its own eggs using smell. Zebra finch eggs are a single colour and do not have any pattern or spots to facilitate visual recognition. Intraspecific brood parasitism (i.e. where birds lay their eggs in the nests of other birds of the same species) is common. Hence, there is a need for the individual to discriminate their own eggs from others. Females were tested in an enclosure in which they were presented with the odour from their own egg and that of an egg from another clutch. Fans blew air over the eggs and into the enclosure. The eggs were hidden from view by being placed in gauze bags preventing any form of visual cue being used. The time that the female spent near the odour of each egg was recorded over a 5-min period. Tests were performed on days 3 and 10 after clutch completion (average hatching time 14 days) with eggs of the same age, i.e. days 3 and 10. Females showed no preference when tested 3 days after clutch completion, but exhibited a strong preference at 10 days after clutch completion for the odour of their own egg (Fig. 15.4). The cues used may be those derived from the developing embryo and permeating the egg and/or those deposited on the egg by the parents during incubation. The delay in recognition may reflect either that time is required for the female to learn the odour of her own egg and/or that the eggs do not present a sufficiently discriminable odour stimulus early in incubation.

The importance of recognizing young is highlighted by the finding that most animals seem to be able to recognize and discriminate their own young from others. Whilst convict cichlids – a popular freshwater aquarium fish species – are not able to recognize their own eggs, once hatched, the parents recognize their own offspring and discriminate them from the young of other cichlids (Myberg,

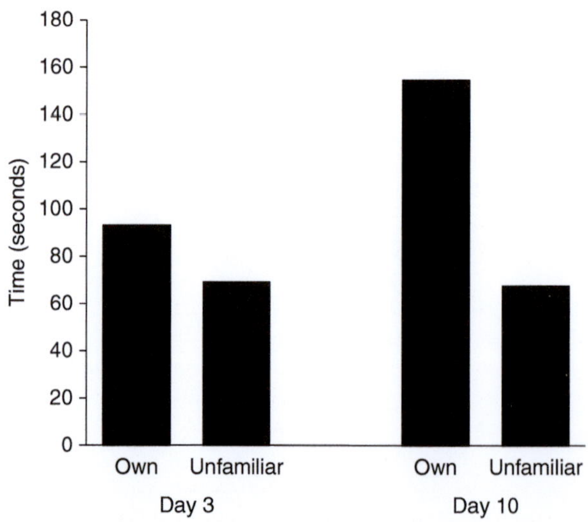

Fig. 15.4. The median time (in seconds) spent by female zebra finches near the odour of their own and an unfamiliar egg at 3 days and 10 days following completion of their clutch. (Redrawn from Golüke *et al.*, 2016.)

1964). Female cats appear to be able to recognize their own young, but those living with other cats may suckle kittens of other group members. Female dogs, when given a simultaneous choice task, spent more time investigating their own pups than unfamiliar pups (Hepper, 1994). Given the potential risk to young from other adults it is not surprising that the young quickly come to discriminate their own parents from other adults. For dogs, the ability to recognize their mother is still present after 2 years of separation (Hepper, 1994). New-borns of many species live for the first part of their lives in a nest with possibly their mother, father and siblings. If removed from the nest, individuals will orient and move back to the nest guided by odour cues. In kittens, the nest odour reduces distress vocalizations, indicating that the kitten both recognizes the odour of its nest and also that it has an appeasing effect, which may reduce separation stress. Zebra finches similarly recognize their own nest through odour (Caspers and Krause, 2011).

There are significant advantages if animals are able to recognize their kin (Hepper, 1986a) and, again, this ability seems widespread amongst companion animals. For example, dogs (Hepper, 1986b), fish (Wisenden and Dye, 2009) and birds (Krause *et al.*, 2012) all exhibit evidence of recognizing and responding preferentially to their kin (see Hepper, 1986a). Animals also produce individually identifiable odour cues that can be detected and discriminated by members of the same species and by members of other species. For example, rats can recognize individual cats by their odours (Staples *et al.*, 2008) and mice can recognize jungle fowl (Karlsson *et al.*, 2010). Moreover, dog owners are able to recognize their own pets from their odour alone (Wells and Hepper, 2000).

Many companion animals live with other animals or regularly meet the same animals, e.g. dog walkers meet up to walk their pets. New dogs joining the group will be subject to increased investigation by the other dogs compared with familiar dogs. However, as the new dog becomes familiar, the level of investigation decreases. Dogs, cats, tortoises and fish all recognize familiar individuals and display more investigation of unfamiliar individuals (Rose, 1970; Dunbar and Carmichael, 1981; Thunken *et al.*, 2009; Nakabayashi *et al.*, 2012). Introducing a new pet into a household with existing animals can present many problems, including increased aggression. However, exposing the new and existing animals to the odour of each other prior to physically meeting significantly reduces the amount of aggression shown (Levine *et al.*, 2005).

Thus, companion animals use odours to facilitate parents recognizing their offspring and vice versa, recognizing kin and individuals, and to discriminate between familiar and unfamiliar individuals and group members. It is worth noting that in order to use odours in social recognition, animals that meet each other have to investigate the odours of the other animal. Whilst there is no doubt that companion animals have excellent olfactory systems to detect and discriminate odours, it is unknown how far odours emanate from the source of odour production. Observation of social encounters between dogs, for example, indicates that olfactory investigation is not undertaken at a distance, but rather requires virtual contact between the nose and source of the odour. Perhaps because of the risk of serious harm that may arise from two animals in close proximity, social investigation may be considered to be almost ritualistic in nature following a similar pattern across all companion animals (with the possible exception of birds, but this may simply reflect a lack of study). Individuals approach each other head on, sniff nose to nose and then move to the head and down the side of the body until investigating the anal region. Such behaviour may reduce the potential for aggression, whilst ensuring that the important sources of odour information are investigated.

Sexual behaviour

Given the importance of reproduction for the survival of the species, it could be expected that sexually receptive females and sexually mature males signal their status to attract mates. Owners of dogs and cats will be particularly aware of the increased attention paid to females when in oestrus or the frantic escape attempts of males when a sexually receptive female is detected.

In the goldfish, where fertilization occurs externally, the synchronous release of mature sperm and eggs is facilitated by odour signals designed to coordinate behaviour and physiological development within and between sexes. Females release pheromones that induce male courtship behaviour and also influence the male endocrine system, promoting the development of sperm. Female pheromones also act as attractants and males can detect these stimuli at very low concentrations and have evolved specific receptor mechanisms for their detection (Kobayashi *et al.*, 2002). Signals from females in oestrus increase their attractiveness and result in increased investigation by males in dogs (Doty and Dunbar, 1974) and tortoises (Auffenberg, 1977). In cats, investigation of fresh urine leads to a specific behavioural response called flehmen. The upper lip curls up and the mouth is part open, a pose held for up to 30 s. The cat may make direct contact with the urine and a pumping action of the tongue and jaw forces fluid and airborne stimuli into the vomeronasal organ. Adult cats exhibit flehmen, but males exhibit this much more, especially in response to the urine of females, and the incidence increases in males when the female is in oestrus. The actual function of this behaviour, perhaps best known from being displayed by horses (e.g. Marinier *et al.*, 1988), is unknown, but may relate to social or sexual recognition.

Male odour may also be influenced by reproductive status. During the breeding season, chin glands of desert tortoises increase their production such that fluid may be detected flowing from the glands (Auffenberg, 1977). Both males and females respond to the odour. Males exhibit aggressive responses, such as ramming, whereas females head bob in response to the odour (Rose, 1970).

Scent marking

One behaviour common to nearly all companion animals is scent marking. Scent marking can be defined as the deposition of an odour from glandular secretions, urination or defecation. Whilst many animals have unique behaviour patterns that accompany the deposition of these odour marks, and these marks are investigated by conspecifics, the actual role of the scent mark is poorly understood. It has been argued that scent marks may act to indicate territorial ownership, the location of a food cache or the presence of the individual. The mark itself may carry information of identity (e.g. group) or status (e.g. dominance).

Males urine mark more than females in dogs (Pal, 2003) and cats (Feldman, 1994). In both species, males exhibit characteristic behavioural patterns to accompany their urination. In dogs, the raised leg urination posture may act to deposit the urine mark at (dog) nose level, increasing its salience to other dogs. Cats are observed first sniffing a vertical object and then turn, back up with raised tail and spray a stream of urine onto the object. Female cats may exhibit the same behaviour, but the urine stream is a less concentrated more diffuse spray. Male cats increase their spraying when females are in oestrus. Urine marking is not restricted to cats and dogs; fish use urine to mark their territories and when other males are present, fish increase their frequency of urination (Maruska and Fernald, 2012). Scent marking may increase in the presence of other older marks from other individuals; it is speculated that animals over-mark the older scent to replace the mark of the original depositor with that of their own but the reasons for this are unknown.

Tortoises, because of their limited urination, mark more with faeces and often mark in and around burrows. For females, this may be related to ownership, since other females are much less likely to use a burrow that contains faeces from another female. Similarly, males are less likely to use a burrow marked with another male's faeces (Bulova, 1997).

Cats also exhibit scratching behaviour, whereby they mark vertical objects, usually on pathways within their territory rather than at the boundaries of their territory. Given the presence of eccrine glands on their feet, this behaviour may act to deposit odour from these glands, combined with a visual signal. Cats also appear to deposit odour through rubbing of the glands of their face onto objects and other cats. Tomcats may mark their environment by face rubbing, especially in the presence of females, as well as marking objects within their environment. Face rubbing between cats may reduce aggression, and kittens are observed to greet their mother with a tail-up display and face rubbing. Males cheek rub more than females, but females neck rub more. Such head rubbing between conspecifics is not restricted to two cats; cats will rub on their human owners. Although humans cannot detect an odour, this does not mean that the cats cannot. Tortoises similarly use their chin glands to mark objects especially during the breeding season (Auffenberg, 1977).

Scent marking, accompanied by distinctive behavioural patterns, is exhibited by many companion animals, although the function(s) of depositing these odours has yet to be fully elucidated. Its presence suggests an important role, but further research is required to appreciate the variety of functions this may serve for different species for their relationship with conspecifics and other animals.

Unique uses of odours

Some animals display specific uses of olfactory stimuli suited to and adapted for their environment. For some of these, whilst the behaviours can be clearly identified, the function(s) of the behaviour remain(s) elusive.

One example of this is scent rolling by dogs. Dogs will typically pick a particularly malodorous substance, then roll in it in a ritualistic manner, covering particularly their neck and shoulder region. Whilst such behaviour can be easily identified and smelled, its function has yet to be determined. Speculations have included: some form of reinforcement or comfort; a means of learning about new odours; a means to hide their own odour; a means to increase their dominance; or it may be a behaviour or part of a behaviour pattern that has been retained in the dog's genes but serves no function for the dog of today.

Odours may serve a unique role in navigation. It is thought that the ability to home in pigeons is achieved, in part, via the use of olfactory signals. Papi *et al.* (1971) reported that anosmic pigeons were unable to find their way home, suggesting the cue used was olfactory, and, recently, study of the atmosphere has reported finding the presence of odour cues that could be used in homing (Wallraff, 2013; Safi *et al.*, 2016). Outstanding olfactory navigation abilities can also be found in fish, where salmon return to their natal stream or pond after years at sea, not having experienced the odour of the hatching area for many years. It is thought some imprinting-like process that occurs after hatching and before the salmon leave their freshwater environment underlies this ability (Dittman and Quinn, 1996; see also Chapter 4). Whether other species also have similar abilities is unknown.

A final olfactory-guided behaviour, in this case unique to mammals, is that involved in the initiation of suckling by the new-born animal. In altricial mammals, survival immediately after birth is dependent upon their ability to find the nipple and commence suckling, thus receiving the milk essential for their survival. As shown for mice in Chapter 2, odour is also essential for cats to guide the new-born to the nipple and enable it to commence suckling. If the nipples are washed, the new-born kitten has great difficulty in locating them (Raihani *et al.*, 2009).

Summary

Companion animals have a varied repertoire of olfactory-guided behaviours. In some cases, this interacts badly with their role as pets, such as when adult male cats spray around the home, although Feliway (a pheromone from the cheek gland of the cat) can be used to reduce this (see also Chapter 11). The olfactory sense is important for the animal to gain information about its environment, to engage in its normal behaviour and also to leave odours to provide security. This latter aspect is often overlooked. Whilst a pet's bedding may smell bad to the pet owner, it is a familiar reassuring stimulus for the pet. Despite large differences in environment, olfactory apparatus and behaviour, companion animals all live in a varied and – for them – important olfactory environment; one as humans we struggle to understand.

References

Alberts, A.C., Rostal, D.C. and Lance, V.A. (1994) Studies on the chemistry and social significance of chin gland secretions in the desert tortoise, *Gopherus agassizii*. *Herpetological Monographs* 8, 116–124. DOI: 10.2307/1467075

Auffenberg, W. (1977) Display behaviour in tortoises. *American Zoologist* 17, 241–250. DOI: 10.1093/icb/17.1.241

Brown, G.E., Jackson, C.D., Malka, P.H., Jacques, E. and Couturier, M.-A. (2012) Disturbance cues in freshwater prey fishes: does urea function as an 'early warning cue' in juvenile convict cichlids and rainbow trout? *Current Zoology* 58, 250–259. DOI: 10.1093/czoolo/58.2.250

Buck, L. and Axel, R. (1991) A novel multigene family may encode odorant receptors: a molecular basis for odor recognition. *Cell* 65, 175–187. DOI: 10.1016/0092-8674(91)90418-X

Bulova, S.J. (1997) Conspecific chemical cues influence burrow choice by desert tortoises (*Gopherus agassizii*). *Copeia* 4, 802–180. DOI: 10.2307/1447297

Caspers, B.A. and Krause, E.T. (2011) Odour-based natal nest recognition in the zebra finch (*Taeniopygia guttata*), a colony breeding songbird. *Biological Letters* 7, 184–186. DOI: 10.1098/rsbl.2010.0775

Davis, R.G. (1973) Olfactory psychophysical parameters in man, rat, dog, and pigeon. *Journal of Comparative and Physiological Psychology* 85, 221–232. DOI: 10.1037/h0035053

Dittman, A.H. and Quinn, T.P. (1996) Homing in Pacific salmon: mechanisms and ecological basis. *Journal of Experimental Biology* 199, 83–91.

Doty, R. and Dunbar, I. (1974) Attraction of beagles to conspecific urine, vaginal and anal sac secretions. *Physiology and Behaviour* 12, 825–833. DOI: 10.1016/0031-9384(74)90020-1

Dunbar, I. and Carmichael, M. (1981) The response of male dogs to urine from other males. *Behavioral and Neural Biology* 31, 465–470. DOI: 10.1016/S0163-1047(81)91546-6

Feldman, H.N. (1994) Methods of scent marking in the domestic cat. *Canadian Journal of Zoology* 72, 1093–1099. DOI: 10.1139/z94-147

Gloriam, D.E.I., Bjarnadottir, T.K., Yan, Y.L., Postlethwait, J.H., Schiöth, H.B. and Fredriksson, R. (2005) The repertoire of trace amine G-protein-coupled receptors: large expansion in zebrafish. *Molecular Phylogenetics and Evolution* 35, 470–482. DOI: 10.1016/j.ympev.2004.12.003

Golüke, S., Dörrenberg, S., Krause, E.T. and Caspers, B.A. (2016) Female zebra finches smell their eggs. *PLoS ONE*, doi:10.1371/journal.pone.0155513.

Hagelin, J.C. (2007) The citrus-like scent of crested auklets: reviewing the evidence for an avian olfactory ornament. *Journal of Ornithology* 148, 195. DOI: 10.1007/s10336-007-0185-6

Hepper, P.G. (1986a) Kin recognition: functions and mechanisms a review. *Biological Reviews* 61, 63–93. DOI: 10.1111/j.1469-185X.1986.tb00427.x

Hepper, P.G. (1986b) Sibling recognition in the domestic dog. *Animal Behaviour* 34, 288–289. DOI: 10.1016/0003-3472(86)90035-7

Hepper, P.G. (1994) Long-term retention of kinship recognition established during infancy in the domestic dog. *Behavioural Processes* 33, 3–14. DOI: 10.1016/0376-6357(94)90056-6

Hepper, P.G. (1995) Human fetal 'olfactory' learning. *International Journal of Prenatal and Perinatal Psychology and Medicine* 7, 147–151.

Hepper, P.G. and Waldman, B. (1992) Embryonic olfactory learning in frogs. *Quarterly Journal of Experimental Psychology B* 44B, 179–197.

Hepper, P.G. and Wells, D.L. (2005) How many footsteps do dogs need to determine the direction of an odour trail? *Chemical Senses* 30, 291–298. DOI: 10.1093/chemse/bji023

Hepper, P.G., Wells, D.L., Millsopp, S., Kraehenbuehl, S.A., Lyn, S.A. and Mauroux, O. (2012) Prenatal and early sucking influences on dietary preference in newborn, weaning and young adult cats. *Chemical Senses* 37, 755–766. DOI: 10.1093/chemse/bjs062

Karlsson, A.C., Jensen, P., Elgland, M., Laur, K., Fyrner, T., Konradsson, P. and Laska, M. (2010) Red junglefowl have individual body odors. *Journal of Experimental Biology*, 213, 1619–1624. DOI: 10.1242/jeb.040279

Kobayashi, M., Sorenson, P.W. and Stacey, N.E. (2002) Hormonal and pheromonal control of spawning in goldfish. *Fish Physiology and Biochemistry* 26, 71–84. DOI: 10.1023/A:1023375931734

Krause, E.T., Krüger, O., Kohlmeier, P. and Caspers, B.A. (2012) Olfactory kin recognition in a songbird. *Biological Letters* 8, 327–329. DOI: 10.1098/rsbl.2011.1093

Krestel, D., Passe, D., Smith, J.C. and Jonsson, L. (1984) Behavioral determination of olfactory thresholds to amyl acetate in dogs. *Neuroscience and Biobehavioural Reviews* 8, 169–174. DOI: 10.1016/0149-7634(84)90037-X

Kruzhalov, N.B. and Boiko, V.P. (1987) Electrical reactions of the olfactory lining and olfactory bulb of the tortoise *Testudo graeca* to certain natural and artificial scents. *Zhurnal Evolyutsionnoi Biokhimii i Fiziologii* 23, 617–623.

Lawson, M.J., Craven, B.A., Paterson, E.G. and Settles, G.S. (2012) A computational study of odorant transport and deposition in the canine nasal cavity: implications for olfaction. *Chemical Senses* 37, 553–566. DOI: 10.1093/chemse/bjs039

Levesque, H.M., Scaffidi, D., Polkinghorne, C.N. and Sorensen, P.W. (2011) A multi-component species identifying pheromone in the goldfish. *Journal of Chemical Ecology* 37, 219–227. DOI: 10.1007/s10886-011-9907-6

Levine, E., Perry, P., Scarlett, J. and Houpt, K.A. (2005) Intercat aggression in households following the introduction of a new cat. *Applied Animal Behaviour Science* 90, 325–336. DOI: 10.1016/j.applanim.2004.07.006

Marinier, S.L., Alexander, A.J. and Waring, G.H. (1988) Flehmen behaviour in the domestic horse: discrimination of conspecific odours. *Applied Animal Behaviour Science* 19, 227–237. DOI: 10.1016/0168-1591(88)90003-2

Maruska, K.P. and Fernald, R.D. (2012) Contextual chemosensory urine signalling in an African cichlid fish. *Journal of Experimental Biology* 215, 68–74. DOI: 10.1242/jeb.062794

Montague, M.J., Li, G., Gandolfi, B., Khan, R., Aken, B.L. *et al.* (2014) Comparative analysis of the domestic cat genome reveals genetic signatures underlying feline biology and domestication. *Proceedings of the National Academy of Sciences USA* 111, 17230–17235. DOI: 10.1073/pnas.1410083111

Myberg, A.A. (1964) An analysis of the preferential care of eggs and young by adult cichlid fishes. *Zeitschrift für Tierpsychologie* 21, 53–98. DOI: 10.1111/j.1439-0310.1964.tb01187.x

Nakabayashi, M., Yamaoka, R. and Nakashima, Y. (2012) Do faecal odours enable domestic cats (*Felis catus*) to distinguish familiarity of donors? *Journal of Ethology* 30, 325–329. DOI: 10.1007/s10164-011-0321-x

Nei, M., Niimura, Y. and Nozawa, M. (2008) The evolution of animal chemosensory receptor gene repertoires: roles of chance and necessity. *Nature Reviews Genetics* 9, 951–963. DOI: 10.1038/nrg2480

Niimura, Y. (2014) Olfactory receptor genes: evolution. *eLS.* DOI: 10.1002/9780470015902.a0020789.pub2.

Office of National Statistics (2016) Population estimates for UK, England and Wales, Scotland and Northern Ireland: mid-2015. Available at: www.ons.gov.uk/peoplepopulationandcommunity/populationandmigration/populationestimates/bulletins/annualmidyearpopulationestimates/latest (accessed 23 January 2017).

Pageat, P. and Gaultier, E. (2003) Current research in canine and feline pheromones. *Veterinary Clinics. Small Animal Practice* 33, 187–211. DOI: 10.1016/S0195-5616(02)00128-6

Pal, S.K. (2003) Urine marking by free-ranging dogs (*Canis familiaris*) in relation to sex, season, place and posture. *Applied Animal Behaviour Science* 80, 45–59. DOI: 10.1016/S0168-1591(02)00178-8

Papi, F., Fiore, L., Fiaschi, V. and Benvenuti, S. (1971) The influence of olfactory nerve section on the homing capacity of carrier pigeons. *Monitore Zoologico Italiano* 5, 265–267.

Pet Food Manufacturers Association (2016) Pet population 2016. Available at: www.pfma.org.uk/pet-population-2016 (accessed 23 January 2017).

PODS (2012) U.S. pet ownership and demographics sourcebook. Available at: https://www.avma.org/KB/Resources/Statistics/Pages/Market-research-statistics-US-pet-ownership.aspx (accessed 23 January 2017).

Raihani, G., González, D., Arteaga, L. and Hudson, R. (2009) Olfactory guidance of nipple attachment and sucking in kittens of the domestic cat: inborn and learned responses. *Developmental Psychobiology* 51, 662–671. DOI: 10.1002/dev.20401

Rajchard, J. (2010) Biologically active substances of bird skin: a review. *Veterinarni Medicina* 55 413–421. http://agriculturejournals.cz/publicFiles/27934.pdf

Romanes, G.J. (1887) Experiments on the sense of smell in dogs. *Nature* July 21, 273–274.

Rose, F.L. (1970) Tortoise chin gland fatty acid composition: behavioral significance. *Comparative Biochemistry and Physiology* 32, 577–580. DOI: 10.1016/0010-406x(70)90475-5

Safi, K., Gagliardo, A., Wikelski, M. and Kranstauber, B. (2016) How displaced migratory birds could use volatile atmospheric compounds to find their migratory corridor: a test using a particle dispersion model. *Frontiers in Behavioral Neuroscience* 10, art. 175. DOI: 10.3389/fnbeh.2016.00175

Sneddon, H.A. (2002) The effects of embryonic chemosensation in vertebrates: a comparative study. PhD thesis, Queens University of Belfast.

Sneddon, H., Hadden, R. and Hepper, P.G. (1998) Chemosensory learning in the chick embryo. *Physiology and Behaviour* 64, 133–139. DOI: 10.1016/S0031-9384(98)00037-7

Sneddon, H., Hepper, P.G., and Manolis, C. (2001) Embryonic chemosensory learning in the saltwater crocodile *Crocodylus porosus*. In: Grigg, G.C., Seebacher, F. and Franklin, C.E. (eds) *Crocodilian Biology and Evolution*. Surrey Beatty, Chipping Norton (NSW), pp. 378–382.

Staples, L.G., Hunt, G.E., van Nieuwenhuijzen, P.S. and McGregor, I.S. (2008) Rats discriminate individual cats by their odour: possible involvement of the accessory olfactory system. *Neuroscience and Biobehavioural Reviews* 32 1209–1217. DOI: 10.1016/j.neubiorev.2008.05.011

Steiger, S.S., Fidler, A.E., Valcu, M. and Kempanaers, B. (2008) Avian olfactory receptor gene repertoires: evidence of a well-developed sense of smell in birds? *Proceedings of the Royal Society B: Biological Sciences* 275, 2309–2317. DOI: 10.1098/rspb.2008.0607

Stoddart, D.M. (1980) *The Ecology of Vertebrate Olfaction*. Chapman and Hall, Andover, UK.

Thunken, T., Waltschyk, N., Bakker, T.C. and Kullman, H. (2009) Olfactory self-recognition in a cichlid fish. *Animal Cognition* 12, 717–724. DOI: 10.1007/s10071-009-0231-2

Walker, D.B., Walker, J.C., Cavnar, P.J., Taylor, J.L., Pickel, D.H., Hall, S.B. and Suarez, J.C. (2006) Naturalistic quantification of canine olfactory sensitivity. *Applied Animal Behaviour Science* 97, 241–254. DOI: 10.1016/j.applanim.2005.07.009

Wallraff, H.G. (2013) Ratios among atmospheric trace gases together with winds imply exploitable information for bird navigation: a model elucidating experimental results. *Biogeosciences* 10, 6929–6943. DOI: 10.5194/bg-10-6929-2013

Wells, D.L. and Hepper, P.G. (2000) The discrimination of dog odours by humans. *Perception* 29, 111–115. DOI: 10.1068/p2938

Wells, D.L. and Hepper, P.G. (2003) Directional tracking in the domestic dog, *Canis familiaris*. *Applied Animal Behaviour Science* 84, 297–305. DOI: 10.1016/j.applanim.2003.08.009

Wells, D.L. and Hepper, P.G. (2006) Prenatal olfactory learning in the domestic dog. *Animal Behaviour* 72, 681–686. DOI: 10.1016/j.anbehav.2005.12.008

Wenzel, B.M. (1968) Olfactory prowess of the kiwi. *Nature* 220, 1133–1134. DOI: 10.1038/2201133a0

Wisenden, B.D. and Sargent, R.C. (1997) Antipredator behavior and suppressed aggression by convict cichlids in response to injury-released chemical cues of conspecifics but not to those of an allopatric heterospecific. *Ethology* 103, 283–291. DOI: 10.1111/j.1439-0310

Wisenden, B.D. and Dye, T.P. (2009) Young convict cichlids use visual information to update olfactory homing cues. *Behavioral Ecology and Sociobiology* 63, 443–449. DOI: 10.1007/s00265-008-0678-1

Zucchi, R., Chiellini, G., Scanlan, T.S. and Grandy, D.K. (2006) Trace amine-associated receptors and their ligands. *British Journal of Pharmacology* 149, 967–978. DOI: 10.1038/sj.bjp.0706948

Index